On the *Future* of Species

Praise for *On the Future of Species*

'A thoughtful and troubling reflection on the future of life, human and otherwise, prepared by a notable expert on what life used to be.' Antonio Damasio, Dornsife Professor of Neuroscience, Psychology and Philosophy, University of Southern California, author of *Descartes' Error*

'A visionary and exhilarating exploration of biology's next great frontier. Woolfson has written a work of astonishing scope and imagination, charting the convergence of artificial intelligence and genome synthesis. He takes readers from the foundations of molecular genetics to the threshold of a world in which we may author entirely new species, whilst reimagining human health. Bold, lucid, and deeply original, *On the Future of Species* is essential reading for anyone interested in the destiny of life on Earth and beyond.' Tim Coulson, Professorial Fellow of Jesus College and joint Head of Biology at the University of Oxford, author of *A Little History of Everything*

'The book we need right now; a clear-eyed, comprehensive look at how AI, gene editing, and synthetic biology are converging to give humanity unprecedented power over its own evolution. Part history, part forecast, entirely compelling. Woolfson guides us through the landmarks of molecular biology with the assurance of an expert and the enthusiasm of a born storyteller. He brings a rare combination of industry experience and literary skill to one of the most consequential questions of our time. Essential reading for anyone seeking to understand where genomic technologies are taking us.' Tom Ellis, Professor of Synthetic Genome Engineering, Imperial College London

'A brilliantly crafted, sweeping exposition with profound insights essential for any citizen of the twenty-first century.' Tim White, Professor of Integrative Biology and Director of the Human Evolution Research Center, University of California, Berkeley

'Darwin wrote *On the Origin of Species* in 1859. With the emergence of large language of life models – artificial biological intelligence – it was the right time to envision *On the Future of Species*. Woolfson delivers that in this brilliant book, a very thoughtful and thorough assessment of the profound implications of editing and rewriting our code.' Eric Topol, Chair of the Department of Translational Medicine, Scripps Research Institute, author of *Super Agers*

'A terrific read and will be interesting to a very broad audience. I particularly enjoyed the blend of engagingly narrated scientific history with forecasting the future. Woolfson is absolutely correct that we are embarking on a new era in which humankind will be able to create completely new kinds of biology that evolution has never explored. This is both tremendously exciting and rather scary!' Patrick Maxwell, Regius Professor of Physic and Head of the School of Clinical Medicine, University of Cambridge

'Scientists already know how to read the language of DNA and to edit it. Woolfson details the new frontier – writing new DNA scripts from scratch.' Thomas R. Cech, Nobel Prize in Chemistry laureate, author of *The Catalyst*.

'A fascinating read exploring the principles of biology applied to medicine from a different lens. I thoroughly recommend for all audiences, from students to healthcare professionals of all backgrounds.' Mumtaz Patel, President of the Royal College of Physicians

'A dazzling and prescient dive into the future of species, Woolfson explores the emergence of Artificial Biological Intelligence, capable of unravelling the genome's regulatory code. Our genomes ... and those of all creatures ... are riddled with 'spaghetti code' hopelessly convoluted by accidents of evolutionary history, demanding an ABI approach that unlocks a brave new world of biological possibility.' Jef Boeke, Founding Director of the Institute for Systems Biology at NYU Langone Medical Center and the leader of the Sc2.0 Synthetic Yeast Genome International Consortium

'Woolfson explores the question that many biologists are asking but few can answer: where will AI-supercharged bioscience lead us?

Authoritatively defines what a "post-Darwinian" future could and should look like. A hopeful guide to the "Brave New World" of AI-powered bioscience.' Tom Ireland, editor of *The Biologist* and author of *The Good Virus*

'Every great leap in human progress begins with a new way of seeing. For centuries, we've viewed biology as something to be studied. Woolfson wants us to see it as something to be authored. . . . This book shows us the next frontier, where the code of life meets the code of machines, and foresight becomes the most essential human trait.' Amy Webb, author of *The Big Nine* and *The Genesis Machine*

'Woolfson explores the profound opportunities and challenges that arise when we turn evolution upside down – beginning not with random mutation but with deliberate selection. Woolfson's book prepares readers for the scientific, moral and societal dilemmas that will accompany such a transformation and concludes by outlining a framework – a scaffold for a manifesto – that humanity will need to deliberate as synthetic biology evolves.' John-Arne Røttingen, CEO of the Wellcome Trust

'The past two centuries have seen two profound shifts in our understanding of evolutionary biology – the nineteenth-century realisation that evolution exists and the gradual unpicking through the twentieth and twenty-first centuries of the genomic mechanisms that control the development of living organisms. Woolfson sets out in this book the fascinating notion that we could one day – perhaps quite soon – see a third revolution in which we directly tap into this genetic code to author entirely new biological species. The implications are profound and I thoroughly encourage anyone with an interest in the natural world to read this important book.' Doug Gurr, Director of the Natural History Museum, London

'Offers a lucid and prescient account of the transition from natural evolution to intentional genome design. As synthetic genomics moves from editing to whole-genome rewriting, this book provides an indispensable guide to the possibilities, responsibilities, and profound choices ahead.' Patrick Yizhi Cai, Chair Professor in Synthetic Genomics, the University of Manchester and International Co-ordinator of the Sc2.0 Synthetic Yeast Genome International Consortium

'Woolfson's lucid prose describes fascinating details about the hundreds of scientists whose work allows us to speed-read genomes now, and to write genomes soon. ... What will new designer species do for us, and to us? This book provides essential background for urgently needed deep discussions.' Randolph M. Nesse, Founding Director of the Center for Evolution and Medicine at Arizona State University, author of *Why We Get Sick: The New Science of Darwinian Medicine*

'*On the Future of Species* explains how we got to understand the chemistry of life, setting the scene for a mind-blowing exploration of the opportunities opened up by this understanding for tackling diseases and possibly improving the human body. But at the same time, it offers wise council about the dangers involved in such tinkering, and points up the need for regulation. It is both timely and authoritative, and deserves to be widely read.' John Gribbin, author of *Science: A History*, *In Search of the Multiverse* and *Six Impossible Things*

'A forward-looking, detailed history, about the developments in the new field of synthetic biology ... Woolfson discusses key ethical steps that society must agree on to move forward. J. Craig Venter, Founder and CEO of the J. Craig Venter Institute and leader of the teams that first sequenced the human genome and made the first synthetic genome and species

'In *On the Origin of Species*, Charles Darwin presented artificial selection as a "magician's wand," enabling expert human breeders to conjure into existence whatever new living forms they desired. According to Woolfson, a vastly more powerful successor now in the works is artificial biological intelligence ... *On the Future of Species* is at once a history of these remarkable developments, a report from the scientific and technological frontlines, and an attempt to chart a path through the morally complex territory ahead. A fascinating and important book.' Gregory Radick, author of *Disputed Inheritance: The Battle over Mendel and the Future of Biology*

'We stand at a threshold where, we may in the future be able to synthesise entirely new genetic scripts and create completely new forms of life. This book is a fascinating look at the possibilities and what their consequences might be.' Venki Ramakrishnan, Nobel Prize in Chemistry laureate, author of *Gene Machine*

On the *Future* of Species

AUTHORING LIFE BY MEANS OF ARTIFICIAL BIOLOGICAL INTELLIGENCE

ADRIAN WOOLFSON

BLOOMSBURY PUBLISHING
LONDON • OXFORD • NEW YORK • NEW DELHI • SYDNEY

BLOOMSBURY PUBLISHING
Bloomsbury Publishing Plc
50 Bedford Square, London, WC1B 3DP, UK
Bloomsbury Publishing Ireland Limited,
29 Earlsfort Terrace, Dublin 2, D02 AY28, Ireland

First published in Great Britain 2026

A catalogue record for this book is available from the British Library

ISBN: HB: 978-1-5266-7097-7; TPB: 978-1-5266-7096-0;
EBOOK: 978-1-5266-7098-4; EPDF: 978-1-5266-7101-1

2 4 6 8 10 9 7 5 3 1

Typeset by Six Red Marbles India
Printed and bound in Great Britain by Clays Ltd, Elcograf S.p.A

MIX
Paper | Supporting responsible forestry
FSC www.fsc.org FSC® C018072

To find out more about our authors and books visit www.bloomsbury.com
and sign up for our newsletters
For product-safety-related questions contact productsafety@bloomsbury.com

To Sheila, Gerald and Margherita

Am not I
A fly like thee?
Or art not thou
A man like me?

— WILLIAM BLAKE, 1794

In them all the wealth and the variety of hereditary transmissions can find expression just as all the words and concepts of all languages can find expression in twenty-four to thirty alphabetic letters.

— FRIEDRICH MIESCHER, 1892

Man, the imperfect librarian, may be the product of chance or of malevolent demiurgi; the universe, with its elegant endowment of shelves, of enigmatic volumes, of inexhaustible stairways for the traveler and latrines for the seated librarian, can only be the work of a god.

— JORGE LUIS BORGES, 1941

For I saw before me in dark contours the beginning of a grammar of biology.

— ERWIN CHARGAFF, 1971

If we are to explain how things are, we must be able to explain how things might have been, or must be, or couldn't be.

— DANIEL DENNETT, 1995

Mathematics is the art of the perfect. Physics is the art of the optimal, but biology is the art of the satisfactory.

— SYDNEY BRENNER, 2003

Contents

Preface

Be all the bruisers cull'd from all St. Giles',
That art and nature may compare their styles
– LORD BYRON, 1812

On the evening of 1 July 1858, during a heatwave that had turned the banks of the River Thames into a stinking morass, a small group of individuals gathered at Burlington House in Piccadilly for a hastily convened special meeting of the Linnean Society of London. They were not there to seek refuge from the heat, but rather to witness the presentation of three papers on the mechanism of evolution that would fundamentally reshape humanity's understanding of its place in the world. The ideas of the authors, the explorer-naturalist Charles Darwin and adventurer Alfred Russel Wallace – neither of whom were present at the meeting – would end up changing our view of life and its origins for ever.

Although shocking to the Victorian sensibility, their message was disarmingly simple. Living species – from the humblest microbes to tulips, beetles, hyenas, and humans – were not the fixed, flawless, and inevitable handiwork of an instantaneous act of divine Creation. They were, instead, the ephemeral products of a slow and relentless process of evolution by natural selection. Chance events and random variations were the driving forces behind species alteration, while differential reproduction was the engine of transformation. It was heredity, rather than God, that determined the nature of living things. Evolution by natural selection served as the indifferent crafter of their gradual modification, lacking any plan or intention. This process of natural engineering, which has defined the history of life on Earth for the approximate four billion years since its inception, might have continued indefinitely.

But this previously unbreakable evolutionary narrative is, instead, about to be turned on its head in a manner that Darwin and Wallace could never have foreseen. This is because the convergence of artificial intelligence (AI) – which promises to reveal the rules of the generative grammar of life – with technologies for constructing the genomes of species on demand has made it possible to contemplate designing new species from scratch. It may also enable us to address one of the major unsolved problems of biology: how genomes specify the complex structures of higher organisms.

We are, as a result, at the threshold of a new era – one in which the blind, directionless process of evolution by natural selection may be augmented, perhaps even surpassed, by artificial evolution. This deliberate, predictive process of artificial species generation – what we might call 'artivolution' – will allow for the intentional, human- and machine-inspired design of novel species, potentially unlocking a generative biology revolution.

For now, our capabilities are rudimentary. We are like children learning to write. We have managed to piece together the genomes of a few simple organisms – viruses, bacteria and a yeast – but we will soon be able to construct those of more complex creatures, including, in the first instance, a multicellular moss. Eventually, we might be able to write out the genome of our own species. Along the way, we may resurrect prehistoric species like the woolly mammoth and more recently extinct creatures like the dodo, great auk and Morant's blue butterfly. We might also – should we choose to do so – be able to write out the genomes of extinct human species. Even more intriguingly, artivolution may allow us to go beyond the simple mimicry of nature's designs. We may, as a result, be able to transcend nature's imagination to create species that have never previously existed.

With the assistance of AI, which has the potential to decode life's generative grammar, and the agency of a chemical printing press capable of rendering the genome sequences of species as if they were the texts of books, our ability to manipulate life's structures could become virtually limitless. Free from the constraints of chance and natural selection, we would no longer need to reference nature's blueprints. We could instead begin to narrate new designs – equipped with the pen, paper and creativity necessary to rewrite life's story. In so doing, we would become the authors of species.

It is not difficult to imagine a future in which any possible organism could be constructed, and where our understanding of human biology becomes so refined that we may comprehend most, if not all, aspects of its functioning. This would revolutionise how we approach human dysfunction, potentially making disease a thing of the past. The realisation of this vision will be driven by the development of what we may call Artificial Biological Intelligence (ABI) – the ability to predict, generate, construct and boot up the genomes of all possible species.

The eventual attainment of ABI, and the ability it conveys to interpret and author genetic sequences, will represent a critical milestone in the history of the biological sciences. The prospect of its realisation is likely to become biology's North Star. Its attainment, to various degrees, will signal the moment when biology comes of age, transitioning from a descriptive science into a predictive engineering discipline. The announcement, in September 2025, of the construction of the first ever AI-designed synthetic virus is the harbinger of this emerging generative biology future.

Were artificial genome design and construction to contribute significantly to species generation, Darwinian evolution by natural selection would no longer be the sole originator of biodiversity. Nature's menagerie – generated by trial-and-error methods of species creation – would, at that point, be complemented by artificial species, designed and synthesised with human intent. Over time, heredity might become marginalised. The ability to master biology's language and to design and construct living organisms according to rational principles will shake the very foundations of the concept of a species.

While offering significant potential utility when firewalled from the wild, the introduction of synthetic species into natural ecosystems presents substantial risks. The continuous evolution of natural species complicates our ability to predict their interactions with artificial species. Synthetic organisms may, furthermore, employ entirely new chemistries that transcend the natural genetic alphabet. These may introduce unfamiliar and unpredictable behaviours into living systems.

Following the potential historic pivot to the design and synthetic construction of artificial life – an inflexion that would alter the nature of life on Earth in ways hardly imaginable – the period of evolutionary history during which natural selection was the sole driver of species generation would reach its conclusion. Natural heredity would be marginalised. The attainment of ABI would represent the apotheosis of humankind's understanding of its own nature and might allow us to contemplate rewriting our own genome.

This has the potential to greatly extend the human healthspan and lifespan, and to eradicate the diseases that have afflicted humanity for millennia. It might also provide opportunities to modify other aspects of human nature. Rewriting the human genome may, in fact, be the only practical way to prevent and cure many devastating diseases. This is because most human diseases arise from small contributions that originate from multiple different components. Yet to do so would raise profound ethical issues and risk impinging on the fabric of human nature itself.

Following the potential attainment of ABI, future organisms – perhaps even alternative versions of humankind – could become programmable. Designed on computer screens through the use of natural and artificial intelligence, their blueprints would be encoded in digital biochemical software. Unlike gene editing, which modifies pre-existing genetic material, this new mode of species generation would allow life to go beyond nature's narratives and to be reimagined. Organisms could be engineered to order, converted into living machines with multiple potential utilities.

We are entering a new age of generative biology, defined by a technological revolution unlike anything that has preceded it – in both scale and significance. It stands poised to fundamentally redefine our relationship with natural biology, and to impact nearly every aspect of human existence – from food security and climate change to healthcare, information storage, energy production, biomaterials, biocomputing and ecosystem preservation.

The convergence of AI-informed genome design and synthetic genomics offers immense opportunities for humankind. It also has the potential to cause catastrophic harm. The attainment of ABI will present an existential challenge to humanity, as the ability to rewrite our own genetic code could fundamentally alter – or erase – human nature as we know it and undermine human autonomy. ABI is also vulnerable to misuse – both intentional and unintentional – and may introduce unanticipated bias into the process of species design.

The writing and rewriting of the genomes of species may, however, paradoxically serve as a means to safeguard humanity from the potentially even greater existential risks posed by the anticipated realisation of Artificial Super Intelligence (ASI) – a hypothetical, but likely attainable form of artificial intelligence that may eventually surpass natural human intelligence.

With the unprecedented power that ABI might potentially convey comes great responsibility. The prospect of the emergence of ABI compels us to confront profound questions about humankind's future. What kind of world do we wish to create and live in, and at what point does human nature cease to be 'natural'? How, furthermore, can we ensure that these technologies would be used safely, ethically and wisely?

This book invites readers to explore the new frontier of generative biology and the potential of ABI to design and create new species. It offers a glimpse of the potentially disruptive power of these technologies, as well as the challenges, risks and ethical dilemmas that we will be forced to address as we navigate their extraordinary possibilities. Facilitating this critical debate will help ensure that, if realised, ABI will impact society in a beneficial manner. Given the exponential rate at which this field is developing, I have attempted, where possible, to explore the timeless principles that would be expected to endure as our knowledge continues to expand.

My aim is to provide non-specialist readers with the background necessary to engage with these questions and form their own opinions about humankind's future. To that end, I have included an outline manifesto for life, which highlights the significance of ABI as both a unifying principle and a landmark in human progress. It also suggests some guiding principles that may help ensure that, if attained, ABI is used responsibly and for the common good.

Any such manifesto must, by necessity, be configured in a manner that delivers justice, well-being, and social benefit equitably to all humankind, while preserving the diversity of natural species and safeguarding the essential features of authentic human nature – including free will and individual liberty.

We are, as a species, about to embark on an extraordinary adventure as we journey into life's unknown future. It is my hope that this book will serve both as a trusted guide and a valued companion, equally well suited to the cautious traveller and to the enthusiastic advocate eager to navigate its uncharted landscapes.

Introduction

> The free, exploring mind of the individual human is the most valuable thing in the world. And this I would fight for: the freedom of the mind to take any direction it wishes, undirected.
>
> – JOHN STEINBECK, 1952

The son of an illiterate Lithuanian cobbler, Sydney Brenner possessed a sparkling intellect that he applied liberally to almost any topic. He was one of a handful of individuals working in Cambridge who pioneered molecular biology – a field of science that would revolutionise our understanding of living organisms. This select group believed that the intricate structures and behaviours of living things could be computed from their molecular components. Complex wholes could be understood from the functioning of their individual parts.

Brenner's discoveries helped establish the relationship between genes and proteins and played a key role in deciphering the genetic code – the set of instructions translating the one-dimensional building-block sequences of DNA into the one-dimensional amino acid sequences of proteins. This near-universal code defined the basic grammar of the language of genes. Molecular genetics rapidly became both a powerful scientific method and a compelling linguistic framework for articulating the nature of biological systems.

Having helped solve one of molecular biology's most significant puzzles, Brenner turned his relentless curiosity to an even greater challenge – how do genes build humans and other living things? He aimed to uncover the molecular basis of biological complexity.

To do that, he would need to learn the language of genomes. If he were successful, it would facilitate the comprehension of life as we know it. It would also open up possibilities beyond natural biology.

Unlike inanimate objects, every organism carries an internal description of itself within its genome. Genomes are, in this sense, compressed molecular descriptions of living beings. Yet the most pressing question remained unsolved: how the one-dimensional genetic sequences give rise to the three-dimensional features of multicellular organisms.

To explore this, and recognising that the biological principles underlying the patterning of simple organisms could illuminate the generative principles of more complex ones, Brenner selected the most straightforward animal he could find – a tiny and apparently inconsequential nematode worm called *Caenorhabditis elegans*. Transparent and just one millimetre long, these creatures inhabited the two-dimensional world of laboratory agar plates, where they grazed on bacteria and reproduced in only three days.

By studying naturally occurring mutants, Brenner showed how a worm builds its brain from just 302 neurons. Remarkably, many of the genes that shaped the worm's neural architecture had counterparts in humans. This suggested that, to some extent, worms and humans constructed their brains using a similar genetic toolkit. The insights gleaned by Brenner from his work on this unassuming creature eventually earned him a Nobel Prize.

I was fortunate to attend one of Brenner's lectures at the Cambridge University School of Clinical Medicine. He bristled with energy and enthusiasm. As he delivered his talk, 'The Human Gene Kit', to a captivated audience, he drew out each word with his characteristic irony, wit, and contrarianism. Delivered just two years before the launch of the Human Genome Project on 1 October 1990, the presentation was both pragmatic and provocative.

After extolling the virtues of sequencing the human genome, Brenner unexpectedly changed tack. He began to argue against commencing the project. It would, he reasoned, be an inelegant and wasteful exercise, requiring vast factories of sequencing machines, years of labour, and billions of dollars. The technology, he insisted, was not at that time sufficient for such an undertaking. Its efficiency needed to improve by a factor of at least 1,000. He then surprised the audience by proposing that the project be delayed until a faster and more efficient method became available.

His foresight, as usual, proved prescient. The publicly funded Human Genome Project led by Francis Collins at the National Institutes of Health (NIH) in Bethesda, Maryland, ultimately took thirteen years to complete and cost around three billion dollars. It proceeded despite his objections. Yet, notwithstanding his reservations about its timing, Brenner never doubted the project's significance. To illustrate this, he offered a simple analogy.

Genomes, he suggested, were analogous to Airfix model aeroplane kits, comprising plastic pieces attached to a frame, which are removed and assembled according to the manufacturer's instructions. Sequencing a genome, he suggested, was akin to producing an inventory of the kit's components. Once the parts list – the 'human gene kit' – had been compiled, the mystery of human life would be revealed. Organisms could at that point be understood by mapping their complexity back to their components. The causality would have nowhere to hide. Without this knowledge, the human genome would continue to remain an incomprehensible black box.

One of the surprises revealed following the completion of the Human Genome Project was that, rather than the anticipated 100,000 genes, the human genome contained just 20,000 – a paltry number and half as many as a loose-leaf lettuce. Even the seemingly insignificant water flea, *Daphnia pulex*, had managed to muster 30,000. The human genome's list of protein-encoding genes was shockingly brief. This suggested that a component of human complexity might lie elsewhere – embedded not in genes, but in the genome's enigmatic non-coding regions.

Identifying the parts of the kit was one thing, but knowing how to assemble them, quite another. Unlike the manufacturers of model aeroplane kits, evolution had not been so kind as to supply genomes with accompanying instruction manuals. The information required to build and operate a human – like other species – was instead scrawled incomprehensibly across their genome sequences. While some of this information was associated with protein-encoding genes, much of it was not. The human instruction manual – the blueprint of life – was located not just in the parts, but also in the non-coding 'genomic frame'.

As a result of the convergence of two transformative technologies – AI-informed genome design and genome synthesis – we now stand at the brink of defining the elusive instruction manuals of organisms from the coding and non-coding sequences of their genomes.

While effortlessly deciphered by cells, the complex layers of higher-order narrative encrypted in combinations of the four-lettered A, T, C and G chemical vocabulary of biological languages, remains elusive to humans. To decode the human genome's instruction manual – the information detailing our assembly and operation – would require sophisticated new approaches. Brenner understood that sequencing the human genome was the necessary first step in defining this 'code within a code'. But the key question was how this could be done most efficiently.

Brenner viewed the painstaking task of sequencing a genome as being similar to paying income tax – if there were a legitimate way to avoid it, one should endeavour to do so. He identified a vertebrate species – the Japanese pufferfish, *Fugu rubripes* – that had a gene repertoire similar to that of humans. Its genome, however, was 7.5 times smaller than that of humans and just four times larger than that of *Caenorhabditis elegans*. Stripped of its excess baggage, this creature's compact genome was a compelling surrogate for the human genome, offering a potential shortcut to unravelling the mysteries of vertebrate existence.

The reason pufferfish genomes were so much smaller than those of humans was due to their relative lack of non-coding sequences – dismissed by Brenner as genetic 'junk'. In failing to acknowledge the significance of non-coding DNA to human genome function, Brenner made one of the few miscalculations of his career, but perhaps his most consequential. Today, we know that non-coding sequences play a key role in regulating genomes. While Brenner's concept of a 'gene kit' emphasised the importance of the genetic parts, it did not account for the instruction manual necessary to assemble and operate them, which is largely inscribed in the non-coding sequences of the genome.

Although non-coding DNA doesn't encode proteins, it makes RNA molecules that are chemically similar to DNA. But whereas the 'messenger' RNA molecules made by genes act as templates for building proteins, the 'regulatory' RNAs produced by non-coding

DNA help define the grammar of DNA's language. They do this by coordinating gene activity. Other types of RNA made by non-coding DNA form structural scaffolds. The 'ribosomal' RNA found in ribosomes, for example, is a core component of these micromachines, which decipher the information in messenger RNAs and translate it into proteins.

While the number and types of genes in the human and pufferfish genomes are strikingly similar, what distinguishes them is not simply their inventory of genetic components, but the way in which their respective gene activities are orchestrated by the RNA molecules produced by their non-coding DNA sequences.

Protein-encoding genes specify the molecular machinery for life. They build physical structures and direct the biochemical reactions that help organise organisms into their individual natures. The regulatory sequences within non-coding regions, on the other hand, form a critical part of the human genome's instruction manual. Without them, we might have ended up as pufferfish, spending our lives submerged in seagrass beds and coral reefs rather than walking upright and pondering our origins.

But what was once dismissed as 'junk DNA' turned out to be a treasure trove of information. Its cryptic sequences detail key components of the blueprints of species. Through the elaboration of non-coding DNA and the invention of a relatively small number of new genes, natural evolutionary engineering was able to transform the *Fugu* gene kit into one that can build a human being.

Brenner proposed a thought experiment illustrating the challenges involved in pinpointing the genetic sequences responsible for species differences. He imagined two closely related hominids walking into a doctor's office. One looked like you or me. The other, afflicted by a peculiar condition, 'looked like a chimpanzee'. Distressed by its inconvenient predicament, the chimpanzee asked the doctor which genes would need to be altered in order to address its terrible condition and restore its nature to that of a human.

This question cuts to the core of biology's greatest enigma: the sequence changes responsible for transforming one species into another. Are we able, for example, to identify the sequences that differentiate a chimpanzee from a human? If so, we would have cracked the secret of life's language. This might enable us to specify the changes that enabled modern human characteristics to be derived from the last common ancestor of humans and chimpanzees.

But constructing the analysis in this way risks oversimplification. It assumes that the uniqueness of humans arises from genes that chimpanzees lack. The notion that each characteristic of a species is controlled by a single gene is, furthermore, in many cases overly simplistic. The reality is more complicated. Consider language. While it may be tempting to imagine that there is a single 'language gene' lurking somewhere within the human genome – thoughtlessly omitted by evolution from the genome of chimpanzees – Brenner suggested that the hypothetical 'language gene' might actually be present in both species.

Language is not a single trait but an emergent phenomenon. It required the coordinated evolution of multiple anatomical structures, including the tongue, pharynx, larynx, vocal cords, and brain. No single gene could impart the capacity for speech. However, once speech had evolved, it's easy to see how a unitary event in the chimpanzee genome could have suppressed it. Perhaps, Brenner quipped, chimpanzees had at some point in their past decided that speaking could get them into trouble, and so evolved a 'language suppressor' gene.

While some traits are governed by single genes, many arise as a result of activity within networks of interacting genes. Individual genes cannot account for complex phenomena like intelligence and consciousness. Nor can the presence of a single faulty gene explain complex diseases, which are caused by small contributions from collections of genes. To understand how species are built, we must learn how genomes generate such phenomena through the coordinated activity of multiple genes. It is not single genes, but rather 'genetic symphonies' that give rise to much of life's complexity.

During the First World War, an English soldier called Hugh Lofting was stationed in Flanders, Belgium. He was badly wounded by shrapnel from a grenade and was sent home to recover. While in the trenches, he wrote a series of illustrated letters to his children. The news of the war had been 'too horrible or too dull' to share, and he became struck by the role that animals played in warfare, filling his letters with fictional animal stories. Observing horses at first hand in wartime, he saw that they suffered and were exposed to the same dangers as soldiers. But unlike humans, when wounded, they received little care. Lofting believed that to address this, and perform surgery on injured horses most optimally, it would be necessary to acquire 'a knowledge of horse language'.

His letters and reflections led, in 1920, to the publication of Lofting's first book, *The Story of Doctor Dolittle*. It recounted the tale of one Dr John Dolittle, a compassionate and eccentric country physician. Preferring the company of animals over that of humans, he learned to communicate with them using their own languages, conversing with creatures as varied as ducks, monkeys, owls – and even a fictional beast known as a pushmi-pullyu. He was taught animal languages by a parrot called Polynesia, who revealed that each species had its own unique way of speaking. If only humans would bother to pay sufficient attention, they would learn to understand them.

While Lofting's tale was a whimsical fantasy, Sydney Brenner realised that it might actually be possible to learn to comprehend – and to 'speak' – the language of animal genomes. This universal molecular dialect would encompass the genetic languages of all biological entities on Earth, from the simplest viruses and bacteria to plants and animals. A kind of biological Esperanto. Were this possible, we would be able to become Dr Dolittles of biology. To achieve this, however, much like learning any foreign language, from German to Bahasa Indonesia, we would need to discern the biological equivalents of phonology, vocabulary, grammar, and semantics. Only then would we be able to unlock the secrets of genomes. At that point, biology's defining theme would shift towards how to use the generative power conveyed by the mastery of the language of genomes for the benefit of humankind.

Rather than producing spoken words or sentences, articulations in the language of genomes manifest as genetic texts – sequences describing the traits of living organisms. Alterations to these can change an organism's characteristics. This may result in beneficial features or disease, depending on whether the language is constructed grammatically or ungrammatically. The genome's grammar embodies the rules governing how the scripted components of organisms behave. One important aspect of this is syntax – the rules governing the structure of genetic 'sentences'. Like spoken language, the meaning of genetic messages is context-dependent. The same protein may, for example, have different roles depending on the setting in which it finds itself.

Yet, decoding a linguistic system as complex as the language of genomes lies far beyond the capabilities of philologists and language scholars alone. It requires the use of artificial intelligence, which can identify patterns in vast datasets that are otherwise invisible to the human mind. AI can help construct models of how the genome operates. It can also teach us how to author new genetic scripts.

The combination of AI and genome synthesis will eventually allow the language of DNA to be fully comprehended and, in so doing, will convey the ability to converse with the genomes of all species – from daisies and humans to honeybees and tortoises. Like Lofting's Dr Dolittle, who learned to communicate with animals through careful observation, we will need to listen attentively to an exhaustive and eclectic cacophony of genetic 'sounds' to unpick their underlying logical patterns.

We may then one day learn how to build mermaids and fairies, and to define the boundaries of what the language of natural biology can and cannot articulate. Sydney Brenner imagined how a student in 2070 might be tasked with describing 'the genetic programme to make a centaur', or to 'prove the impossibility of a genetic programme that would make a centaur'. The mastery of life's biological languages would transform biology into a mechanistic discipline, capable of deterministic and predictive outcomes – something currently confined to the realm of physics.

But to comprehend the language of human beings, it is not the human genome that we need to understand so much as 'humanity's genome'. Brenner argued that it was necessary to examine 'ordinary people – with no hypothesis – in large numbers'. Sequencing the genomes of a broad swathe of humanity would provide a basis for comprehending how genomes generate both the commonalities of human traits and individual diversity.

We continue to accumulate vast quantities of data, yet our ability to interpret and translate it into actionable knowledge remains a challenge. The language of humanity's genome is embedded within the totality of its sequences. We need to find a way to extract its meaning – a task to which artificial intelligence is ideally suited. Much in the same way that DNA sequencing helped us to learn the language of genes, AI promises to be instrumental in teaching us the language of genomes.

The key distinction between biology and physics is that living organisms contain encoded information. If we can learn how the information encoded in DNA's molecular language is converted into the characteristics of living organisms, we will have solved biology's central problem – how genome sequences map to the form and function of organisms. The underlying framework for this analysis is based on the notion that living organisms are machines operating through biological computation. As we will see, biological machines work in ways that are distinct from human-made machines, but the metaphor is nevertheless useful.

It was René Descartes, the seventeenth-century French philosopher, scientist and mathematician, who first suggested in *Discours de la méthode* (Discourse on the Method, 1637) that the behaviour of non-human animals could be understood as being mechanical. He concluded that animals were essentially 'animal machines' (*bête-machines*): complex assemblies of mechanical parts, and devoid of mind.

In *Meditationes de prima philosophia* (Meditations on First Philosophy, 1641), Descartes extended this idea to humans, who, like animals, were perceived as being machine-like in their design. However, he added in a crucial distinction – an immaterial soul or mind that was annexed to the human body through an organ in the brain called the pineal gland. Human machines were thus a composite of body and mind, which were distinct, but closely entwined. The pineal formed the point of connection between mind and matter. The human machine was a fusion of body and soul. Descartes believed that this insight paved the way for a new and 'infallible' form of medicine. 'Human life' could, furthermore, 'be prolonged' if we were able to define 'the appropriate art' for understanding human nature.

A little over a century later, in 1747, Julien Offray de La Mettrie, in *L'homme machine* (Man a Machine), refined Descartes' ideas to contrive the modern, scientific view of human nature. He rejected the notion of an immaterial soul, asserting instead that all aspects of human behaviour, including thought and consciousness, were the products of the human machine. The mind was simply an extension of the body's mechanistic function.

Were the human body constructed like human-made machines – say, a plane or a television, for example – engineering it would be straightforward, as would treating disease. But while some of the parts of humans resemble the 'orthogonal' components of human-made machines – in which changes to one do not impact the others and each is independently repairable or replaceable – this is the exception rather than the rule. Most functions in the human body, and their corresponding components, are in fact interdependent, overlapping and interconnected.

The future of generative biology will likely rely on efforts that make biological systems more modular and predictable, so as to more closely resemble human-made machines. This will involve reducing their complexity, streamlining their designs, and disentangling any interdependencies between their components to increase their engineerability. While the human body-as-a-machine remains a useful metaphor, biology deviates from textbook engineering principles in many ways.

The messy architecture of natural designs remains one of the most substantial obstacles to medical progress and our ability to redesign nature. Even the simplest viral genomes contain gratuitously intricate features serving no obvious function. To understand how biological systems came to be so complex, we must turn to the second theoretical foundation underpinning our efforts to reconfigure nature: the notion that cells are biological computers.

On 8 June 1954, the mathematician Alan Turing was found unresponsive by his housekeeper in Manchester, with an apple laced with cyanide by his bedside. Turing had played a pivotal role at Bletchley Park during the Second World War, leading the team that cracked the German Enigma code and helped alter the course of the war. Yet it is to an earlier personal tragedy that we must turn to trace the roots of

Turing's other monumental contribution – the idea that a machine could be capable of thought. The untimely death of Turing's closest childhood friend Christopher Morcom – a brilliant pianist and mathematician – contributed to his early fascination with cryptography and the possibility of creating an electronic mind.

Turing's ideas on computing machines began to crystallise in his 1936 paper, 'On Computable Numbers, with an Application to the Entscheidungsproblem'. In it, he described a theoretical model of computation – later known as the Turing machine – that would form the inspiration and logical basis for the modern, general-purpose, programmable computer.

Turing envisaged an infinite tape, digitally encoded with a set of instructions that functioned as memory – what we now recognise as software. The tape was fed into a machine – which today we would call hardware – that systematically scanned it, reading and writing symbols according to a predefined set of rules. In its original form, each Turing machine was built for a specific task, its operational rules hardwired into its design. To perform a different computation, one would need to construct an entirely new machine, tailored to a new set of rules.

In 1947, Turing expanded his ideas, proposing a more versatile version of his original machine, which became known as the Universal Turing Machine. Unlike the original, task-specific version, the universal machine was general-purpose, and capable of executing any computation that could be described by the instructions encoded on the input tape. In essence, it could compute anything that any other machine could compute.

Inspired by Turing's work, the mathematician and engineer John von Neumann transformed the abstract concept of universal computation into a tangible reality. In his 1948 paper 'The Theory of Self-Reproducing Automata', von Neumann introduced the idea of a universal constructor. This was a self-reproducing machine, the prototype of an actual physical device, which was capable of reading a blueprint of itself stored as a set of instructions describing its design, and then using these to build a physical copy of itself. This self-replicating machine could also construct any other machine describable in code. Notably, while the instructions described the machine's function, they could not themselves perform that function, drawing a sharp distinction between function and execution.

The work of Turing and von Neumann laid the foundations for understanding biological systems as programmable computing machines. As Sydney Brenner aptly observed, biological systems are a special kind of computing device. 'The best examples of Turing's and von Neumann's machines' he remarked, 'are to be found in biology.' Biology, he argued, is essentially '(very low energy) physics with computation'. The genomic 'code-scripts', – the digital sequences of nucleotide bases – function as programmes or tapes that symbolically describe the nature of all living organisms. These genomic tapes, Brenner asserted, are 'at the core of everything'. The machinery of the cell, in turn, performs algorithmic processes on these genetic tapes, executing the operations necessary for life.

The idea that biological systems encode information – with genomes functioning like computer programmes to facilitate molecular computing – was first implied, though not explicitly stated, by Francis Crick in his 1957 lecture 'On Protein Synthesis', delivered at University College London. The French biochemist François Jacob later recalled Crick – 'tall, florid', with 'long sideburns' – speaking with 'evident pleasure and volubly, as if he were afraid he would not have enough time to get everything out'. In his lecture, Crick described the one-way flow of genetic information from DNA to the amino acid sequences of proteins, drawing a parallel with how a machine reads and executes a programme. The amino acid sequences of proteins could not be back-translated into the nucleic acid sequences of DNA. The fact that synthesised amino acid sequences can 'do almost anything', and take on an almost infinite number of forms, suggests that cells possess a capacity akin to that of a Universal Turing Machine.

While computer programmes are written in binary code, DNA sequences utilise a quaternary code – comprising sequences of four nucleotide bases – that spells out the genetic instructions for constructing and operating living organisms. These nucleotides, the chemical 'letters' of the biological alphabet, are strung together into long chains that read like the typographical text of a printed book.

The metaphor of biological systems as computers was further refined by François Jacob and Jacques Monod in their 1961 paper on the *lac* operon, a region of bacterial DNA that controls lactose metabolism in *Escherichia coli* (*E. coli*). They described how regulatory genes function like molecular switches that deploy Boolean logic – a set of logical rules connecting true or false conditions. This laid the groundwork for understanding genetic circuits in terms of conditional logic gates, much like those used in electronic devices such as transistors. For example, gene A would activate only if both gene B and gene C were switched on, while gene D would activate if either gene B or gene C were on.

Yet just as the metaphor of biological systems as machines has its limits, so too does that of biological systems as computers, and genomes as programmes. While remaining a useful analogy for modelling their behaviour, biological computing differs significantly from the type of computation that occurs in human-made digital machines. For one thing, in biological systems there is no clear separation between software and hardware. The nucleotide sequences of genomes serve as the 'tape', storing programmatic instructions for building and operating living things. But they simultaneously function as molecular machines, whose dynamic three-dimensional structure regulates gene activity. Moreover, while computers operate digitally, cells rely on analogue-like molecular gradients that influence events probabilistically. Computer programmes run in a context-independent manner, but genetic programmes are often impacted by environmental factors. And while computers focus their computation in central processing units and process information sequentially, cells perform countless biochemical interactions in parallel.

Perhaps the most fundamental difference, however, is that unless explicitly designed to do so, computer programmes do not spontaneously rewrite their own code. The programmatic information in biological systems, however, evolves. Genetic programmes are subject to continuous mutation, recombination, and a host of other processes that modify their information, leading to constant revisions over time. The principal 'author' of these modifications to natural genetic programmes is evolution by natural selection.

Naturally evolved and artificially engineered programmes both share a fundamental design limitation, in that they have a tendency to degrade over time. But while human coders can modify their programmes substantially and precisely at will, or even start again from scratch, evolution is restricted to tinkering with its genetic programmes. Its rewriting strategy is imprecise and incremental. It cannot, furthermore, entirely reimagine the genetic code script of a species, as the heredity mechanism demands that each iteration is firmly rooted in the past.

This inability to effectively rationalise its designs, combined with the fact that natural modifications to the genetic code occur as the result of chance events rather than by intention, means that biological programmes are often convoluted, clumsy, and inefficient. Evolution optimises for survival rather than elegance. Its reliance on ancient designs imposes fundamental constraints on natural systems. Imagine, for example, if the design of modern computers were constrained by the features of an abacus, or if the characteristics of modern laser printers had to be bolted onto the wooden frame, ink roller, and printing plate of a 1440 Gutenberg press.

Computer programmers have a specific term for code that is tangled, unstructured, outdated, and difficult to maintain. They call it spaghetti code. Such code, marked by complex interdependencies and ad hoc fixes, becomes increasingly difficult to modify, extend and maintain without causing unanticipated effects. In short, it's a mess. The same principles apply to biological systems. Evolutionary adaptations, layered upon pre-existing structures, often result in convoluted workarounds rather than in definitive, streamlined solutions.

Without careful maintenance and consistent adherence to engineering principles, most code written by programmers eventually degrades into spaghetti code. To some extent this results from the behaviour of the individuals tasked with writing and repairing it. Instead of getting to the bottom of a problem and addressing the root cause, programmers are prone to improvisation. These expedient patches, while solving immediate problems, often clash with the parent code, creating new layers of complexity and irregularity. Programmers who inherit such code rarely have the time or inclination to go back and fix it. New features are, additionally, often bolted onto existing code without fully considering the implications, extending programmes far beyond the scope of their original design.

In software development, programmers mitigate the inevitable drift towards spaghetti code through a process known as refactoring. This involves rewriting parts of the code to improve its clarity, efficiency and modularity, without changing its behaviour. It is an essential practice that should ideally be performed at regular intervals.

If code is not regularly refactored, it becomes impenetrable, cumbersome, and, in some instances, incomprehensible. At that point, it may be easier to abandon the code and start again. Deciding when to do this requires significant judgement. Trying to untangle and fix inherited problems can be very time-consuming, especially when a system has long outgrown its original design.

Genomes containing the information of natural species were not programmed by humans but by evolution by natural selection. They are the result of natural engineering. Unlike the code crafted by humans, the nucleotide sequences of the genomic 'tapes' of living organisms cannot be systematically refactored. As a result, genomes carry the burden of approximately four billion years of accumulated mutations – a substantive technical debt that has not yet been paid down.

The consequences of this imperfect natural engineering were recognised as early as 1779 by the philosopher David Hume. In *Dialogues Concerning Natural Religion*, he argued that living creatures, including humans, appeared to have been designed by a 'stupid mechanic'. Rather than reflecting the handiwork of a rational designer, the details of their construction suggested the agency of an unguided process marked by complexity, inefficiency, and the use of rudimentary – and in some cases inelegant – solutions. Evidence of this abounds.

Were we to have the luxury of redesigning the human genome from scratch, optimising it for healthy longevity, it is unlikely that any rewrite aligned with modern engineering practices would resemble the genomic architectures generated by Darwinian evolution. The opportunistic and haphazard manner in which evolution has programmed the genomes of living things, while creating extraordinary variety, complexity, and beauty, has left them replete with flaws and jeopardy.

At a time when most people believed that species were created instantaneously as the result of a singular divine act, it would have been unthinkable to imagine that the design of humans – or any other species – could have been configured otherwise. But today we know that evolution does not follow a predetermined script. Rewind the tape of life and let it play again, and the probability of arriving at anatomically and biochemically modern human beings would be vanishingly small. This challenges the idea of human exceptionalism, as well as the idea that all other known species were inevitable. Life, in all its complexity, is as much the product of contingency as it is of natural selection. Without the asteroid impact that wiped out three-quarters of Earth's species approximately 66 million years ago, it is unlikely that humans would be here at all.

In recent years, advances in gene editing and genome engineering have provided us with the unprecedented ability to rewrite genetic code with intent. Gene therapies have already had a profound impact on medicine and society, with multiple treatments approved to repair or replace faulty genetic components. In time, we may be able to cure all 'monogenic' diseases – those caused by a single faulty gene – of which 8,511 had been defined by 2025.

Yet as the capacity to refactor human biology increases, so too does the realisation of an uncomfortable truth – access to these therapies is not evenly distributed. Nearly half the world's population lives in poverty, unable to afford even the most basic healthcare, let alone cutting-edge genetic medicine treatments. The future of genomic medicine – and who benefits from new therapies – may therefore ultimately be determined by economics, rather than by science or the underlying unmet medical needs.

An even greater challenge awaits when we inevitably reach a biological ceiling in the treatment of genetic disease. This is likely to occur once there are effective therapies avaiable for most or all monogenic diseases. But because the majority of human diseases are polygenic – caused by interactions between multiple genes rather than a single mutation – we will still be left managing most of the iceberg. The complexity of polygenic systems will make straightforward genetic fixes more challenging. At that stage, the consequences of the genome's spaghetti code will become fully apparent. The irrational tangled structure of our genomes will present a formidable barrier to the more extensive type of rewriting that may be required to address polygenic disease.

If we project the aims of clinical medicine to its extreme – its grand ambition to eliminate all diseases – we may be confronted with a stark choice. Would we be willing to rewrite parts, or even the entirety of the human genome to achieve medical cures or to secure protection from disease? And if so, might such alterations inadvertently reshape other aspects of human nature? This raises a deeper question: is the human genome computable? If it is not – or only partially so – there may be no reliable way to predict the impact of complex genetic modifications. In that case, the only way to test a biological programme would be to synthesise the candidate genome and 'run' the resulting organism.

However much we may wish to refactor the human genome according to best engineering practices in order to achieve healthy longevity, natural entanglements are likely to impose substantive limits on the extent to which different systems and subsystems can be untangled. The most important question, then, is not whether we can free ourselves from disease, but the potential risk and cost we are willing to accept to do so.

The consequences of the potential ability to programme and reprogramme biological systems extend far beyond healthcare. If we succeed in learning the language of genomes, we will gain the power to reshape the natural world. Yet even as we acquire this capability, we are simultaneously destroying it – undermining ecosystems and driving countless species to extinction. Teetering on the edge of ecological collapse and irreversible climate change, we must turn away from the destructive technologies of the industrial age and invent and adopt sustainable biological alternatives. Advances in engineering biology could bring enormous benefits to humankind, addressing urgent global issues such as helping to attain global food security and to prevent, or even reverse, environmental degradation.

With the emerging ability to programme genomes, it should be possible to meet most of humanity's needs through biologically inspired designs. To ensure this future is sustainable, we must also protect remaining wilderness, rewild the Earth, and preserve natural species. If we engineer life in a measured and responsible manner, these goals need not be in conflict. Eventually, Darwinian evolution will no longer be the sole force driving genetic programming. Naturally evolved species will be joined by artificial ones, as humans become programmers of life itself. This new artificial world of designed, synthetic organisms could exist alongside, but firewalled from, the natural world – two separate systems, aligned, interdependent, and mutually supportive.

It is a curious quirk of history that the two key technologies driving today's extraordinary biological revolution – AI-augmented genome design and the ability to synthesise entire genomes – can trace their origins to unlikely places: in the case of AI, to the dingy video game arcades of the late 1970s; and in the case of genome synthesis, to a small, impoverished village in British-ruled India.

Indeed, without the public's enthusiasm for video games – particularly the blockbuster *Space Invaders* – we might never have developed graphics processing units (GPUs). Originally created to satisfy gamers' desire for increasingly sophisticated video graphics, GPUs turned out to be perfectly suited for the parallel computations required by modern AI. Their ability to process vast amounts of data made them a foundational platform for the deep learning algorithms that will help AI to decode the language of genomes.

Similarly, without the vision and dogged persistence of the Indian-born American biochemist Har Gobind Khorana, we might never have chemically synthesised the first artificial gene – a breakthrough that laid the groundwork for writing synthetic genomes. Even in 1968, when gene synthesis technology was still in its infancy, Khorana foresaw its potential dangers. 'In the distant future,' he warned, 'when all this comes to pass, the temptation to change our biology will be very strong.' That future is no longer distant.

The question is no longer whether we can redesign and rewrite life, but when – and whether we should. What limits and safeguards should we place on these activities to protect not only humanity and other natural species, but the very fabric of human nature itself: free will, conscience, empathy, morality, and intellectual liberty?

We are living at a remarkable moment in history, where the convergence of AI and synthetic genomics – products of human ingenuity – is poised to close the loop between genetics and culture. Heredity, long the dominant force of biological nature, is now being directly challenged, placing the relevance of biological history at risk.

It is time for humanity to initiate a debate, to define what it means to be human, and to decide how we will move forward safely, ethically and responsibly as a species. In doing so, we must not forget our obligation to protect and preserve Earth's fragile species who will accompany us on this uncertain journey. We cannot truly be human without them.

1
The Language of Life

By manipulating DNA, we create new species at will.

— ARTHUR KORNBERG, 1987

Confined to a sanatorium in the tiny village of Davos, perched among the sweeping valleys and dramatic peaks of the eastern Swiss Alps, the 51-year-old Swiss chemist Friedrich 'Fritz' Miescher, incapable of relaxing even as his life drew rapidly to its premature conclusion, began to investigate the effects of high altitude on red blood cell counts. But his weakening body soon failed him, and on 26 August 1895, Miescher succumbed to disseminated tuberculosis. He left behind a disordered collection of unpublished laboratory notes and unfinished manuscripts that failed to meet his exacting standards. These scattered documents, alongside his modest portfolio of published papers, would form the foundations of the science of genetic chemistry: a body of knowledge that would detail the chemical nature of the hereditary material of all life on Earth.

Miescher's devotion to science bordered on obsession. He reportedly almost missed his own wedding due to his desire to complete an experiment. When his laboratory ran out of glass vessels, he brought in his Sèvres porcelain dinnerware. One can only imagine his frustration as he witnessed his life being undermined by the invisible genetic instructions carried within the microorganisms multiplying inside his cells, long before he could uncover the generative secrets of the remarkable chemical substance he had discovered.

Miescher himself was an unlikely scientific hero. Hard of hearing, myopic, fidgety, insecure, restless and introverted, his life had focused on a single problem of profound significance: the chemical nature of the cell, and in particular the chemistry of the tiny organelle known as the nucleus, found in the cells of every complex organism. The German biologist Ernst Haeckel had, in his 1866 book *Generelle Morphologie der Organismen* (General Morphology of Organisms), correctly speculated that the nucleus might contain the hereditary material – but no one had proved it.

Miescher was convinced that it was the methods of chemistry – or rather the cell staining and microscopy practices of his contemporaries – that would ultimately unravel the basis of heredity. Chemistry, in his view, was more fundamental than cell structure; it was the key to life itself. Yet the precise nature of the hereditary material had proved elusive.

After completing medical school in Basel, and encouraged by his enthusiastic uncle, the eminent anatomist Wilhelm His, Miescher persuaded Felix Hoppe-Seyler – the rising star and pioneer of the nascent field of chemical biology, which we today call biochemistry – to take him on as a research student. Working in Hoppe-Seyler's renowned *Schlosslaboratorium*, housed in the medieval fortress of Schloss Hohentübingen, overlooking the Neckar River in Tübingen, Germany, Miescher entered a rarified intellectual hub, which held the view that chemistry was essential to life processes. In the autumn of 1868, he embarked on his quest to determine the chemical composition of the nucleus. Like many others, he assumed the nucleus would contain a novel type of protein – at the time, the only class of biomolecule thought to have the necessary complexity to serve as the basis of heredity.

Using pus extracted from bandages gathered at a local hospital clinic, Miescher devised a method for separating the nuclei of white blood cells from the surrounding cytoplasm. Washed from the pus, the cells formed a sediment that could easily be separated from the residue. After collecting the cells, the nuclei were then isolated by digesting the surrounding connective tissue with enzymes from a pig's stomach, which destroyed the remaining proteins. They were then extracted with acid and alkali. To his surprise, the addition of acid to the purified nuclei produced a precipitate. This dissolved in alkali and contained large amounts of phosphorous, but no sulphur, the chemical hallmark of proteins.

In a letter to his uncle dated 26 February 1869, Miescher noted that he had decided to name this unusual nuclear material – whose composition was unlike any previously identified chemical substance – nuclein. In just a few months, and with minimal formal training, he had single-handedly isolated the hereditary material of life. Today we know it as deoxyribonucleic acid, or DNA – the molecule in which the biological information of all species is written. The repository of life's encoded secrets. The paper detailing the results, '*Ueber die chemische Zusammensetzung der Eiterzellen*' (On the Chemical Composition of Pus Cells), was published in 1871.

Unfortunately, Miescher's idea that the nucleus housed a unique chemical substance was met with derision and vigorously contested. In a letter to Hoppe-Seyler in 1872, Miescher expressed his regret at publishing his findings and exposing himself to 'the mob'. Discouraged by the unfavourable reception, he retreated to the comfort of his home town of Basel, and turned his attention to a less controversial subject: the transformations occurring in the bodies of salmon as they migrated up the Rhine from the ocean. Yet he did not entirely abandon his work on the hereditary material. He continued to refine his methods for isolating DNA, though he chose not to publish further on the mysterious nuclear substance he had discovered.

Although he stopped short of concluding that the substance he had identified was the hereditary material – still believing that only proteins had the necessary complexity to perform such a function – Miescher was the first to isolate and describe DNA. By demonstrating that it could not, like proteins, pass through a parchment filter, he also became the first to characterise it. He showed that the size and complexity of nuclein far exceeded that of proteins. DNA molecules were gigantic.

Having unwittingly discovered the chemical nature of the hereditary substance, Miescher felt compelled to speculate on how the hereditary information of each species might be uniquely represented within its chemical structures. In doing so, he made another significant, although at the time wholly unappreciated, contribution to the history of genetics. This was detailed in a letter to his uncle written in December 1892, more than two decades after his discovery of DNA and just a few years prior to his death.

In the letter, Miescher conjectured that the information of living things was encoded in chemical structures, much as alphabetic letter combinations spell out the meanings of words. 'Just as all the words and concepts of all languages can find expression in twenty-four to thirty alphabetic letters,' he wrote, so too could the 'wealth and variety of hereditary transmissions' also 'find expression'. A finite number of components in 'chemical structures' could, he suggested – like the letters and words of a language – be arranged in endless combinations specifying an infinite number of biological utterances. He had inferred that genetic information was encoded.

Living organisms, he proposed, were governed by a chemical grammar analogous to that of written languages. This generative grammar specified the meaning of the biological instructions coded within chemical text. The ability to represent biological information in chemical notation – like the ghosts materialising to Ebenezer Scrooge in Charles Dickens's *A Christmas Carol* (1843) – conferred upon the heredity mechanism the capacity to encode the information of all past, present and future organisms. If the grammar governing this chemical language of life could be comprehended, then humankind might one day author its own genetic narratives.

Miescher's discovery was, in this sense, more than just a milestone in our understanding of the chemistry of heredity. It was the opening chapter of a story that would culminate in Artificial Biological Intelligence: the ability to use AI and genome-writing technologies to design and construct natural and artificial synthetic genomes.

On the night of 17 April 1941 – nearly half a century after Miescher's death in Davos – Frederick Griffith, the newly appointed director of the Emergency Public Health Laboratory Service, met a tragic end. A German air raid struck his building in Russell Square, London. The work performed by this reclusive scientist, who had previously enjoyed walking his dog on the Sussex Downs and skiing in the Alps, would play a pivotal role in bringing the chemistry of the hereditary material to the attention of the wider scientific community, who – for the greater part – remained unaware of Miescher's pioneering discoveries.

Griffith's painstaking, decades-long cataloguing of different strains of *Streptococcus pneumoniae* – the commonest bacterial cause of lobar pneumonia – provided the perfect training for the experiment that would secure his place in scientific history. His 1928 paper 'The Significance of Pneumococcal Types' described a startling observation: a harmless strain of *S. pneumoniae* could be transformed into a virulent one simply by mixing it with heat-inactivated virulent bacteria and injecting the combination into healthy mice.

This remarkable transformation implied that a chemical factor within the dead bacteria had reprogrammed the live strain, permanently altering its nature. At the time, the profound implications of this for understanding the chemical basis of heredity were barely appreciated. Yet while Griffith's work was motivated by the need to treat lobar pneumonia – an often fatal disease in the pre-antibiotic era – his experiment revealed something extraordinary and of general significance. It suggested that the hereditary material functioned not just as a passive archive of biological information, but as dynamic programme that was transferable between cells, and capable of altering the behaviour of their biological machinery beyond the point of its origin.

The next major breakthrough came from an unlikely source. Socially charming, articulate and sharp-witted, Oswald Avery – the son of a Baptist missionary to New York's Bowery district, an area once notorious for its flophouses, saloons and brothels – was, like Griffith, captivated by pneumococci. As with many of his peers, Avery was initially sceptical of the so-called 'Griffith experiment', even after other laboratories had confirmed the result. Yet Avery realised that, if genuine, the phenomenon offered an unprecedented opportunity to define the chemical basis of the mysterious 'transforming principle'.

Working at the Rockefeller Institute in Manhattan alongside his colleagues Colin MacLeod and Maclyn McCarty, Avery made a startling revelation: the chemical basis of the transforming principle was DNA – not protein.

Their 1944 paper should have shaken biology to its core. Instead, it was met with incredulity. The prevailing view continued to insist that only proteins had the structural complexity necessary to carry hereditary information. DNA, in contrast – with its repetitive chain of just four different chemical nucleotides – appeared far too simple to encode life's vast complexity. Miescher's early vision of biological cryptography – the idea that life's information was written in code – had been forgotten.

It wasn't until 1951, when the quietly spoken Alfred 'Al' Hershey and his knitting-obsessed assistant, Martha Chase, completed their experiments at the Cold Spring Harbor Laboratory on Long Island, that the scientific community began to be convinced. Using bacteriophage T2, a virus that infects *E. coli*, they devised an ingenious approach that enabled them to track the separate movements of DNA and protein during viral infection. The simple structure of bacteriophage T2 consisted of DNA packaged inside a protein coat. One or the other had to be responsible for the virus's ability to reprogramme bacterial cells following infection. All they needed to do was trace where each component ended up.

By tagging the DNA and proteins with distinct radioactive isotopes – phosphorus for DNA and sulphur for proteins – Hershey and Chase were able to determine which of the two components entered the bacterial cells following infection. They mixed the virus particles with bacteria, and once the viruses had attached to the bacteria, a blender was used to sheer off the viral shells attached to the outside of the infected bacteria. The mixture was then centrifuged to separate the viral shells from the bacteria, and each was then separately analysed.

The results were striking. The radioactive tag recovered from the viral shells was found to be rich in sulphur, but contained little or no phosphorus. As sulphur is found in proteins but not DNA, this indicated that the viruses had injected their DNA into the bacteria, leaving their protein coats behind. Meanwhile, the presence of radioactive phosphorous inside the bacteria with only a trace of radioactive sulphur confirmed that it was viral DNA rather than protein that had been injected into the bacterial host cells, as phosphorous is found in DNA but not in proteins.

Hershey and Chase cautiously concluded that DNA 'had some function' in the growth of phage within bacteria. But the broader implications were clear. The Hershey–Chase experiments provided compelling evidence that it was DNA – rather than protein – that formed the chemical basis of the hereditary material. But it was still far from clear that DNA formed the basis of the genetic material for all living things.

It was the elucidation of DNA's three-dimensional structure that confirmed its likely role as the carrier of hereditary information and suggested how this information was faithfully replicated. This discovery instigated a sustained effort to confirm DNA's role as the hereditary material, and to determine the mechanism by which the chemical information in DNA coded for the amino acid sequences of proteins. This landmark achievement was the result of the combined efforts of an eclectic and sometimes confrontational group of larger-than-life figures who adopted complementary approaches to address the same problem. Scattered between London and Cambridge, they transformed biology into an informational science, and in so doing, established the foundations of modern genetics.

In the aftermath of the Second World War, Maurice Wilkins – who had helped fractionate uranium for the Manhattan Project at Los Alamos – found himself at a loose end. He soon joined John Randall's laboratory at the University of St Andrews in Scotland. Randall, also a wartime veteran, had helped invent airborne radar, a technology that profoundly shaped the Allied war effort. When Randall moved to King's College London, he brought Wilkins with him. There, Wilkins became part of an interdisciplinary band of scientists that was referred to by some of their Cambridge colleagues at the time as 'Randall's Circus'. This eclectic and at times dysfunctional group of individuals dedicated itself to applying the methods of physics to biology's greatest challenges, including the enigma of heredity – whose secrets, Randall believed, were locked within the structure of DNA.

Inspired by the pioneering work of William Astbury and Florence Bell performed at the University of Leeds in the late 1930s – whose X-ray photographs first revealed that DNA had a regular, ordered structure – Maurice Wilkins decided to try his own hand at photographing DNA. X-ray crystallography, at the time a powerful new technique, allowed scientists to infer the positions of individual atoms in complex molecules by analysing the diffraction patterns produced as the X-rays scattered off them. Astbury and Bell's early photographs had already hinted that DNA's nucleotides were strung together like beads on a necklace, stacked 'on top of one another like a great pile of plates'.

With characteristic wartime resourcefulness, Wilkins cobbled together an X-ray camera from surplus military parts. Alongside his

PhD student Raymond Gosling, he set about capturing images of DNA using Astbury and Bell's pioneering method. The DNA fibres Gosling worked with – 'drawn out like a filament of spider's web' by Wilkins's artful use of a paper clip – had been provided by visiting Swiss biochemist Rudolf Signer, who purified them from a calf thymus.

After several frustrating attempts, in 1950 Gosling finally produced a DNA fibre preparation of DNA that yielded a striking X-ray image. Soon after, another team member, Alec Stokes, managed – while on the train home from work – to deduce from the criss-crossing and concentric black markings that DNA must have a helical structure.

It seemed only a matter of time before the precise structure of DNA would be revealed. Yet despite such early triumphs, John Randall remained sceptical. To his eye, the fuzzy images produced by Gosling and Wilkins were inadequate for the mammoth task of determining DNA's architecture. Nor was he convinced that they had the necessary skills to interpret them. If the structure of DNA was to be solved, a team with deeper expertise would be required.

Without informing Wilkins, John Randall made a pivotal decision. In January 1951, he recruited the gifted crystallographer Rosalind Franklin to the laboratory, reassigning the DNA project to her and placing Raymond Gosling under her supervision. Unsurprisingly, this arrangement generated significant tension between Wilkins and Franklin, which was exacerbated by the fact that King's did not allow female faculty members to dine with their male colleagues at lunchtime – an important forum for scientific exchange.

Franklin and Gosling swiftly set to work, producing a beautiful series of high-resolution X-ray photographs. Franklin meanwhile immersed herself in the mathematical challenge of mapping the obscure markings in the diffraction patterns onto the positions of individual DNA atoms. She soon showed that DNA existed in one of two interchangeable forms, labelled 'A' and 'B'. While the A form yielded sharper photographic images, it was the partially-crystalline hydrated B form that was biologically relevant. Although blurred by water molecules, which distorted the DNA, the images from the B DNA proved much easier to interpret.

Among the photographs taken by Franklin, Photo 51, captured in May 1952, turned out to be especially clear and consequential. It helped her to conclude that the B form of DNA was helical. Her notebooks indicate that by February 1953 she had inferreed that the A form was likely helical also. The data suggested that biological information was encoded in the unique ordering of the four nucleotide bases. Franklin noted that 'an infinite variety of nucleotide sequences' would make it 'possible to explain the biological specificity of DNA'. If this were indeed the case, then Miescher's suggestion, more than half a century earlier, that biological information was encoded in combinations of chemical structures would have been vindicated. Biology would be revealed as being a form of cryptography.

Meanwhile, in the cold, dank Cavendish Laboratory in Cambridge, the 'ambitious but shy' 23-year-old American upstart James 'Jim' Watson, who had arrived in Cambridge in the autumn of 1951, found himself intellectually paired with Francis Crick, the 35-year-old son of a shoemaker. Crick had spent much of the war devising ways to infer the shapes of German ship hulls from the wakes spotted using aerial reconnaissance photographs. Together they realised that the determination of DNA's three-dimensional structure was biology's holy grail. Solving this problem would reveal, definitively, how life's hereditary information is both represented in DNA and replicated.

Watson's interest had been piqued a year earlier, in 1950, when he visited a scientific meeting in Naples. It was there that he chanced upon one of Wilkins's X-ray photographs of DNA. This was an event that would change the course of history. He immediately applied to work with Wilkins, but was rejected.

Although they lacked direct access to the King's College X-ray data, Watson and Crick believed they could determine DNA's structure by building scale models grounded in chemical principles. But despite labouring for weeks attempting to fit their homemade metal and cardboard cut-out models of DNA into plausible configurations, they were unable to identify a workable structure. Lacking access to relevant experimental data to test and refine their models, they were working in the dark.

Two pivotal events changed everything. The first came in January 1953, when Wilkins showed Watson a copy of Photo 51 during his visit to King's College London. While by that time Watson was likely already aware – through his regular updates from Wilkins – that the B form of DNA formed some type of helix, he had, only days before, seen Linus Pauling's incorrect rendition of DNA's structure.

Franklin's photograph likely gave him a jolt, making him realise that life's most compelling conundrum had yet to be solved. The second was in February 1953, when Max Perutz shared a King's College scientific progress report with Watson and Crick that included Franklin's crucial X-ray crystallography data. This demonstrated, among other things, that DNA's two strands ran in opposite directions. These developments, which occurred without Franklin's knowledge or consent, proved instrumental in Watson and Crick's construction of the double helical model of DNA.

Following a six week period of intense activity between January and March 1953, Watson and Crick were able to work out the principle of base pairing – namely that an A could only pair with a T, and a C only with a G – and finally solved the structure of the B form of DNA. They concluded that it formed a double helix. Their model explained how genetic information was stored and replicated, and provided compelling evidence that DNA was the hereditary material. They had solved 'the secret of life', and defined the informational basis of species. The elucidation of the double helical structure of DNA initiated a biological revolution.

Although initially failing to adequately acknowledge Franklin's essential contribution, in 1954 Watson and Crick admitted that without Franklin's data it 'would have been most unlikely, if not impossible' for them to have succeeded. Their landmark paper announcing DNA's double helical structure was published in *Nature* on 25 April 1953, a month before the coronation of Queen Elizabeth II. Yet despite the important insights that their work provided into DNA's structure, many scientists continued to remain unconvinced that DNA was the hereditary material.

Five years later, on 16 April 1958, Rosalind Franklin died of ovarian cancer at the age of just thirty-seven. Four years after her death, in 1962, James Watson, Francis Crick and Maurice Wilkins were awarded the Nobel Prize in Physiology or Medicine, for their discovery of the molecular structure of DNA.

Following the elucidation of DNA's structure, the next significant challenge was to understand the molecular details of how DNA was synthesised and whether this could be performed artificially under defined conditions. To do so, it would be necessary to determine how DNA sequences were replicated. This final piece of the heredity puzzle would prove critical, setting the stage for biology's generative and synthetic future. It was solved through the determination and imagination of Arthur Kornberg – the son of a sewing-machine operator in the sweatshops of New York's Lower East Side, who later opened a small hardware store.

Growing up on the streets of Brooklyn in the 1930s, Kornberg lived in a one-room, windowless apartment, and had little contact with nature. Unlike the beetle-obsessed Victorian naturalists of the nineteenth century, he found himself hunting not for exotic insects, but for the colourful covers of discarded matchbooks. These offered fleeting glimpses into cosmopolitan worlds far beyond his immediate environment. Following a stint as a navy doctor, this childhood fascination transformed into what he later called a 'love of enzymes'. Kornberg possessed the extraordinary foresight to realise that chemistry formed the basis of a 'universal language' of life – a language capable of conjuring up infinite biological possibility.

In Kornberg's mind, DNA was 'like a tape recording'. It carried a message containing 'specific instructions for a job to be done'. The information in these instructions could be copied and deployed repeatedly 'elsewhere in time and space'. This gave genetic information a timeless mathematical quality. While many sceptics – even following the publication of Watson and Crick's paper – continued to question DNA's central role as the hereditary material, Kornberg fully embraced the double helix model. And he understood, crucially, that it suggested the mechanism by which DNA copied itself.

Convinced that the chemical reactions underpinning heredity could be artificially recreated in a test tube, Kornberg embarked on a quest to identify the enzyme responsible for synthesising DNA. Like the sewing machines his father once operated, this enzyme would stitch together DNA's nucleotide components, one by one, into the predefined sequences that formed the informational 'cloth' of life. If this enzyme could be identified, it might be possible to author the DNA sequences of species deliberately, specifically and at will.

Each nucleotide – the basic building block of DNA – is composed of three elements: a sugar, a phosphate and a nucleotide base.

Individual nucleotides are chemically welded together to form the long strings of nucleotides from which DNA strands are built. Each sugar molecule links to one of the four nucleotide bases, while the sugars themselves form a rigid structural backbone – a chemical spine – that holds the strand together. The four bases – adenine, cytosine, guanine and thymine – attach to the sugars, and their precise ordering, like the arrangement of alphabetic letters in written language, is what makes the genomes of each species unique. Yet while the English-language alphabet has twenty-six typographical letters, the language of DNA uses just four chemical characters.

The two strands that make up DNA molecules wind around one another in opposite directions, to form a double helix. The bases on each strand pair up through weak chemical bonds – adenine (A) with thymine (T), and cytosine (C) with guanine (G). Watson and Crick's model elegantly suggested how DNA replicated itself: the two strands unzip, and each acts as a template for the synthesis of its complement. Kornberg realised that there had to be an enzyme mediating this process, and set out to find it.

In 1956, just two years after arriving at Washington University in St Louis, Missouri, and at the age of thirty-seven, Kornberg announced that he had succeeded. He named the enzyme that he had isolated 'DNA polymerase'. Exactly as he had predicted, this enzyme enabled the synthesis of DNA outside living cells by sequentially joining nucleotides together. The order of nucleotides in the newly synthesised DNA was determined by the template strand it was provided with. Like a tailor following a paper pattern, the enzyme faithfully made an exact complementary copy.

Kornberg made his breakthrough using a cell extract prepared from the bacterium *E. coli*. The synthesised DNA was chemically indistinguishable from natural DNA. He was awarded the Nobel Prize in Physiology or Medicine in 1959 for his discovery of the mechanism of biological DNA synthesis. The next challenge was to see whether this insight into natural DNA synthesis could be used to make artificial DNA.

Just over a decade later, in 1967, Kornberg and his colleagues Mehran Goulian and Robert Sinsheimer announced that they had achieved precisely that. Using the DNA polymerase that Kornberg had isolated, they chemically synthesised the small genome of a virus called phi-X174. A reference copy of the virus's genome served as the template. Remarkably, the artificially synthesised virus was biologically active – it reproduced itself across two generations and, as Kornberg noted, 'could have continued to reproduce itself indefinitely'.

Kornberg had shown that DNA polymerase could serve as the basis of a machine for artificially synthesising DNA. But because it relied on the pre-existing information in a DNA template to direct the synthesis of the new DNA, it could not – at least at that moment – provide a way to write bespoke DNA sequences of DNA at will. The tailor, in other words, was still obliged to copy existing patterns rather than fashioning new ones.

Although the newly synthesised viral genome was an exact copy of the natural one, Kornberg's results hinted at the possibility of synthesising artificial genomes with defined sequences of choice. It marked the conceptual birth of a new kind of 'engineering biology'. Rather than describing life as it existed, Kornberg's work raised the future prospect of going beyond imitation and authoring new life. For the first time in life's history, it was possible to imagine the synthesis of entirely new living entities, designed according to predefined specifications. But to do so, a method would be required that could synthesise DNA without the need for a template.

On 16 December 1967, President Lyndon Johnson announced that Kornberg's team had 'unlocked a fundamental secret of life'. They hadn't, as some media reports claimed, quite 'made life in a test tube' – a virus is not a living organism, relying as it does on host cells to replicate – but they had proved that DNA sequences could be artificially synthesised outside a cell.

An editorial in *The New York Times* speculated that, as a result of the discovery, it might soon be possible 'to make an exact copy of a genius such as an Einstein'. Kornberg himself, however, had more modest aspirations. He proposed that the method might be used to create artificial DNA sequences of therapeutic significance – for example, attaching a piece of 'harmless viral DNA', which could then be 'used as a vehicle for delivering' a 'gene into the cells of a patient' to correct an inherited defect. This idea would later form the foundations of what became known as gene therapy.

But Kornberg's vision extended far beyond simply modifying viral DNA sequences to treat human diseases. He understood that heredity was no longer the exclusive mechanism by which genomes could be created, and that the ability to synthesise genomes with a predefined informational structure would give humanity the unprecedented power to create new species. The foundations of the age of genome design and synthesis had been established. Humankind had begun to scribble and was beginning to learn how to write the language of life.

2

The Library of Species

> Present in every human being are two desires, a desire to know the truth about the primary world ... and the desire to make new secondary worlds of our own.
>
> – W. H. AUDEN, 1968

It was my first day at the MRC Laboratory of Molecular Biology in Cambridge. Also known as the LMB, the laboratory was famous for having made some of the most important discoveries in the history of molecular biology. It had produced an unprecedented number of Nobel Prize winners as well. I made my way up to the Department of Protein and Nucleic Acid Chemistry on the fifth floor and knocked on the office door of my PhD supervisor, the Argentinian biochemist César Milstein.

Dressed in a colourful woollen jerkin, he looked at me with surprise, and then with warmth and enthusiasm. He offered me a cup of maté – the traditional Argentinian tea-like drink – and began to brew it carefully. Almost as an afterthought, he asked me what I planned to work on, and following some discussion remained silent for what felt like an eternity. He then stood up, smiled, and said 'CD1'. And in that moment it was decided. I would work on CD1.

CD1 was at that time a molecule of unknown function found at the surface of certain cells. It had been discovered in Milstein's laboratory using the monoclonal antibody technology he had developed with his postdoctoral student Georges Köhler. César informed me that due to space limitations I would need to conduct my experiments on the surface of a refrigerator. His technician, John Jarvis – a jovial local who had grown up alongside Syd Barrett, the enigmatic former frontman of Pink Floyd – then appeared, and guided me to a small fridge protruding from under the bench. This meagre rectangular space would end up being the home for my scientific investigations for well over a year.

Later, as we ascended the staircase to the dilapidated canteen on the top floor, César regaled me with stories about the enzymes aldehyde dehydrogenase and phosphoglucomutase that he had once worked on in his former life in Buenos Aires. The canteen offered panoramic views of Cambridge's skyline and was filled with scientific chatter. Teatime at the LMB had become an institution. It was the occasion on which laboratory members converged to discuss their experiments and seek out the advice and opinions of their colleagues.

As I stood in line, a mischievous, ginger-haired woman called Joy poured some tea into a pale green ceramic cup for me. César and I found a table, and after a few minutes were joined by a short, immaculately dressed man, with a slight frame and Austrian accent, who introduced himself as Max Perutz. Standing at the table – due to a longstanding back problem – he unwrapped some sandwiches from a Tupperware container, and began to eat them slowly and methodically. Before long, we were joined by a softly spoken and unassuming man called Fred Sanger.

Between the three of them, César, Max and Fred had won a total of four Nobel Prizes, with Fred claiming an unprecedented two. Each had left an indelible mark on the history of medicine and biochemistry. Together with their colleagues at the LMB, they had pioneered the field of molecular biology. Under Perutz's visionary leadership, the scientists at the LMB had unveiled many of life's most fundamental molecular secrets and developed methods to explore them.

One of these involved determining the positions of atoms in biological molecules through the analysis of X-ray diffraction patterns. X-ray crystallography transformed the invisible, nanoscale world of molecules – measured in billionths of a metre increments – into something visible. It rendered the molecules of life, once confined to the realm of the unknowable, as vivid and comprehensible as the objects on a kitchen table. X-ray diffraction became the lens through which the hidden molecular world was animated. As a result, protein molecules – with a structural capacity reminiscent of miniature cathedrals, bridges and Ferris wheels – could now be perceived in the macroworld. This novel way of probing the invisible underbelly of living systems led to a profound change in our understanding of how life works at the molecular scale.

Perutz, an avid mountaineer, found inspiration not just in the laboratory, but in the glacial expanses of the Jungfraujoch in the Swiss Alps. During his vacations he studied how delicate snow crystals transformed into the dense, massive ice crystals of glaciers. This fascination with structural transitions led him to conceive of a fantastic journey into the hidden atomic landscapes of haemoglobin.

Haemoglobin, an immense and complex protein comprising more than 10,000 atoms, acts as a molecular machine, ferrying oxygen and carbon dioxide between the lungs and the body's tissues. It was a formidable problem; no one had yet solved the structure of such a large protein molecule. Yet by 1937, he had produced the first X-ray photographs of haemoglobin – providing the first tentative glimpses into its enigmatic world.

Despite being told he was crazy to take on this task, by 1960 he had solved its structure. Seeing it for the first time, he described the experience as 'like reaching the top of a mountain after a very hard climb and falling in love at the same time'. The intense joy that comes when 'nature reveals its great secrets', he believed, was something unique to science.

While Perutz was scaling the heights of molecular structure, Milstein, guided by the advice of his mentor Fred Sanger, had shifted his attention to antibody molecules. With the help of Georges Köhler, Milstein revolutionised medicine by demonstrating that it was possible to generate antibodies of a single, defined specificity. These so-called 'monoclonal' antibodies could precisely recognise proteins at the surface of cells, opening up a new world of medical applications – including the targeted attack of cancer cells. The work of Köhler and Milstein also laid the foundation for what later became the field of antibody engineering.

In the later part of his 1984 Nobel Lecture, 'From the Structure of Antibodies to the Diversification of the Immune Response', Milstein looked beyond his immediate achievements to imagine biology's future. He foresaw the convergence of molecular biology with technologies able to extract insights from vast datasets and speculated about the 'exhilarating possibility' that it might one day be possible to predict the three-dimensional structures of proteins directly from their amino acid sequences.

This would enable science to move beyond nature and allow the design of artificial antibodies from first principles. Milstein had, with remarkable prescience, anticipated the transformative impact that AI would have on biology through its power to design new biological molecules. An accompanying technology able to write out those designs and turn them into actual DNA sequences would be required to realise his vision.

Fred Sanger was a doctor's son and a Quaker. By the time of my teatime encounter with him he had been retired for more than a decade, spending most of his time gardening at his house in the nearby village of Swaffham Bulbeck. He described himself as both an 'inveterate sequencer' and as 'a chap who messed about in the lab'. His calling had been to decode life's molecular languages – a pursuit that paid off spectacularly when he became the first person in history to determine the complete amino acid sequence of a protein.

The protein he sequenced was insulin, a hormone that regulates blood sugar. Sanger meticulously mapped the order of its fifty-one amino acids, demonstrating that a protein of a given type had a unique amino acid sequence. This was an immensely consequential finding, as it implied that DNA sequences determined the ordering of amino acids, thereby forging a link between the genetic code script and the machinery of life.

Building on this seminal result, Sanger understood that following Watson and Crick's elucidation of the double helical structure of DNA, the next step was to establish a method for determining the order of the nucleotides in DNA's information-rich sequences. If DNA was the source code of 'living matter', then the unique arrangement of the four different nucleotides along its backbone likely endowed each species with its unique characteristics. This was achieved, in part, by specifying the amino acid sequences of proteins. Sanger realised that he would need to work out a way to read DNA sequences. For a time, he admitted, he 'didn't see any hope of doing it'. But with his characteristic spirit of adventure, he decided, regardless, to give it a go.

In 1975, along with his technician Alan Coulson, Sanger unveiled a method for sequencing DNA. Using DNA polymerase – the enzyme discovered by Arthur Kornberg – he showed that he could synthesise a series of radioactively labelled fragments of different sizes from the DNA strand under investigation. These were then fractionated by running them on a gel in an electric field – a method called gel electrophoresis. This enabled single-nucleotide resolution of the fragments, which were then stained and visualised as distinct bands.

For the first time in life's history, the information of heredity – encoded in the precise order of nucleotides – could be determined. Sanger had enabled genes to be analysed at the chemical level. Suddenly the digital text of life could be read like the letters, words and sentences of a book. At that time, the rudiments of the language of biology had not yet been defined, so the meaning of specific nucleotide sequences remained a mystery. Sanger's method functioned like a telescope, illuminating the previously invisible letters of DNA's alphabet like brilliant stars in the night-time sky. It offered the tantalising prospect of reading and deciphering the enigmatic nucleotide sequence texts that determine the nature of species.

Sanger went on to refine and automate his sequencing technique. That same year he demonstrated its power by sequencing the genome of bacteriophage phi-X174, one of the simplest known biological entities. At just 5,400 nucleotides, its diminutive genome – a single-stranded circle of DNA coding for just nine proteins – was minuscule compared to the approximate 3 billion nucleotides of the human genome. Sanger had triumphed once again, coaxing the modest virus to relinquish life's most coveted secrets. Like Alice who fell down a rabbit hole in Lewis Carroll's *Alice's Adventures in Wonderland* (1865), he had opened up a portal into the infinite landscapes of a fabulous DNA sequence world.

It was ironic that this self-effacing lover of sailing and nature – who was not inclined to wander too far from his garden in a quiet Cambridgeshire village, known principally for its thirteenth-century church and annual Gilbert and Sullivan productions – had developed a method capable of unlocking a timeless space larger than the universe itself. Indeed, the logical space of all possible DNA sequences is so vast that it could not be accommodated within the observable universe. Its construction would swiftly consume every particle in existence. Sanger had hacked into the invisible universe of biological information.

Yet despite this astonishing success, the prospect of sequencing an entire human genome remained, at that time, far beyond technological reach. The method was not yet scalable. But Sanger, always pragmatic, saw a way to demonstrate the feasibility of sequencing larger genomes. He turned to mitochondria – the tiny, energy-producing organelles located within the cytoplasm of human cells. These intracellular powerhouses contain their own tiny genomes, just 16,569 nucleotides long. Sequencing a mitochondrial genome, Sanger reasoned, would serve as a surrogate for the larger genomes of complex species, including those of humans. He entrusted the project to his research assistant Bart Barrell. Their findings, published in 1981, paved the way for what would eventually become the Human Genome Project.

Sanger's dideoxy method went on to be used to sequence the genomes of a multitude of different species, including bacteria, yeast and nematode worms. But his crowning glory came on 26 June 2000 – twenty-three years after the publication of the dideoxy sequencing method – when J. Craig Venter and Francis S. Collins stood in the East Wing of the White House and announced the completion of a draft human genome sequence. While the public consortium led by Francis Collins had taken close to thirteen years to sequence the human genome at a cost of around 3 billion dollars, the private consortium led by Venter – using a method called 'whole-genome shotgun sequencing' – had managed to complete a draft sequence in just nine months, and for a paltry 100 million dollars. President Bill Clinton marked the moment by proclaiming that 'humankind was learning the language in which God created life'. The human genome had finally relinquished its most fundamental secret.

The sequencing method was not limited to living species. It could also be used to extract information from the genomes of extinct organisms. Allan Wilson, a biochemist at Berkeley, California, showed that it was possible to sequence DNA from a salt-preserved quagga specimen. This subspecies of zebra, which was native to South Africa, was hunted to extinction in 1883. It was also possible to sequence DNA retrieved from the teeth of a 1.65-million-year-old mammoth and from the 'fossil genome' of a Neanderthal man.

Given that it was possible to recover information from life's past and present, the question of whether it might also be possible to travel into life's potential future inevitably emerged. If existing and extinct species are specified by their genome sequences, then it follows that huge numbers of unrealised genome sequences must have been 'passed over' by history. These represent the untapped portion of biological possibility. Some of these previously unrealised genome sequences are readily accessible, but encode information incompatible with the generation of a living species. Others lie so far afield within the genetic landscape that they are functionally inaccessible to natural evolutionary processes. None of these unrealised species have ever existed, and therefore no physical DNA sequences exist from them that might offer insights into their potential nature.

Were we to explore the vast expanses of the library of species that is the DNA sequence 'Wonderland', in an attempt to make it knowable, we would need to adopt a new approach. The literary imaginations and philosophical reflections of Jorge Luis Borges and Daniel Dennett are instructive in this regard as they offer a tangible vision of this infinite combinatorial landscape.

Jorge Luis Borges, the reclusive Argentinian librarian, author, polymath and expert on Anglo-Saxon literature, was born in 1899. Raised in the Palermo district on the shabby northern outskirts of Buenos Aires, Borges crafted miniaturist poetry and short stories that conjured up fantastic, metaphysical worlds.

The 'chief event' of his life was, by his own account, his childhood fixation with his father's library – a room with glass-fronted shelves, containing thousands of volumes. Here, books serve as a metaphor for the genomes of living things, each detailing the unique narrative of a different species. But in the case of genomes, the story is written not in letters, words and paragraphs, but in the nucleotide sequences from which DNA is made. In this imaginary 'biological library', the genomes of species are stacked side by side, just as books are housed together on library shelves.

Borges was mesmerised by his father's library, seeking the solace of solitude among its treasures. He rarely strayed far, except on his visits to see Bengala, a famed tiger at the Palermo zoo, celebrated for its pronounced longevity. The library – and the imaginary worlds it invoked in his mind – was his playground. Lacking childhood friends, he invented two imaginary companions to accompany him on his journeys into the fantastical worlds he created. While these mental excursions were often exhilarating, they could also be terrifying.

This early preoccupation with his father's library likely inspired Borges's celebrated story, 'The Library of Babel'. Published in 1941, as part of his collection *The Garden of Forking Paths*, it is regarded as one of the most influential works of twentieth-century literature. In it Borges explores the mathematical concept of infinity through the metaphor of a library of 'indefinite and perhaps infinite' size – a space containing every book that could ever be written. Only a mathematically minuscule fraction of this 'unending' library's potential books, Borges noted, is ever realised in the actual world, where it become subject to the slow erosion of time. Such a library, he remarked, 'can only be the work of a god'.

The books in this hypothetical 'Borgesian library' contain every possible allowable arrangement of twenty-five alphabetic characters. The library is thus 'complete' in its coverage of literary possibility, realisable with its constrained alphabet – timeless, eternal, 'useless' and 'incorruptible'. Everything that can be written with these twenty-five letters already pre-exists as an endless array of typographical texts, randomly distributed across the library's infinite expanses. But the library lacks structure, and most of its books are 'formless and chaotic'. They contain combinations of letters amounting to gibberish, and convey no comprehensible information. There are, after all, far more ways to be meaningless than to be coherent.

Borges's decision to use a constrained alphabet of just twenty-five letters, rather than the conventional twenty-six letters of the English alphabet or twenty-seven of the Spanish, reflects a deliberate choice to create an abstract and alien linguistic system that invokes unfamiliar, alternative worlds.

Scattered at sparse intervals within this vast sea of nonsense and irrelevance are rare seams of books that contain coherent meaning. Some are volumes that exist in our own world, while others deviate from the originals. Still others are unrecognisable, retaining only a few ghostly sentences from the parental editions. Intermingled with endless tomes of noise, the library contains books on every imaginable subject including: manuals, comic books, weather reports, commentaries and opinions, as well as biographies detailing your life and mine that project into both the past and future. It includes humanity's complete literary canon from Lucretius to Dickens and Murakami.

To help identify meaningful works, the library is populated by librarians charged with the tireless task of retrieving them. They wander across the infinite expanse, testing their predictions, while searching for coherence and significance. The vast majority of meaningful texts have never been and will never be realised. They are destined to remain silent, irrelevant and untouched, gathering invisible mathematical dust for eternity.

While acknowledging that humans are 'imperfect' librarians, Borges speculated that it might nevertheless be possible to formulate 'a general theory of the library' – a framework that would reveal 'the formless and chaotic nature of almost all the books'.

The metaphor of the Library of Babel applies to biological possibility and captures its staggering scope. The challenge of formulating a predictive theory of the library of biological possibility – a framework that could help identify meaningful regions within its endless landscapes – has so far eluded us.

Yet the ability to synthesise genomes makes the prospect of constructing and exploring defined regions of this library of all possible genome sequences a tangible, physical reality. The key challenge, however, lies in identifying the rare islands of possible biological meaning within this expansive search space – the elusive regions that contain descriptions of potentially viable species. To reach them, we will need a method for charting this mathematical jungle that can reliably predict where coherent biological territory might be located. Without such a system, the task of retrieving a coherent book from the sea of typographical nonsense would be like looking for a microscopic needle in an endless haystack.

Unlike Borges's books, which are drafted in combinations of alphabetic letters, the language of biology utilises a much simpler alphabet, comprising combinations of the four nucleotide bases of DNA. The philosopher Daniel Dennett formulated a version of this nucleotide sequence rendition of the Library of Babel, which he called 'The Library of Mendel'. But I prefer to think of the imaginary library of all possible DNA nucleotide sequences as 'Fred's

Library' – in honour of Fred Sanger, who provided us with physical access to these otherwise inaccessible informational landscapes.

Fred's Library contains the complete collection of every possible genome of every possible length. It is a wondrous genetic landscape or 'genomeverse' containing every conceivable genome sequence. Instead of housing literary works like Borges's fictional library, it contains the genetic blueprints of every organism that has ever existed – ants, stick insects, dandelions and pterosaurs, to name but a few. It contains not only actual organisms, but hypothetical ones. Unicorns, giant humans with feet the size of a cricket bat, immortal humans, miniature humans, humans resistant to cancer, humans the size of fleas, flying pigs, singing plants, talking fruits and photosynthesising ants.

It remains to be seen whether the logical 'truth' of these genomes – their capacity to compute viable organisms – can be discerned from their DNA sequences alone. The vast majority are false genomes, masquerading as legitimate blueprints for life. They lack the essential organisational structure necessary for life. Within this genomeverse, the regions encoding viable species exist as rare and occasional mathematical islands, scattered across endless oceans of irrelevance. The sheer scale of possibility ensures that most will never be realised; there simply isn't enough time, space or matter in the universe to permit it.

To find plausible entities, sequences that compute viable living organisms, one would need to traverse vast sections of wilderness, populated by irrelevancy and gibberish. Along the way, one would encounter the twisted, fragmentary imperfections of impossible creatures – the secret sea of unrealisable beasts – that creep, crawl, whirr and whizz aimlessly through the twisted landscapes of mathematical implausibility.

And then, of course, there is the tedium of the endless repetitions. For much of the traveller's journey through the library, boredom would reign supreme. Here, nonsense is the norm; the plausible, the reasonable are 'an almost miraculous exception'.

Nature has devised a machine capable of exploring a tiny corner of Fred's Library. This engine – Darwinian evolution by natural selection – has allowed life to breach the library's outermost edges. It is a biological time machine capable of traversing the present, but it is also the generator of the past. Once a genome sequence emerges through evolution, it becomes subject to the passage of time. The history of life on Earth – encompassing all past and present species – is its improbable and magnificent legacy.

Yet there is a fundamental difference between human-made structures and biological species. Human creations arise from deliberate design, shaped by conscious intent. Nature's complexity, in contrast, though it may appear purposeful, is the outcome of a blind and indifferent evolutionary process. Evolution operates without foresight, driven not by intention but by chance mutations and the pressures of environmental selection.

At the heart of evolution by natural selection lies the generation of mutations: mistakes, damage and recombination events that tweak or shuffle DNA sequences. In the microbial world, DNA may also be exchanged across species. Although statistically random, mutations can cluster at genomic 'hotspot' regions, with certain types of changes being more likely than others. Evolution thrives on these accumulated accidents. Without them, natural travel across DNA sequence space would not be possible, and life would remain confined to a narrow corner of Fred's Library, unable to explore its wondrous landscapes.

Natural DNA damage results from ultraviolet light, chemical exposures and the inevitable errors introduced during genome replication by the DNA polymerase enzyme. Mutation rates themselves are finely tuned by evolution: too high, and the genomic information risks melting away; too low, and evolution will lack the raw material at the core of its motor.

Chance, too, has played a decisive role in shaping evolution's long and meandering exploration. Take the Cretaceous–Paleogene extinction, for example, the most recent of the so-called 'big five' mass extinction events, which occurred around 66 million years ago. It wiped out roughly three-quarters of Earth's species, including the non-avian dinosaurs, and is thought to have been caused by the impact of an asteroid that was over ten kilometres in diameter. When it crashed into the Yucatán Peninsula, the Chicxulub asteroid released an amount of energy equivalent to 10,000 times the world's nuclear arsenal, plunging the planet into a catastrophic global winter. Without this cosmic event, the evolutionary trajectory of life on Earth – including the emergence of humans – would have been profoundly different.

The unpredictable nature of mutations, combined with the random calamities and contingencies of deep time, ensures that if the tape of life were rewound and replayed, we would almost certainly not end up with humans. Our origins reflect the agency of both chance and natural selection. Natural life cannot escape its history.

The human imagination – restless, insatiable and forever searching out the possibility of an improved future – has, across history, intuited the potential for alternative worlds and different ways of being. Long before modern genetic science, Neolithic farmers pioneered selective breeding methods, crafting the first genetic time machine capable of exploring life's future. It allowed humans to imagine how species might be, rather than passively accepting them as they were. The domestication of wild species occurred simultaneously in several far-flung geographical regions across the world. By selecting for favourable traits in plants and animals, early agriculturalists unlocked a new frontier of genome exploration, unwittingly laying the groundwork for modern biotechnology. Even in its rudimentary incarnation, the ability to manipulate genomes set the stage for the emergence of advanced civilisations – which would, over time, invent increasingly sophisticated methods for modifying genetic information.

One of the greatest triumphs of Neolithic selective breeding occurred some 10,000 years ago in Mexico's Balsas River Basin, where a wild grass called teosinte was gradually transformed into maize. Through generations of selective breeding, farmers turned an unremarkable, tough-seeded, wild grass into one of humanity's most essential staple crops. By the time European colonisers arrived in the fifteenth century, maize had become the dominant food production system across the Americas. Today, it accounts for roughly one-fifth of all the calories consumed annually by humans.

Recent studies have shown that this transformation was primarily driven by small genetic changes, often in regulatory genes – the master switches that achieve outsized effects by orchestrating development. Slight adjustments to these critical control points can give rise to entirely new features. In the case of maize, these innovations reshaped teosinte's architecture, transforming its stony-sheathed seeds into cobs with hundreds of exposed kernels. Interbreeding between different teosinte strains also played a pivotal role. By 6,000 years ago, teosinte was only partially domesticated. But once it reached the Mexican highlands, farmers crossed it with local highland strains. This infusion of new genes supercharged the genome of 'second-wave' maize, endowing it with traits that completed its domestication.

The Victorians took selective breeding to extravagant new heights, channelling their fascination with unnatural selection into reimagining the nature of dogs. Queen Victoria herself kept a King Charles spaniel called Dash. Yet this 'passion for dogs' had a darker underside. The distinctive spots of Dalmatians, for example, introduced vulnerabilities that left the breed prone to debilitating urinary tract diseases.

Selective breeding is, however, an imperfect time machine. It can only explore adjacent genetic territories, inching forward step by tentative step. Its navigation of genetic space is performed, moreover, without maps. It has no notion of direction, of where it is heading, or of what it might find when it arrives there. To build more sophisticated time machines capable of venturing deeper into the alien worlds scattered across Fred's Library would require further leaps of the human imagination.

The visionary seventeenth-century English philosopher Francis Bacon imagined just such a quantum leap. Burdened by debt, tangled in the ambiguities of his sexuality and beset by a host of other inconveniences – including a brief incarceration in the Tower of London – his impulse to transform himself could not have been greater. Forced into retirement in 1621, Bacon turned inward, reflecting on human nature and the personal flaws that had contributed to his downfall. But being of a practical disposition, he was not content to languish in philosophical consolations. He aimed to formulate something more immediate and actionable: a method for addressing human shortcomings more directly and definitively.

The ideas he jotted down in his notebooks in the final years of his life were published in 1627 – the year following his death – in the form of the utopian novel *New Atlantis*. With uncanny prescience,

Bacon outlined a systematic approach for reconfiguring the nature of living things – a vision that extended far beyond the slow, fumbling methods of selective breeding. His treatise was nothing less than a manifesto for life – a systematic plan for the generation of artificial species. The knowledge needed to manipulate the nature of living things would, he proposed, be achieved through the reverse engineering of species. The systematic dissection of their components would reveal the secrets of their construction. This, in turn, would allow humankind to forward engineer entirely new species for the betterment of society.

In Bacon's imagined, fantastical world, the nature of living things was no longer fixed, immutable or ordained. Instead, it was highly malleable. The structures of life could be reshaped through artificial means, altering species in 'colour, shape' and 'activity', even to the point of dwarfing them. Human minds could, similarly, be transformed to become 'free of pollution and foulness' and devoid of selfish and lustful drives. Lacking any concept of genetics and unaware of the physical basis of heredity, Bacon nonetheless – 400 years ahead of his time – anticipated the possibility of synthetic genome design and construction.

Future life, Bacon envisioned, would be governed by the principles of a new philosophy devoted to the optimisation of life's structures. These rules would be defined through experimentation on all manner of 'beasts and birds', performed to 'take light what may be wrought upon the body of man'. Such experiments, he believed, would allow the products of interventions to be defined in advance, enabling the nature of living things to be modified in a predictable, purposeful manner. Bacon concluded that the essence of a species was determined by an underlying generative grammar – a 'language of life' that, if properly decoded through the study of nature's mistakes and creations, could be harnessed to reconfigure nature's creations.

In conceiving of the possibility for the rules-based design of species, Bacon sketched out an agenda in which the 'secret motions of things' would be laid bare, and the 'works and creatures of God' improved beyond their inherited form. While revolutionary, Bacon's utopian vision lacked the means for its realisation. Bacon's attempts to actualise this utopian world were, however, prematurely curtailed by an unfortunate incident. While travelling through snow-lined

London in the winter of 1626, he dismounted from his coach in Highgate to perform an impromptu experiment. He caught a 'chill' and died soon after of pneumonia.

It would take another three centuries before the first true prototype of genome engineering emerged. In 1927 the New Yorker and geneticist Hermann Joseph Muller introduced the idea in a lecture titled 'The Problem of Genetic Modification', delivered to a packed auditorium in Berlin. Frustrated by what he called the 'sluggish' rate at which nature introduced mutations into genomes, Muller set out to find a way to augment this artificially.

Muller regaled the enthralled Berlin audience with the story of how, back at the University of Texas, he had bred fruit flies in a refrigerator to shield them from the intense Texan heat. Using a machine of his own design, he had then proceeded to irradiate them with a high dose of X-rays, which boosted their mutation rate a hundredfold. While selective breeding merely shifted the frequencies of naturally occurring mutations, Muller was able to introduce artificial mutations directly into the genomes of species. More strikingly, the mutations proved heritable – suggesting that they had been permanently fixed into the genetic code script.

Muller was awarded the 1946 Nobel Prize in Physiology or Medicine for his discovery. Yet while he had produced the prototype of a molecular time machine capable of leaping a little further into the future than selective breeding, his method remained constrained by its starting material. It could not, moreover, direct mutations to specific sites within the genome. For all its ingenuity, Muller's technology was unable to approximate to even a fraction of Bacon's grand four-century-old vision.

It wasn't until 1972 that a decisive new step was taken. Like Arthur Kornberg, Paul Berg was raised in the Coney Island neighbourhood of Brooklyn, even attending the same school a few years later. His family were Russian immigrants and his father a furrier. When Kornberg established the Department of Biochemistry at the Stanford Medical School in Palo Alto, California, in 1959, he invited Berg to join as a founding member.

Berg is sometimes referred to as 'the father of genetic engineering', and his contribution to the history of biochemistry was nothing short of remarkable. He developed methods for combining DNA from different species and in so doing ushered in the age of recombinant DNA. Berg demonstrated that DNA from three different biological entities – two viruses and a bacterium – could be spliced together into a single artificial 'recombinant' DNA molecule.

The chimeric DNA was assembled by constructing artificial 'tails' that were stuck onto the ends of each piece of DNA. Berg's technology functioned as a kind of word processor for DNA, allowing genetic sequences from disparate origins to be spliced together with precision. This new ability to create artificial DNA molecules enabled scientists to engineer organisms in ways previously unimaginable. Berg was awarded the Nobel Prize in Chemistry for his discoveries in 1980.

While effective, Berg's method was intricate and laborious, requiring no fewer than six different enzymes. But soon, a simpler alternative emerged that allowed DNA fragments to be joined together in a single step, using just two enzymes. This breakthrough was conceived by Herbert Boyer and Stanley Cohen in early 1972, during an evening stroll near Honolulu's Waikīkī Beach. The paper they published in 1973 with their colleagues Annie Chang and Robert Helling marked the dawn of a new kind of genetics that blended biology with the precision of engineering.

Bacteria often harbour small circular DNA fragments called plasmids. These self-replicating loops of genetic material are distinct from bacterial chromosomes. Boyer and Cohen cut these with an enzyme called EcoRI, which was first identified in Boyer's lab in 1970. EcoRI cleaves DNA at specific nucleotide sequences producing 'sticky ends' – complementary overhangs that naturally snap back together. Using the same enzyme, they then excised antibiotic resistance genes from a different plasmid. Because the sticky ends from both DNA fragments were complementary, they could be seamlessly joined together using a second enzyme, DNA ligase.

When introduced into bacterial cells, this artificial hybrid construct conferred antibiotic resistance on the recipient, marking the first successful cloning of a gene. This proof of concept laid the foundations for the democratisation of genetic engineering and the subsequent biotechnology revolution.

It was not at that time known whether DNA from one species could be combined with that of another to create functional hybrid DNA. Many scientists suspected that there were 'natural barriers created across evolution' that would block the sharing of genes between species. But in a series of follow-up experiments, Boyer and Cohen demonstrated otherwise. They created recombinant DNA molecules that fused bacterial genes with those of fruit flies and African clawed frogs. Remarkably, these hybrid genes functioned inside their bacterial hosts, which synthesised biologically active hybrid proteins.

With this achievement, Boyer and Cohen showed it was possible to breach the evolutionary barrier separating species, which had existed for billions of years. In the new world of Berg, Boyer and Cohen's recombinant technology, DNA from different species could be effortlessly stitched together, transcending the natural boundaries imposed by evolution. Scientists could now isolate genes from any organism, rearrange them into novel configurations, and then introduce them into 'microbes, plants, and animals, including humans'.

This revolutionary technology also paved the way for gene therapy: the insertion of corrective genes into human patients to repair genetic errors. By 1996, recombinant methods had enabled the construction of the first true gene-editing tools, comprising restriction enzymes, 'programmed' by fusing them to DNA-binding sequences called zinc fingers. For the first time ever, scientists could introduce precise cuts at specific genomic locations, a level of precision previously unimaginable.

Yet despite these astonishing advances, biology still lacked access to the full expanse of Fred's Library – the vast conceptual space containing all possible DNA sequences. Engineering the DNA that evolution had already provided could only take scientists so far. To unlock the library's boundless potential, it would be necessary to learn how to synthesise DNA from scratch.

As it turned out, someone had already commenced work on this formidable challenge, and had even completed the sketch of a prototype that hinted at the possibility of transcending heredity and venturing into uncharted genetic landscapes.

Towering over nearly everyone around him, impeccably dressed, quick-witted, impulsive and an accomplished tennis player, the

Glaswegian biochemist Alexander Robertus Todd cut a formidable figure. He was brilliant, erudite, and had a remarkable memory; his office is reputed to have looked more like a stately home than a laboratory. The son of a tramway worker, Todd had ascended swiftly through the ranks of British academia. In 1938, at just thirty, he was made head of the prestigious Chemistry Department at the University of Manchester. It was there in 1939 that he began to develop an interest in nucleotides, the information-encoding units of DNA. By 1944, he was head of the Chemistry Department at Cambridge University.

Among Todd's long-standing passions was the study of naturally occurring coloured pigments. One day, he noticed that when the bean aphid *Aphis fabae* was squashed, the pale yellow fluid seeping from its body slowly turned red. Intrigued by the chemistry underlying this phenomenon, Todd dispatched members of his laboratory to the fields along the River Cam to collect specimens. With the help of his enthusiastic co-workers, the chemical details of the phenomenon were soon unravelled.

It was while working on pigments in flowers as a graduate student in Oxford that Todd had begun to appreciate the importance of synthesis – not merely as a technical tool, but as a means of opening up 'wider vistas regarding the significance and function of biologically important compounds'.

Todd's main experimental interest soon turned to the structure and chemistry of vitamins, particularly vitamin B_1 – also known as thiamine – whose absence causes the debilitating disease beriberi. As thiamine played a key role in the enzymes reactions catalysing nucleotide synthesis, Todd realised he would need to study DNA more closely. Beyond helping him to understand vitamins, the chemistry of DNA held the secret to life's most fundamental biological processes. If he succeeded in copying nature's methods and could synthesise nucleotides and the bonds between them, he would be able to fashion the 'keys to life itself'. The best way to understand life was to learn how to synthesise it.

In 1955, Todd and his colleague A. M. Michelson achieved a historic milestone. They had, for the first time ever, chemically synthesised a minute fragment of artificial DNA. It was tiny – just two nucleotides long and far smaller than the tiniest genome – but they had shown it was possible to construct DNA of a predefined sequence from scratch. Crucially, the synthetic material was indistinguishable from natural DNA.

This new capacity to chemically synthesise DNA was nothing short of revolutionary. It offered the potential to traverse the entire landscape of DNA sequence space. Life would no longer be tethered to pre-existing genetic material. Humankind had finally acquired the rudiments of a molecular time machine capable of materialising instantaneously anywhere within Fred's Library. Significant improvements would be required, but the foundations were in place. Todd had established the foundations of the science of synthetic genomics. He was awarded the Nobel Prize in Chemistry for his work on nucleic acid synthesis in 1957.

While the ability to synthesise DNA greatly expanded our access to Fred's Library, journeys into its infinite landscapes would require more than just a molecular time machine; they demanded a map. Without having some familiarity with the library's geography, the odds of arriving at a meaningful destination within its mathematical expanse were infinitesimally small. What was needed was a guidebook that could identify potential destinations, and highlight regions of particular interest – much as a Michelin Guide might recommend hotels and restaurants worthy of a detour.

In *The First Voyage Around the World*, published in 1524, the Venetian explorer Antonio Pigafetta chronicled his three-year circumnavigation as part of Ferdinand Magellan's Spanish expedition. Of the original 240 men who set out, only 18 returned. Pigafetta's detailed maps and observations charted previously unrecorded regions – the vastness of the Pacific and the narrow straits that would later bear Magellan's name. Fred's Library still awaits its own Magellan: a patient and meticulous cartographer to begin the momentous task of mapping out the contours of its landscapes.

But unlike Earth's finite geography, the DNA sequence landscapes of Fred's Library are infinite. The navigation of its vast, uncharted mathematical territories and oceans are, furthermore, equally fraught with dangers, pitfalls and uncertainties. We do not yet know enough about its structure to attempt the cartography of even its smallest regions. However, without such a map, the sheer immensity of this space makes any meaningful journey difficult to contemplate.

The instruments required for this cartographic task are not the sextants, chronometers and compasses used for geographical exploration, but rather AI, massive biological datasets, and the ability to construct synthetic DNA sequences at scale. Combined, this strong iteration of artificial biological intelligence will eventually allow us to chart DNA sequence space. The modern-day Pigafetta circumnavigating Fred's Library will set sail not in a ship, but at the helm of an AI-enabled digital computer.

The potential destinations within the metaverse of potential genomes possess a concrete reality, as tangible in their own way as the Eiffel Tower, Times Square or the Albert Memorial. Each genome sequence in this endless expanse is, literally, a piece of informational real estate. A destination to be visited, and whose nature may be defined without physically having to go there.

Arthur L. Samuel's epiphany was to realise that computing machines could be configured to acquire artificial forms of intelligence, not by rote instruction, but by letting them experiment and learn. As one of the pioneers of this type of machine learning, he laid the groundwork for genetic cartography.

In the 1940s, while working on the design of one of the earliest computers at the University of Illinois, Samuel imagined writing a checkers (draughts in the UK) programme capable of beating a world champion. He left Illinois in 1949 to join IBM, and using an IBM 7094 computer – one of the most powerful machines at that time – developed The Samuel Checkers-playing programme. This was one of the first examples of non-numerical computation empowered by artificial intelligence. It went beyond simple number crunching, as it was able to learn from its interactions.

Rather than programming the computer with the 500 quintillion scenarios required to play proficiently, Samuel provided it with an artificial 'neural network' architecture, allowing the machine to learn from the games it had played previously. His innovation paid off. In 1962, Samuel shocked the world when his programme secured a historic victory against a professional checkers player from Stamford, Connecticut, called Robert Nealey, who went on to become the state champion. For the first time ever, a computer had defeated a human opponent in a game that demanded intelligence and strategic thinking.

Samuel's achievement laid the foundation for an approach to AI that would eventually inspire the development of a type of artificial neural network called a large language model, or LLM. Although initially developed to generate the text of written languages, LLMs turned out to be highly adept at identifying hidden patterns in biological datasets. This is because the nucleotide sequences of genomes are structured much like the sequences of letters in the words of natural languages. In conjunction with huge genomic databases and the ability to write out the sequences of genomes at scale, they hold the key to defining the generative rules of biology.

3
Constructing Synthetic Genomes

> Our ability to manipulate the information content of nucleic acids, which we equate with their sequence, is dependent on our ability to put together the information by chemical synthesis.
>
> – HAR GOBIND KHORANA, 1968

In early November 2016, David Evans, a virologist from the University of Alberta in Edmonton, Canada, stepped onto the podium at the Eighteenth Meeting of the WHO's Advisory Committee on Variola Virus Research, held in Geneva, Switzerland. The attendees of the closed-door meeting had already been stunned by a presentation on short fragments of variola virus – the causative agent of small-pox – retrieved from a seventeenth-century Lithuanian mummy. In the ensuing discussion it had been revealed that had the virus been 'intact, and well preserved', it might still have been 'virulent and life-threatening' to humans. Thankfully, it was not. Since the formal declaration of its global eradication on 26 October 1979, smallpox – a virus estimated to have killed up to 500 million people in the nineteenth and twentieth centuries – has remained extinct.

Against this backdrop, the title of Evans's talk, 'Recreation of Horsepox Virus Using Synthetic Biology', could not have been more controversial. Evans described how he and his colleagues Ryan Noyce and Seth Lederman had synthesised the horsepox virus – a close relative of smallpox – entirely from scratch. They had brought a virus back to life that no longer existed in nature. Astonishingly, this feat was achieved in just six months, and at a cost of only $100,000.

By painstakingly assembling ten mail-order segments of chemically synthesised DNA, they had created a 212,000-nucleotide-long genome that produced a fully functioning virus that was capable of infecting cells and replicating.

The implications were immediately obvious. If a horsepox virus could be chemically synthesised, identical methods could be used to recreate smallpox. Due to the similarity between horsepox and smallpox, Evans had inadvertently produced an instruction manual for synthesising the causative agent of one of the deadliest diseases known to humankind. Critics contended that the work had 'crossed an important rubicon in the field of biosecurity' and 'brought the world one step closer to the re-emergence of smallpox as a threat to global health security'. Some believed the work was reckless and should not have been performed at all. Evans had, they argued, made the world 'more vulnerable to smallpox'.

The WHO Advisory Committee, however, took a more sanguine view, noting that most technologies are 'dual-use' – capable of both good and harm. This was true of genome synthesis, much as it had been true 'for more basic technologies like fire'. The committee concluded that 'the historical record has clearly demonstrated that society gains far more than it loses by harnessing and building on such scientific technologies'.

Evans's work, nevertheless, highlighted a glaring regulatory gap. Remarkably, neither state nor federal legislation had addressed the issue. The science of synthetic biology was evidently advancing far faster than the systems designed to oversee it.

As a result of the democratisation of scientific knowledge, the reduced cost of experimentation and the ever-increasing ease of performing sophisticated procedures, the publication of this work generated considerable potential risk. The information and methods Evans used to construct the virus were now in the public domain. Evans defended his work, arguing that, despite the dangers of publicising the relatively simple methods by which pathogenic viruses could be chemically synthesised, he felt it was important that the world acknowledge 'the fact that you can do this', and begin addressing the associated biosafety and biosecurity concerns.

While the synthesis of artificial smallpox viruses had once been a matter of theoretical concern, Evans had shown that it was now a real possibility. Advances in synthetic biology and genome design, he noted, would soon make it possible to synthesise viruses that had 'never existed in nature' – viruses that could potentially prove to be as pathogenic as smallpox.

Although a virus is not technically a living thing – it depends on a host cell to replicate – it is nonetheless a biological entity. It serves as a surrogate for all biological species, both living and extinct, as well as for those that have never before been realised. By demonstrating the possibility of resurrecting an extinct virus, Evans's work demonstrated that no organism can ever be considered truly extinct. The ability to recreate them – at least at the genome level – remains in perpetuity, irrespective of their physical existence. The unsettling fact is that the potential for smallpox, for example, will continue to endure in perpetuity. The information of the virus exists in a timeless mathematical form.

Through the rematerialisation of an extinct virus, Evans had illustrated the tangible reality of Fred's Library. What had once been a theoretical concept was now anchored in material fact. If a virus could be resurrected, then so too, at least in principle, could a dodo, a sabre-toothed salmon with its tusk-like protrusions, and the genomes of countless species that have never previously existed. A tiny crack had opened up in the lid of the Pandora's box that is Fred's Library, offering a fleeting glimpse of its extraordinary contents.

Yet, the 212,000-nucleotide-long genome of the horsepox virus was far larger than the tiny bits of DNA synthesised by Alexander Todd in his Cambridge laboratory in 1955, which comprised just two nucleotides. To transform Todd's rudimentary technology into a more sophisticated molecular time machine, capable of the types of feats achieved by Evans, would require a touch of genius. This came in the form of an Indian scientist called Har Gobind Khorana.

A singular figure who appreciated the elegance of chemistry, Khorana would, alongside his mentor Todd, become one of the pioneers of synthetic biology. He taught the world how to write genetic sequences.

Born in 1922 in a small village called Raipur, in the Punjab region of British-held India, Har Gobind Khorana was the youngest of five children. His father, a *patwari* – an agricultural taxation clerk – worked for the British government. The family were among the few in their impoverished village who were literate. Khorana received his early education under a tree, until his father helped establish a schoolhouse.

As a child, he would sit on the steps of the local post office, transcribing letters for the illiterate villagers. In the mornings, he scoured the village for a house with a smoking chimney to collect a burning ember to light the family's breakfast fire. These humble beginnings instilled in him a frugality that would define his life. To conserve his lead pencils and minimise paper use, he taught himself to write in a characteristic microscript – his cramped, disciplined handwriting an early symbol of his purposeful mind.

After studying chemistry at the University of the Punjab, in Lahore, Khorana moved to post-war Liverpool for his doctorate, graduating in 1948. He hoped to find a position in Switzerland, but was unable to secure funding. Undeterred, he went anyway, securing a place in the Zurich laboratory of the Croatian-Swiss chemist Vladimir Prelog.

His experiments floundered, but this apparent misfortune turned out to be the making of his future success – and, ultimately, the synthetic biology revolution he helped create. With time free from the bench, Khorana immersed himself in the German organic chemistry literature in the library. Among these texts, he discovered a 1932 paper on a chemical called carbodiimide. His memory of this paper would later prove critical to his innovations in DNA synthesis.

After briefly returning to India, Khorana's family pulled together the funds for his passage back to England. In 1949, he joined Todd's efforts to unpick the chemical structure of nucleic acids. Todd made him appreciate the profound importance of chemical synthesis for understanding the nature of the hereditary material. In 1952 – three years before Todd published his famous paper on the synthesis of dinucleotides – Khorana moved to Vancouver to establish his own laboratory.

Guided by the German-born pharmacologist Otto Loewi's maxim that 'we must be modest except in our aims', Khorana resolved to work only on big problems. Accordingly, he set himself the task of becoming the first person in history to chemically synthesise an artificial gene. He realised that if a technology could be developed to synthesise existing genes, it would also be possible to synthesise the sequences of genes that had never previously existed in nature. This, he understood, would mark a unique moment in the history of life: the creation of biological information independent of a hereditary process. Such authored genetic information would not be constrained by life's prior history. The synthetic creation of genetic material, Khorana believed, would liberate life from the constraints of evolution – enabling the crafting of sequences of any imaginable composition.

At the University of Wisconsin-Madison, Khorana committed himself to developing a new method for synthesising nucleic acids. At a time when the very idea of creating an artificial gene seemed fantastical – and scientists were still barely able to join two nucleotides together – Khorana envisaged something extraordinary. His plan was especially audacious, as his attempts would precede the development of Sanger's DNA sequencing method by nearly two decades. There would, as a result, be no easy way to verify the sequences of any artificial genes he succeeded in making.

To achieve this monumental task, Khorana formulated an entirely new synthesis strategy, which became known as the phosphodiester method. He began by dividing the sequence of the artificial gene into manageable parts. He then developed methods for synthesising short DNA fragments corresponding to these sequences – oligonucleotides – that were far longer than anything previously achieved. Finally, he devised ways to assemble these fragments into a single, continuous piece of DNA representing an artificial gene.

A derivative of the chemical compound carbodiimide – dicyclohexylcarbodiimide (DCC), which he had first encountered in Switzerland – proved critical. The library time following his failed experiments in Zurich had finally paid off. Acting as a coupling agent, DCC enabled the formation of the natural phosphodiester bonds that link nucleotides into DNA sequences. With this tool in hand, Khorana was able to synthesise DNA sequences of up to ten or twelve nucleotides long, of any composition. These small building blocks became the foundation of his synthetic genes.

Khorana's strategy was both meticulous and elegant. The single-stranded pieces of DNA were paired together to form small double-stranded DNA fragments. Each fragment had single-stranded overhangs, which ensured that the complementary fragments connected to one another in the correct order. Once aligned, the short, double-stranded interlocking fragments were chemically joined using the newly discovered enzyme DNA ligase, which effectively 'welded' the DNA pieces together. To enhance the enzyme's efficiency, Khorana added a phosphate group to the DNA ends with another enzyme, T4 polynucleotide kinase. By blending chemical precision with enzymic assembly, Khorana developed a hybrid method that fundamentally reshaped the field of molecular biology.

His innovation had immediate applications. By synthesising short DNA sequences with defined repeating patterns, he was swiftly able to decipher the genetic code – identifying how each of the sixty-four different nucleotide triplets, or codons, mapped onto the specific amino acids in proteins, or served to start or stop their incorporation. This work, for which he was awarded the 1968 Nobel Prize in Physiology or Medicine, illuminated one of biology's most fundamental mysteries. Yet for Khorana, the work on the genetic code was both a distraction and a stepping stone. His principal goal remained unchanged: the artificial synthesis of genes that would enable the creation of entirely new biological information.

Khorana was well aware of the profound implications of his work. Were it possible to 'manipulate the information content of nucleic acids' with precision, biology could be rewritten in a manner that transcended nature's renditions. But he had run up against a significant challenge. In the early 1960s, no protein-encoding DNA molecules had been sequenced. There were, consequently, no gene sequences for him to copy. Without a blueprint, it would be impossible to construct an artificial gene.

While many genes code for proteins, others code for a different type of nucleic acid, called RNA. The only gene at that time that had been definitively linked to an encoded product was one that coded for an RNA. It was clear that this was the gene he needed to synthesise. Like DNA, RNA is made up of sequences of nucleotides, each comprising a sugar, one of four nucleotide bases and a phosphate group. But RNA differs from DNA in several critical ways. The sugar in RNA is ribose, not deoxyribose; the base thymine (T) is replaced by uracil (U); and RNA typically comprises just a single strand, whereas DNA forms a double helix.

A fortuitous breakthrough came in 1965, when the Cornell biochemist Roger Holley published the sequence of a seventy-seven-nucleotide non-coding gene that produced a special type of RNA called alanine transfer RNA (tRNA). This molecule, essential for protein synthesis, ferries the amino acid alanine to ribosomes – the cellular factories that assemble proteins according to the instructions in another type of RNA, called messenger RNA (mRNA). With the published sequence of alanine tRNA in hand, Khorana finally had his template.

By 1970, after a decade of painstaking effort, Khorana completed the synthesis of artificial alanine tRNA – humanity's first ever artificial gene – using off-the-shelf chemicals. Anticipating the transformative implications of this achievement, he noted that 'the same principle could be used eventually to add the synthetic gene to other genomes'. He had developed a universal method for generating and rewriting the information of living things.

But this prototype artificial gene lacked the regulatory elements necessary for its expression in living cells. To address this, Khorana synthesised a bacterial gene complete with the so-called promoter and terminator sequences – the genetic 'start' and 'stop' signals that regulate gene expression. This gene, known as tyrosine suppressor tRNA, had been discovered at the LMB in Cambridge.

To synthesise the tyrosine suppressor tRNA gene and its regulatory elements – which, at 207 nucleotides, was almost three times the size of the alanine tRNA gene – Khorana needed to assemble twenty-six distinct DNA fragments. Once constructed, the synthetic gene was inserted into a defective virus called bacteriophage lambda, which lacked a functional tyrosine suppressor tRNA gene. The mutant virus, previously unable to proliferate in bacteria, came back to life when provided with Khorana's synthetic gene. By 1976, Khorana had shown that artificial DNA could not only mimic natural genes, but also function within living systems.

The following year, these new methods enabled Herb Boyer to synthesise the first ever artificial human gene, somatostatin – a hormone that regulates growth and insulin production – and express it in bacterial cells. Shortly afterwards, in 1978, Keiichi Itakura at the City of Hope National Medical Center in California refined Khorana's methods to produce artificial human insulin genes, ushering in the age of biotechnology. His improvements reduced the time required for gene synthesis from several years to just a few months. Genentech, the pioneering biotechnology company, adopted this approach to manufacture synthetic insulin, which was soon used to treat human patients with diabetes. It became the first genetically engineered drug to be produced at scale and to receive approval from the US Food and Drug Administration (FDA).

Khorana's demonstration that it was possible to synthesise artificial genes suggested that the same approach could, in principle, be used to synthesise entire genomes. But while revolutionary, Khorana's method had several limitations. Unlike chemical reactions controlled exclusively by enzymes – which were highly specific, recognising just one type of substrate – the non-enzymatic reactions he employed were harder to control and prone to error. Khorana addressed this by using 'protecting' reagents, but he was unable to shield the phosphate groups in the nucleotides. This led to unwanted 'branching' reactions at the phosphate, linking nucleotides together in unintended ways. A multi-step purification process was required to remove these contaminants.

To make matters worse, as the length of the oligonucleotide increased, the amount of branching increased dramatically, imposing a limit on how long the oligonucleotides could be. And because the reactions were performed in solution, the oligonucleotides had to be purified after each step. Khorana's method was, as a result, slow and laborious.

Recognising these challenges, Robert 'Bob' Letsinger proposed synthesising DNA on a solid support, rather than in solution. This simple but profound innovation laid the groundwork for the high-throughput DNA synthesis methods that would revolutionise the field for decades to come. One of the pioneers of modern DNA synthesis, Letsinger had spent his wartime years at MIT developing methods for producing synthetic rubber. He arrived at the Northwestern University Chemistry Department in 1946 aged twenty-five, his young family being accommodated in a tin hut.

Using what he referred to as a 'popcorn' polymer, Letsinger developed a new chemical method for synthesising oligonucleotides. Among other things, this solved the problem of the unprotected phosphate groups that had plagued Khorana's earlier work. Known as phosphotriester chemistry, Letsinger's method was faster, more efficient and more user-friendly. But it wasn't without flaws. One significant drawback was the swelling that occurred in the organic polymer supports.

These limitations were eventually addressed by Marvin Caruthers, Letsigner's former graduate student, who would go on to introduce significant new innovations into the science of DNA synthesis. When Caruthers was interviewed for a position at Northwestern University, Robert Letsinger shared an audacious vision: he planned to synthesise large biological molecules on a polystyrene support. Caruthers joined Letsinger's team and helped develop the phosphotriester method – an elegant, if imperfect step forward. He subsequently joined Khorana's laboratory, becoming part of the artificial gene project. It was there that he became fascinated with the question of how genes were regulated by their surrounding sequences. To probe these interactions meaningfully, he realised it would be necessary to develop more efficient ways to build DNA.

Caruthers took a faculty post at the University of Colorado Boulder, and in 1976, a graduate student called Mark Matteucci joined his laboratory. Matteucci set about adapting Letsinger's chemical synthesis process, turning to a more efficient solid support: a porous silica glass matrix. Two years later, a postdoctoral fellow, Serge Beaucage, further refined Letsinger's chemistry, culminating in a modified version called the phosphoramidite method – a technical triumph.

With this new method, incorporating both the improved support and refined chemistry, everything changed. Oligonucleotides – twenty to thirty nucleotides long – could now be routinely synthesised in under a day, a process that had previously taken months of painstaking effort. The yield of correctly formed bonds, furthermore, approached 98 per cent. The new method set unprecedented standards, going far beyond anything that had previously been possible. Caruthers and his team had achieved the unthinkable: a rapid, accurate, high-yield and near-flawless synthesis process that democratised DNA synthesis. It remains the basis for oligonucleotide synthesis today. That same chemistry also became the foundation for the production of the first automated DNA synthesiser, manufactured by Applied Biosystems – a company Caruthers helped establish. As a result of these technological advances, it finally became possible to start to think about the feasibility of constructing small genomes.

One eventual application of this technology was the synthesis of the horsepox virus genome by David Evans. Yet the first fully artificial genome – synthesised without a DNA template – belonged to a far smaller, but no less consequential entity.

In a landmark moment for synthetic genomics, on 9 August 2002 Eckard Wimmer and his colleagues, Jeronimo Cello and Aniko Paul, at the State University of New York at Stony Brook, announced that it was 'possible to synthesize an infectious agent' in a test tube, using only chemical means and 'solely following instructions from a written sequence'. They had chemically synthesised the entire genome of the poliovirus.

This feat touched on a profound philosophical question about the nature of life itself, since there is some debate about whether viruses are living organisms, given they are dependent on host cells to replicate their genomes. Wimmer, however, took a more practical view. He saw viruses as 'entities that alternate between non-living and living phases'. Before infecting a host cell, a virus was simply a non-living chemical – inert and lifeless. Once inside, the 'chemical' virus metamorphised into something that was alive. By that logic, Wimmer's team had done something extraordinary: they had rationally designed and synthesised the genome of a living organism.

In the accompanying paper, Wimmer described how his team had painstakingly assembled the tiny, 7,741-nucleotide-long genome of the poliovirus from scratch, using mail-order DNA fragments. At the time, it was the largest piece of artificial DNA ever constructed.

Poliovirus – the agent behind the devastating disease polio – primarily affects young children, attacking the nervous system and resulting in irreversible paralysis. It is often fatal. Following the triumph of the smallpox eradication campaign, international health agencies coordinated to eradicate polio, but eliminating it proved much harder than smallpox, and polio continues to persist in economically disadvantaged regions around the world.

Because the genome of polio is made of RNA, once the DNA version of the genome had been made, it needed to be converted into RNA using an enzyme called RNA polymerase. The lab-built virus, once assembled, behaved much like its natural counterpart.

When administered to mice it caused paralysis within a week. The artificial virus was near-identical to that of the natural one, differing only in nineteen 'watermark' sequences that had been inserted into its genome to flag its synthetic origin. The watermark sequences, however, had a detrimental effect on the artificial virus. As a result, up to 10,000 times more virus was required to kill the mice compared to the natural version. Aside from this, its behaviour was virtually indistinguishable from that of the natural virus.

Published less than a year after the attacks on the World Trade Center in New York and the US Mail anthrax attacks in Florida and New York, it is hardly surprising that Wimmer's paper generated controversy. Although funded by the Defense Advanced Research Projects Agency (DARPA) as part of a biowarfare countermeasure programme, Wimmer's work underscored the potential dangers of synthetic biology if placed in the wrong hands – and raised urgent questions about the need for regulation. Wimmer himself maintained that in publishing the paper he was acting responsibly, drawing public attention to the biowarfare risk society faced in the absence of appropriate oversight.

The genome sequences of numerous pathogenic viruses – including HIV, SARS and Ebola – were already freely available online. All that was required to construct a lethal virus was a computer, mail-order DNA, sufficient funding to cover synthesis and assembly, and basic laboratory equipment. Smallpox, with its devastating history, had already been named as one of the pathogenic viruses that might, in principle, one day be synthesised. But with a genome twenty-four times larger than that of poliovirus, most experts considered that a remote prospect. Evans's synthesis of the horsepox genome, which is twenty-eight times larger than poliovirus – just sixteen years later in 2018 – upended that assumption, demonstrating the unprecedented speed with which the field of synthetic genomics was advancing. This was driven by the development of new methods and innovations in automation.

In the years that followed, the pace of synthetic genomics continued to accelerate. While it had taken Wimmer three years to complete the synthesis of the 7,741-nucleotide poliovirus genome, just eighteen months later, in 2003, Craig Venter, Hamilton Smith, Clyde Hutchison and Cynthia Pfannkoch at the Institute for

Biological Energy Alternatives in Rockville, Maryland, announced that they had synthesised the 5,386-nucleotide genome of the phiX174 virus. They had completed this in just two weeks. Their assembly method – polymerase cycle assembly (PCA) – used a single pool of chemically synthesised oligonucleotides, and proved faster and more efficient than anything available at that time.

By 2005, two years after Venter's breakthrough, another striking development emerged. A group led by Adolfo García-Sastre at the Mount Sinai School of Medicine, New York, described how they had resurrected the highly virulent flu virus responsible for the 1918 Spanish flu pandemic, the deadliest pandemic in human history. The so-called 'Spanish flu' was responsible for the death of around 50 million people worldwide. The reconstructed sequence of the extinct virus – once ominously described as 'perhaps the most effective bioweapons agent now known' – was pieced together from fragments of DNA extracted from the body of an Alaskan woman, buried in permafrost since 1918.

The authors openly acknowledged that there could be no absolute guarantee that the virus would not at some point escape – either accidentally or by design. History, after all, offered some cautionary tales. The SARS virus, for example, had accidentally escaped from laboratories in Singapore in 2003, and in Beijing the following year. Nor could they prevent the knowledge that they had published from being co-opted by a state-run bioweapons programme or rogue actor. But they argued that the potential benefits far outweighed the risks. Their work opened up new avenues for understanding and preventing viral disease, was important for public health and, they maintained, had been conducted using rigorous safety protocols.

On 24 January 2008, Hamilton Smith, Daniel Gibson, Craig Venter and Clyde Hutchison at the J. Craig Venter Institute announced another major milestone in synthetic biology. They had chemically synthesised the largest artificial genome ever created by humankind: an artificial version of the genome of the bacterium *Mycoplasma genitalium*.

At 582,970 nucleotides long, this synthetic bacterial genome was nearly three times the size of the horsepox genome. The team constructed the artificial genome sequence using 101 DNA 'cassettes', each between 5,000 and 7,000 nucleotides, spanning the entire genome. The synthesis of these chemical building blocks had been outsourced to commercial DNA synthesis companies. Each cassette overlapped with its neighbouring sequences, ensuring the correct order during assembly. The cassettes were combined into progressively larger fragments in five stages. In the final step, four large fragments – each around 144,000 nucleotides – were joined using the recombination machinery of yeast.

The resulting artificial genome was near-identical to its natural counterpart, distinguished only by several unique 'watermark' sequences, deliberately inserted to clearly identify the synthetic version. A gene called MG408 was, additionally, disrupted to prevent the bacteria from attaching to mammalian cells, thereby reducing its potential pathogenicity in humans. The choice of *Mycoplasma genitalium* was deliberate, as this simple organism operated with just a few hundred genes and had one of the smallest genomes of any known independently replicating cell that could be efficiently grown in the laboratory. It was an ideal candidate for synthetic genome experimentation.

The team had succeeded in authoring, from scratch, the instruction manual for a simple biological machine. This achievement, Venter declared, marked the beginning of 'a new design phase of biology', one in which the components of living things could be purposefully designed and constructed to address specific problems. Although only a small number of changes had been introduced into the natural organism's sequence, the potential for more ambitious engineering was self-evident.

The synthesis of this tiny genome marked a paradigm shift in biology. The ability to synthesise genomes introduced a powerful new type of 'hammer' for cracking organisms open – not by dismantling them, but by designing and building them – to understand how life works. This new paradigm disrupted the long-standing Baconian model of scientific enquiry, which had sought to understand nature by breaking it apart. Synthetic life, in contrast, promised to reveal the generative rules of biology through its construction, offering the potential to design entirely new organisms with specific outcomes and purposes in mind.

The next step was even more ambitious: the creation of a synthetic organism. The implications were staggering. Synthetic genomes, authored by humans and containing nucleotide sequences that had never before existed in nature, would function like computer software, reprogramming living cells. For the first time in history, life would have the capacity to bypass the constraints of natural evolutionary programming. Its future trajectory would no longer be shaped solely by nature, but by human ingenuity. It would mark the beginning of the end of heredity's hegemony, and represent a profound shift in humankind's relationship with nature.

Creating a synthetic organism, however, required one crucial demonstration – proof that an artificial genome could animate a living cell. To achieve this, it would be necessary to determine whether a synthetic genome could be transplanted into the cell of another species and successfully 'booted up'. The cell's native genome would need to be removed and replaced with a synthetic one. If successful, this would accomplish something unprecedented in the history of biology: the transformation of one species into another. It would also mark the creation of the world's first human-made species – a form of life unbounded by ancestry.

To prepare for the transplantation of a synthetic genome into a natural cell, it was first necessary to show that the natural genome of one species could be transplanted into the cell of another, and take over its metabolic processes. In 2007, in an extraordinary experiment, Venter and his team showed that this was indeed possible. The 1.1-million-nucleotide-long genome of the bacterial species *Mycoplasma mycoides* was transplanted into a recipient species *Mycoplasma capricolum*, replacing its native genome. The donor genome did not have to be destroyed as the cell divided once the new genome was inserted. As only transplanted cells carried an antibiotic resistance gene, they could be isolated using antibiotic selection.

The result, as Venter described it, was 'a clean change of one bacterial species into another'. Once it had incorporated the donor genome, the host cell assumed the characteristics of the donor species. The molecular machinery of one species could, it turned out, be converted to run on a completely different operating system. As predicted, the transplanted genome reprogrammed the cell like a piece of software uploaded into computing machine hardware. The next question was whether the synthetic genome would behave the same way in its new context.

In 2010, Venter's team provided the answer. They successfully transplanted a synthetic *Mycoplasma mycoides* genome into *Mycoplasma capricolum*. The synthetic genome was identical to the natural one, except for the introduction of watermark sequences. The host cell acted as little more than a 'chassis' for the synthetic genome's programme, which directed it to produce an entirely new set of proteins. They named this new species – the world's first synthetic organism – *Mycoplasma mycoides JCVI Syn 1.0*, referring to it as 'the first self-replicating species on the planet whose parent is a computer'.

Unsurprisingly, the breakthrough provoked significant debate. While hailed as a tremendous breakthrough, the work raised profound ethical questions and carried extraordinary – and unknowable – implications for society's future. Venter's team acknowledged that there were 'philosophical issues that have broad societal and ethical implications'. Yet despite such unknowns, Arthur Caplan, the prominent NYU Grossman School of Medicine bioethicist, provided his blessing, referring to their achievement as 'one of the most important scientific achievements in the history of mankind'.

After fifteen years of tireless effort, Venter had succeeded in building the first artificial species. While the term 'artificial' in this context referred to a genome that had been chemically synthesised, it was essentially just a faithful replica of its natural counterpart. But it raised the possibility that in the future it might be possible not only to replicate genomes, but to author entirely new ones to generate species that had never previously existed.

Humankind's growing mastery of artificial genome synthesis is nothing short of revolutionary. Yet life's fantastic journey into its potential post-Darwinian future – a world where naturally evolved genomes coexist with artificial ones – is only just beginning. If Har Gobind Khorana's genome assembly method was akin to the craftsmanship of a handwritten manuscript, the technologies developed by the pioneers of genome synthesis, like Venter, were reminiscent of Gutenberg's movable-type printing press.

Invented in 1440, the Gutenberg press transformed human communication. It combined the use of movable metal type with a printing press and printer's ink, and was capable of producing pages of text at scale, with remarkable speed and precision. By 1456, Gutenberg had produced his renowned forty-two-line Bible, the first book printed in the Western world. This resulted in an information revolution, enabling the communication of facts and ideas across the world on a scale never previously imaginable. Laboriously hand-copied manuscripts could now be replaced by mass-produced books and pamphlets in multiple different languages, facilitating the rise of literacy and the democratisation of knowledge across the globe.

The development of a 'Gutenberg press' for genome synthesis could similarly transform the life sciences. At present, however, we are still in the pre-technological equivalent of scribes writing laboriously by hand. While current methods can synthesise genomes of increasing complexity – including viruses, bacteria, yeast and eventually a moss – they remain slow, error-prone and prohibitively expensive. Moreover, while the current approach of stitching together small DNA fragments into larger assemblies works reasonably well for simple genomes, it is impractical for the synthesis of complex ones. Even more challenging is the fact that current genome construction methods struggle with difficult sequences – especially highly repetitive sequences and those with a high GC nucleotide content.

Nature, of course, has been assembling DNA for billions of years, using DNA polymerase enzymes to copy genetic templates with extraordinary precision. Were it possible to make a DNA polymerase enzyme that did not require a template, predefined DNA sequences could, in principle, be synthesised to order. Such a universal DNA printing machine would amount to a type of molecular 'Gutenberg press' for genomes. It would enable any nucleotide sequence to be printed out with impunity.

A natural enzyme, terminal deoxynucleotidyl transferase (TdT), which evolved to help generate antibody diversity, has the unique ability to add nucleotides together without a template. Though it has been used to synthesise DNA, its output is currently limited to relatively short DNA fragments. Yet this rudimentary capability could, in theory, be expanded through the use of AI-guided protein design and engineering – potentially allowing it to write out much longer DNA segments, and perhaps, one day, even entire genomes. It may, alternatively, be possible to use AI to reimagine natural DNA polymerases in ways that enable them to write DNA in the absence of pre-existing templates. Until then, we will need to rely on innovations in chemical DNA synthesis and assembly to increase our capacity to write DNA at scale.

However, the synthesis of DNA alone is insufficient. To unlock the full potential of synthetic biology, we must learn not only to build genomes, but to design them as well. While the evolutionary 'programming' of natural DNA sequences has produced a great richness and variety of species, our present-day ability to author artificial genome sequences remains rudimentary – reminiscent of a preschool infant learning to scribble. The synthesis of the simple genomes of viruses and bacteria represents the nursery slopes of genome construction. But they will be the first terrain and testing ground on which the ability to design genomes will demonstrate its future potential.

Yet it is only once we have decoded the universal generative rules of biology that we will be able to establish the versatile principles and generality necessary to author completely new forms of life. Should we choose to do so, we will need to move beyond the mere copying of existing genomes and transform DNA into a predictive engineering material, capable of composing new biological narratives reaching beyond the imagination of nature and history.

4

The Accidental Species

> Physics was once called natural philosophy; perhaps we should call biology natural engineering.
>
> – SYDNEY BRENNER, 2012

In May 1848 the 25-year-old naturalist and 'ardent beetle-hunter' Alfred Russel Wallace realised his long-held dream of exploring the tropics. On arriving in Brazil, he made his way up the Amazon River's northern tributary, the Rio Negro. Lacking wealth and social connections, he was more Indiana Jones than gentleman explorer. To fund his scientific expeditions, he was forced to become an entrepreneur, collecting specimens and shipping them back to his agent in England.

After four years spent documenting an astonishing variety of tropical species, Wallace boarded a freighter, the *Helen*, bound for London. Three weeks into the journey, in mid-Atlantic, the ship's captain entered Wallace's cabin to inform him that the vessel was on fire. Hastily packing two small lifeboats with notebooks and supplies, the crew watched as Wallace's collections – including a menagerie of parrots, monkeys and other live animals – were consumed by flames. After ten harrowing days at sea, they were rescued by a cargo ship, the *Jordeson*, en route from the West Indies to London. Fortunately for the history of science, this self-effacing adventurer – who would, along with Charles Darwin, transform the world by linking the past of species to their present – was soon safely home, gorging on 'beef steaks and damson tart', which he described as 'a paradise for hungry sinners'.

A decade after he had first set out, in February 1858, while 'shivering' from a malaria-induced fever on the remote island of Ternate in the Malay Archipelago, following an expedition to the Far East, Wallace dispatched a letter to Charles Darwin, England's most eminent naturalist, which also contained an essay, entitled 'On the Tendency of Varieties to Depart Indefinitely from the Original Type', outlining Wallace's discovery of evolution by natural selection.

On opening the letter on 3 June 1858, the procrastinating and intensely cautious Darwin – who had been contemplating his own evolutionary ideas for more than two decades but had delayed publication out of concerns for their reception – was shocked to discover that the letter 'contained exactly the same theory' of species formation as his own. In a frantic letter dated 18 June 1858 to his friend, the geologist Charles Lyell, Darwin confessed, 'I never saw a more striking coincidence. If Wallace had my manuscript sketch written out in 1842, he could not have made a better short abstract!'

Challenging the prevailing belief in the immutability of species, Wallace, like Darwin, had invoked a natural selective mechanism to explain their origin. But Darwin had not yet announced or published his own near-identical theory. To secure Darwin's priority, his supporters – Charles Lyell and the botanist Joseph Hooker – swiftly arranged for three papers on the mechanism of evolution to be presented at a special meeting of the Linnean Society, held on the evening of 1 July 1858, at Burlington House in Piccadilly. This meeting, addressing Wallace and Darwin's perspectives on 'The Laws which affect the Production of Varieties, Races and Species', would become one of the most important – as well as one of the most consequential – in the history of science.

Neither Wallace nor Darwin was present to hear their papers read. Wallace had not been invited, and was seriously unwell in Papua New Guinea. Darwin, meanwhile, was mourning the death of his eighteen-month-old son from scarlet fever.

While conceding that these two 'gentlemen' had 'independently and unknown to one another conceived the very same ingenious theory', Lyell and Hooker asserted that Darwin had first used the term 'natural selection' as early as 1844 in an unpublished essay, although the phrase had a different meaning at that time. Furthermore, Darwin had outlined his thoughts on evolution by natural selection in a letter to the prominent Harvard botanist Asa Gray in 1857. By this account, Darwin's articulation of the concept of natural selection had pre-dated Wallace's by more than a decade.

Over time, Wallace's contribution to the discovery of evolution by natural selection was overshadowed by Darwin's. One reason for this was Wallace's reluctance to acknowledge the universal applicability of natural selection. He insisted, for example, that it could not explain the origin of humans. His later involvement with spiritualism and his public defence of spurious supernatural phenomena, such as 'table-tapping', did little to bolster his credibility within the scientific community. Regardless, had Wallace submitted his essay to the journal that had previously published his work, rather than naively sending it to Darwin, the theory of evolution by natural selection might today be known as Wallacian evolution.

Spurred on by Wallace's letter, in which he articulated his convergence on the same theory, Darwin published his bestselling popular account of natural selection, *On the Origin of Species*, on 24 November 1859 – just over a year after the Linnean Society meeting. The book propelled the theory of evolution by natural selection to global awareness and elevated Darwin to iconic status.

The theory of evolution became one of the most controversial scientific ideas of all time, rivalled only, perhaps, by Galileo Galilei's assertion in his 13 March 1610 treatise *Sidereus Nuncius* (Starry Messenger) that the Sun – not the Earth – sits at the centre of the solar system. Humanity, Darwin proposed, was not the designed and preordained product of an instantaneous act of divine Creation, but rather a derivative of apes and simpler life forms.

Darwin's narrative of human history differed substantially from all prior accounts and was considered heretical by many as it removed God's direct involvement in the generation of species – including humankind. The suggestion that life had evolved through a natural, gradualist process appeared to undermine divine agency. It contradicted the biblical narrative detailed in the Book of Genesis, and all other theological accounts of life's origin.

At the core of Darwin and Wallace's 'dangerous idea' was the notion that living organisms have descended from pre-existing ones, incrementally accumulating small changes over vast expanses of geological time. This process amplified and elaborated traits that conferred a reproductive survival advantage, while diminishing the frequency of those that did not. All life on Earth is meshed into a single, sprawling and entangled tapestry of design, deeply rooted in antiquity.

This shared evolutionary history has instilled Earth's species with a remarkable unity of form and function. The French biochemist Jacques Monod highlighted this biochemical commonality, when he stated in 1954 that what holds true for a bacterium is also 'true for an elephant'.

While the detailed nature of different species has changed across evolutionary time, the genetic programmes controlling them have not. They have persisted, essentially unchanged, for over 500 million years. There is, moreover, a fundamental unity between all species at the molecular level. With the exception of some viruses, all known biological species encode their hereditary information using the nucleotide sequences of DNA. They also build their structures from proteins (and in some cases RNA), according to instructions encoded in DNA. Fruit flies use the same family of genes to build their body as humans, while the tentacles of cuttlefish are fashioned using the same genes as those controlling the growth of human arms and spider legs.

One striking demonstration of this cross-species biological unity – and of how biochemical processes are conserved and redeployed – may be found in Edward Marcotte's 2015 study of the baker's yeast *Saccharomyces cerevisiae*. Despite having last shared a common ancestor with humans over a billion years ago and possessing vastly different levels of complexity, yeast and humans share roughly a third of their genes. While the yeast and human versions of these genes are not identical, their sequences and functions are similar, suggesting that they evolved from a common ancestor.

Working at the University of Texas at Austin, Marcotte's team replaced 414 essential yeast genes with their human 'orthologue' equivalents. Astonishingly, nearly half the human genes tested functioned in yeast, integrating perfectly into its metabolic pathways. In some cases, entire pathways were replaceable. The experiment demonstrated that deep evolutionary connections persist between species – and may be maintained across vast durations of time. Yet in some cases the functions of identical genes are not simply conserved, but repurposed.

Light-sensitive proteins called opsins, found in human eyes, have, for example, been redeployed for hearing in flies. In the African clawed frog *Xenopus laevis*, in contrast, an opsin-derived protein called melanopsin helps regulate its body clock. This process of evolutionary redeployment, known as exaptation, suggests both nature's parsimony and its ingenuity. It transforms old tools and repurposes them.

This inherent flexibility of evolution – the ability to use near-identical genetic programmes in novel ways – is facilitated by the modular organisation of anatomical and biochemical structures. The mammalian spine, for instance, is divided into five anatomically distinct units, each functionally independent of the others. This modularity allows for the independent evolution of each section. As a result, in the cat family – which includes species as diverse as the clouded leopard, cheetah, domestic cat and lion – the lumbar spine has been able to diversify in size and shape to accommodate different lifestyles, while other regions have remained unchanged. But not all biological systems have managed to achieve this degree of modularity. And with that lack of flexibility come constraints that introduce trade-offs that may lock suboptimal design features into a species' architecture.

The dung beetle genus *Onthophagus* – encompassing around 2,000 different species – offers a vivid illustration of nature's evolutionary trade-offs. Generously distributed across the globe, these beetles inhabit the droppings of animals as diverse as kangaroos, bison, zebras and toads. Male dung beetles battle fiercely for access to females, who dig elaborate tunnel systems beneath the ground. These branching chambers are used to store compacted dung balls that serve as nurseries for their offspring. Males guard the entrances to the tunnels, evolving outsized horns to help them do so.

This has resulted in an evolutionary arms race, driving the development of successively larger horns. These horns may emerge from one of three anatomical locations: jutting from the front of their face, protruding from the base of their head, or projecting from their thorax. But in evolution, optimisation in one domain often introduces compromises in others – and dung beetles are no exception. The evolution of outsized horns comes at a significant cost. Males with facial horns frequently have diminished antennae. Those with head-based horns tend to have smaller eyes, while those with thoracic horns have a shrunken wing size. These trade-offs reflect the interconnected nature of developmental pathways, which places numerous constraints on biological possibility. To engineer a dung beetle with large horns and fully functional eyes and wings would likely require a rewiring of its genome to decouple these developmental processes.

The theme of trade-offs extends to many other species, including ground spiders of the Gnaphosidae family. These arachnids hunt aggressive prey – such as ants and other spiders – and deploy glue-like silk threads, evolved from ancestral 'anchor' threads, to ensnare them. But this adaptation carries an associated cost. It has left them barely able to secure their webs to the surfaces of their environment. In 2017 Jonas Wolff at Macquarie University in Sydney showed that as hunting silk became more consequential for these spiders, the glands used to produce their anchor threads diminished in size. This has impaired their ability to secure their webs and to spin the 'draglines' that enable them to scoot across the web surface. Evolution's opportunistic tinkering, while making these spiders formidable hunters, has compromised their web-building capabilities.

Rather than reflecting the work of providence or divine craftsmanship, Darwin and Wallace reframed such adaptations within the context of natural generative laws. The 'endless forms most beautiful and most wonderful' that Darwin referenced in the closing section of *On the Origin of Species* were not the purposeful creations of a master designer, but the by-products of a disinterested, natural selection process.

Evolution's creativity reflects a purposeless and unguided process that lacks foresight and is indifferent to its outcomes. Species emerge not as a result of the predetermined 'grand designs' devised by divine agents, but through an interplay of historical contingency, chance mutations and natural selection. Their intricate structures are shaped not by intention, but by the mechanics of physics, chemistry and nature's unique mechanism of engineering through natural selection. Jacques Monod, in his 1970 book *Chance and Necessity*, famously described evolution as the outcome of 'chance caught on a wing'.

The Darwin–Wallace evolutionary theory, as such, ran counter to creationist explanations of life's origins – such as those articulated by the English theologian and philosopher William Paley in his 1802 *Natural Theology; or, Evidences of the Existence and Attributes of the Deity, Collected from the Appearances of Nature*. Paley likened the complexity of an organism to that of a watch, arguing that the intricate nature of organisms implied the handiwork of a 'watchmaker' or 'intelligent designer', operating at the time of Creation. In this view, species originated instantaneously and were fixed and immutable from the moment of their inception.

The root source of the structures found in living systems, as Santa Fe Institute complexity theorist Stuart Kauffman argued in his 1993 *The Origins of Order*, arises 'for free'. This intrinsic order emerges spontaneously in physical and chemical systems, without the assistance of genetic programmes. The information in self-organising systems is generated not by design, but through the operation of the rules of physics and chemistry. Evolution by natural selection has learned to tune and tweak these wells of spontaneously generated order within biological systems, exploiting them to achieve outcomes that would not otherwise be possible.

This 'order for free' provides the feedstock for evolution, which manipulates it through the agency of genes. Genes capture the intrinsic order of physics and chemistry, and help channel it into the unique features of living organisms. While being extensively utilised, this spontaneous emergent order is anticipated – but not directly coded for – by genome sequences. This important component of biological complexity is therefore 'unspecified', meaning its informational content is not genetically encoded.

Examples of such wells of natural order and spontaneous self-organising patterns abound in nature. The hexagonal symmetry of snowflakes is one such example. Their intricate shapes arise as frozen water droplets progressively increase in size through the successive deposition of water molecules onto ice crystal surfaces. The shape of any individual snowflake is determined by the temperature and humidity at which it forms. Despite their remarkable complexity, the structures of snowflakes are 'uncoded', lacking a genetic blueprint that specifies their form. The unique patterns of snowflakes represent 'programme-free' order, or what Kauffman calls 'order for free'. There are an estimated 10 quintillion (10 to the power of 18) individual water molecules in any given snowflake. The probability that any two will share the same morphology is so vanishingly low that, for all practical purposes, no two are ever identical.

The structures of proteins provide another example of spontaneously generated, uncoded order. The primary amino acid sequences of proteins spontaneously 'fold' into the three-dimensional structures required to perform their functions. The protein-folding process is not specified or directed by genes, or any other external programme. The complex ordering of a protein's atoms in three-dimensional space that occurs during the process of folding is unspecified by genetic programmes, coming instead for free.

Rather than being digitally represented in the abstract language of genes, the information required for folding a protein is programmed into the physics and chemistry of the amino acid sequence itself. The potential to form a folded structure is an inherent property of an amino acid chain. With the exception of proteins requiring so-called 'foldase' proteins to guide them into their final structures, the laws of physics and chemistry do everything else. The rules for protein folding are intrinsic – hard-wired into the sequences themselves – and are not represented genetically. Evolution has learned how to manipulate the wells of natural order that drive proteins into their three-dimensional structures and channel them.

While the appearances of biological entities are reminiscent of the structures of human-made artefacts, the manner in which the products of natural and artificial engineering originate could not be more different. Evolution designs its species very differently from mechanical engineers. As Richard Dawkins stated in *The Blind Watchmaker* (1986), in evolution's case the designer – assisted by chance and self-organisation – is natural selection. If life were crafted by a watchmaker, it would be a 'blind watchmaker', as evolution lacks any preconceived design concepts or directionality. It has no agenda, purpose or plan. It opportunistically exploits and incorporates molecular opportunities, transforming them into novel structures. It favours short-term utility over long-term optimality, preferring design solutions that are 'good enough' and readily attainable over those that are optimal. Evolution is content to run with 'whatever works'.

This impromptu, improvisational process, which in his 1977 paper 'Evolution and Tinkering' the biologist François Jacob described as evolutionary 'tinkering', repurposes existing functions to address emergent unmet needs. Jacob defined a tinkerer as an individual that gives 'his materials unexpected functions to produce a new object. From an old bicycle wheel, he makes a roulette; from a broken chair, the cabinet of a radio. Similarly, evolution makes a wing from a leg or a part of an ear from a piece of a jaw.' The new molecular structures and functions generated by evolution reflect an analogous type of retooling of existing components and features. This opportunistic mode of construction results in the various imperfections and suboptimalities observed across the living world. As Darwin himself noted in *On the Origin of Species*, we should not 'marvel if all the contrivances in nature be not, as far as we can judge, absolutely perfect'.

The exuberant helmets of the treehopper *Publilia modesta*, which give it a distinctive otherworldly appearance, provide a striking example of how molecular tinkering generates morphological novelty. These ornate helmets are crafted by genes that once built wings. At some point in evolutionary history, the suppressed wing-building pathway was reactivated by an acquired mutation. Freed from the constraints imposed by flight, and working with the same ancient starting material, evolution reimagined the wing remnants, transforming them into strikingly different forms. Without their spectacular helmets, treehoppers would lose their magnificence, and assume instead the humdrum appearance of miniature cicadas.

It is, however, neither necessary nor inevitable that biological complexity should increase across evolutionary time. While all life traces its origins to single-celled ancestors, some species have retained their simplicity, while others have embraced complexity. The bacteria living in, on and all around us, for example, are reminiscent of those found in mineralised bacterial microfossils dating back more than three billion years. Instead of advancing their complexity toward more elaborate forms, the history of bacteria has been one of treading a relentless lateral trajectory. Being a bacterium has involved a billions-of-years-long rendition of a fugue on an identical, minimalistic theme. Yet, despite their simplicity, bacteria are among the most successful species on Earth, thriving in habitats as diverse as the hot springs of Yellowstone National Park and the permafrost of the North Pole.

It took approximately 4 billion years from the time of life's inception for unicellular organisms to unlock the developmental potential necessary to create multicellular animals. This milestone event occurred in the Precambrian, around 541 million years ago, in a geologically sudden burst of evolutionary creativity known as the Cambrian Explosion. During that time, the body plans of all complex organisms, including those that would eventually give rise to humans, were established.

Humans are unique among Earth's species, in that they have developed an unprecedented capacity for complex culture. This parallel, extra-genetic information system externalises information storage in non-biological substrates and goes far beyond the cultural capabilities of non-human organisms, which include termite mounds, birds' nests and tool use in crows and chimpanzees. Complex culture in humans has given rise to teapots, harpsichords, Punch and Judy shows, phone boxes, digital computers, automobiles and drugs that can effectively treat certain types of cancer. It has also created telescopes capable of peering back in time to distant galaxies, and technologies that can rewrite the genetic code.

Surprisingly, the complexity of a species does not correlate with the size of its genome. The gigantic genome of the New Caledonian fork fern, *Tmesipteris oblanceolata*, native to several Pacific islands, for example, spans an astonishing 160 billion nucleotides, making it around fifty times larger than the human genome. The genome of the sluggish, mottled-brown Neuse River waterdog salamander *Necturus lewisi*, inhabiting two small river drainages in North Carolina, is similarly thirty-eight times larger than that of a human. Even the speckled grasshopper, *Bryodemella tuberculata*, has a genome that is twenty-one times larger than our own. In contrast, the densely packed genome of the Japanese pufferfish, at just 400 million base pairs, is one-eighth the size of the human genome. It is reassuring, in this context, that the size of the human genome exceeds that of the lettuce *Lactuca sativa*, which is 2.5 billion nucleotides long.

Since its inception, around four billion years ago, life's diversity has been shaped by natural evolutionary engineering – each genome being the rewritten progeny of prior generations. Species are not rigid and fixed, but rather fluid islands of continuously shifting genetic variation. Their source code, written in nucleotide sequences, forms what Richard Dawkins, in *River Out of Eden* (1995), called a 'digital river' of descent – a flow of genetic information that traces back to the earliest genomes and courses forward into an uncertain future.

Over time, genomes have expanded, contracted, fused, split, duplicated, reshuffled, and been corrupted by errors. They have also acquired new genetic material by picking up code from viruses and bacteria, and by fusing with bacteria – a process known as endosymbiosis. These mechanisms generate the raw material for evolution, enabling species to diversify through genomic innovation.

A striking example of the evolutionary kinship between species and their commonality of descent lies in the genome of one of the most ancient and primitive animal species: the mud-dwelling starlet sea anemone *Nematostella vectensis*. Native to the East Coast of the United States, it holds a remarkable secret. When a team led by Daniel Rokhsar and Nicholas Putnam at the Department of Energy Genome Institute in Walnut Creek sequenced its genome in 2007, they discovered that it contained nearly half a full set of human genes. Indeed, of its 18,000 protein-encoding genes, an astonishing 7,766 are shared with humans.

The layout of the sea anemone's genes mirrors that of humans, with large clusters having maintained their relative positions for hundreds of millions of years. The sequences of individual sea anemone genes are also similar to those of humans. Remarkably, sea anemones contain versions of 283 human disease-associated genes, including *BRCA2*, which is linked to breast cancer. The *BRCA2* gene in sea anemones is closer to the human version than it is to that of flies or worms, demonstrating the closeness of its kinship with us. Remarkably, this set of genes appears to have originated hundreds of millions of years before the evolution of humans.

The tale of the sea anemone genome illustrates the remarkable genetic unity running across all animal species. The shared genetic features between humans and sea anemones indicate that these species descended from a common ancestor – an ancient creature living around 700 million years ago – which provided the foundational genetic toolkit for generating complex animal form. The 7,766 genes that we share with yeast represent the 'knowable' part of this ancestral gene repertoire. This hypothetical common ancestor of sea anemones and humans, known as the 'eumetazoan ancestor', gave rise to all animals with two-sided symmetry, known as bilaterians. The astonishing diversity of bilaterians – including flies, flatworms, slugs, lobsters, frogs, birds, whales and humans – was accomplished through 'extreme tinkering', which involved the reshuffling and repurposing of the ancestral gene set.

Evolution played with this foundational gene kit to produce strikingly different species. In the case of some complex organisms, like flies and worms, the ancestral kit contracted rather than expanded – a reminder that genome complexity is not linked in any straightforward way to organismal complexity. Moreover, alterations to the way existing genes are regulated may be as consequential for biological innovation as the introduction of new genes.

The connection between genetic and morphological complexity is, however, far from straightforward. The marine species *Trichoplax adhaerens* – whose Latin name translates as 'sticky hairy plate' – is the simplest known free-living animal. Despite barely qualifying as an animal due to its morphological simplicity, this millimetre-long, disc-shaped creature glides around the Red Sea like an amoeba – and yet its genome is surprisingly like our own. It contains an estimated 11,514 genes, many of which are present in humans and other complex animals.

But appearances can be misleading. The purple sea urchin, *Strongylocentrotus purpuratus*, which resembles a pincushion and bears little outward resemblance to humans, has a genome that is just 814 million nucleotides long, and boasts an unexpected 23,500 genes, which is more than humans. In a somewhat embarrassing twist of evolution, we are closer in kin to sea urchins than we are to beetles, flies, crabs and clams. These distant cousins of ours, with their spiny exteriors and streamlined forms, suggest that many of the basic genetic innovations required to build humans were in place millions of years before our origin. Sea urchins and sea anemones are part of our family tree. The profound genetic unity of species underscores the fact that all life is connected by a shared genetic heritage.

The nature of organisms reflects billions of years of evolutionary programming that has been hard-wired into genetic scripts. Yet the organisation of the products of natural evolutionary engineering is fundamentally different from that of human-engineered machines.

One of the most striking differences relates to the design of their components. In engineered machines, components are designed to perform single, well-defined functions. This property, known as 'orthogonality', allows individual parts to be upgraded, repaired, or replaced without disrupting the system as a whole. Biological machines, in contrast, are anti-orthogonal. Their components often perform multiple, overlapping functions, meaning that changes to one component are likely to impact many others.

While this interconnectivity confers robustness – allowing core functions to persist even when the system is damaged – it also introduces constraints. Because of this relative lack of orthogonality and the need to satisfy multiple, often conflicting constraints, optimising biological systems can be extraordinarily challenging. The structures of biological machines are further constrained by the fact that they need to be evolvable. If they were not 'built to change', they would struggle to adapt to their continuously changing environments and would be poor substrates for evolution.

Another key distinction lies in their mechanisms of innovation. Biological machines cannot start anew; they can only build new designs by modifying existing ones. In contrast, human-made machines can be entirely reimagined, unbound by their historical antecedents. Human innovation thrives on this freedom, celebrating it in the ever-evolving designs of rationally engineered machines.

Perhaps the most profound difference between natural and artificial engineering, however, is the outsized role played by chance in the configuration of natural machines. While each component of a human-made machine is optimised in a precise and deliberate manner, evolution relies on chance mutations to provide the raw material for its innovation. The options available to evolving organisms are dictated by the timing and pattern of these mutations, which result from chance events. Evolving a genome is like spinning a roulette wheel in a Las Vegas casino – the outcome of each spin is dictated by probability, yet every wheel carries its own biases. Together, this makes the outcome of any individual spin unpredictable. Existing biology is the result of a long-running gamble in the casino of evolutionary history, shaped by a complex interplay of chance and necessity.

Chance also exerts its influence through the historical circumstances species encounter and the sequence in which events, such as mutations, are encountered. Historically contingent events can have a dramatic impact on the evolution of a species, and the outcomes of evolution are exquisitely sensitive to these details of history. Because of this vulnerability to historical contingency, François Jacob argued that if life's logic and chemistry were recapitulated on another planet, the evolutionary outcomes would likely be unrecognisable. Hypothetical Martians, he speculated, might look nothing like us.

The human genome bears the hallmarks of chance. One of the most surprising discoveries to emerge from the sequencing of the human genome was that approximately half of its nucleotide sequences comprise remnants of ancient viral infections. These sequences are fossilised traces of viruses that infected the germline cells of our primate ancestors millions of years ago. The viral-related sequences in the human genome include: Human Endogenous Retroviruses (HERVs), Long Interspersed Nuclear Elements (LINEs), Short Interspersed Nuclear Elements (SINEs), LTR retrotransposons (non-HERV) and DNA transposons. Together, these account, respectively, for approximately 8 per cent, 20 per cent, 13 per cent, 8 percent and 3 per cent of the human genome. That such a large portion of the human genome has been cobbled together from the remnants of ancient viruses as a result of their chance infection of germ cells over vast expanses of time is truly astonishing.

Many of these viral remnants remain active today, producing RNA. Some of the sequences derived from integrated viruses have been repurposed and help make us human. Genes have flowed from viruses into humans and evolution has opportunistically co-opted some of these, domesticating them and turning them into genes that perform important biological functions. These viral sequences, furthermore, provide a pool of raw material that might one day form the basis of new functions. One former viral gene, *Syncytin*, which was domesticated around 100 million years ago, plays an essential role in the development of the human placenta.

The opportunistic takeover of this repurposed viral gene, however, came with its own jeopardy. It created a vulnerability in the placenta by locking in the need for a receptor called *ASCT2*, which contains a binding site for pathogenic viruses. In an elegant evolutionary twist, another viral sequence, known as *Suppressyn*, was co-opted to produce an antiviral agent that compensates for this vulnerability. Had the genomes of our primate ancestors not by chance been infected by the virus whose sequence was ultimately repurposed to make *Syncytin*, placental mammals might never have originated – and therefore humans might not have evolved. A seemingly insignificant chance viral infection in a primate ancestor's genome initiated a chain of events that shaped our evolutionary outcome.

Consider the loss of tails in primates, another chance event that traces back to a random genetic mutation occurring roughly 25 million years ago. This mutation resulted in the insertion of a 300-nucleotide-long sequence into the non-coding region of the *TBXT* gene, disrupting its role in tail development. The mutation likely led to the loss of tails in the common ancestor of apes and humans, leaving only a vestigial tailbone. Monkeys lack this genetic change and have retained their tails.

The absence of a tail is considered a defining feature of humans. But while it has been argued that the loss was crucial for the evolution of two-legged locomotion and enabled humans' ground-dwelling lifestyle, there is no evidence to suggest that possessing a tail hinders bipedalism. In fact having a tail might even enhance it. Humans might actually have been better suited to walking if they had retained their tails. Furthermore, if this chance mutation had not occurred and been fixed, humans might still have tails today.

The order in which mutational events are introduced into genomes can have a profound impact on future evolutionary possibility. The early addition of one mutation rather than another may shut down large swathes of evolutionary 'opportunity space', while simultaneously opening up new pathways. Some functions are routed into suboptimal solutions and get stuck there. This is the result of a phenomenon called 'epistasis', whereby the 'fitness' of a mutation is a function of the status of the rest of the genome.

A mutation may be beneficial when occurring in the context of one genetic background, but neutral on another. Had mutations been introduced into genomes in a different historical order, alternative evolutionary pathways would have been explored. Founder 'permissive' mutations – often neutral in isolation – are frequently necessary for subsequent mutations invoking evolutionary change. Without the context provided by the 'founder' mutation, subsequent mutations may be neutral or deleterious. Had the initial ordering of mutations been more favourable, they might have encountered more optimal opportunities. But the attainment of near-term adaptive solutions through the fixation of specific mutations may constrain access to genetic landscapes able to furnish optimal solutions to future adaptive problems. Every move in genetic sequence space simultaneously opens up and shuts down vast swathes of sequence opportunity space. As the future always remains unknown, the long-term consequences of any specific movement in DNA sequence space remains unpredictable.

This context dependence of mutations likely renders many aspects of evolution irreversible. The specificity of the cortisol receptor in humans, for example, depends on the presence of seven ancient mutations. But when these mutations were artificially reversed to engineer a more promiscuous protein reminiscent of the ancestral one, the result was a 'dead', non-functioning receptor. Other mutations that had in the intervening time accumulated in the protein had made the original constellation of mutations obsolete.

The impact of a mutation may also change across an organism's lifespan, such that traits favoured early on in life may subsequently become problematic. A phenomenon called senescence, for example, prevents cells from dividing. While this can help prevent cancer in youth by halting unchecked cell division, senescent cells accumulate in tissues, causing a host of complications.

Occasionally, historical accidents force evolution into awkward rabbit holes. In 1995, the biologist Timothy Onasch studied a family of enzymes known as the hydratase-dehydratases. He demonstrated how quickly a quirk of evolutionary history can lock a species into a suboptimal metabolic outcome. Hydratase-dehydratase enzymes, which catalyse the addition or removal of water in metabolic reactions, come in two versions: *syn* and *anti*, reflecting the difference in how their atomic structures are oriented. One might expect evolution to have selected the most optimal versions, and generally that is what has happened. The more efficient *anti* forms are used preferentially, while *syn* pathways are typically marginalised. However, eight hydratase-dehydratases have adopted the less favourable *syn* form. The anomalous use of the chemically inefficient version of the enzyme likely reflects historical contingency rather than an adaptative optimisation.

Alternative molecular histories – the pathways time forgot – can be explored by artificially reconstructing the genome sequence space around the ancestral proteins that gave rise to modern ones. The evolution of protein sequences involves exploring the topology of a regional DNA sequence landscape within Fred's Library. Mapping the distribution of functions across small regions of sequence space reveals how different starting points are linked to distinct destinations and evolutionary opportunities. But in evolution, the point of embarkation is determined by chance. It's like deciding that you want to go on holiday but being unable to choose the destination. Your cruise ship might end up in New York, but you might equally find yourself vacationing in Rio de Janeiro or Monte Carlo.

The distribution of destinations within localised DNA sequence space determines how fruitful an evolutionary journey is likely to be and comprises a fitness 'landscape'. Some landscapes are more navigable than others, containing pathways that allow a sequence to gradually 'improve', via functional intermediates. These types of 'rugged' landscapes are readily traversed by evolutionary processes. Others, in contrast, comprise vast, featureless plains – regions that are near impossible to navigate using gradual evolutionary processes. Evolutionary landscapes are highly degenerate, and many alternative sequences can adequately perform a required function. Yet each represents a distinct embarkation point for subsequent journeys, and is associated with its own unique network of constraints and opportunities.

Fitness maps of specific regions of sequence spaces can be constructed to identify the richness of the possibility within each region, and to determine whether evolution has identified the best of all possible available options, both locally and globally. Movement in a particular direction in sequence space results in a serial chain of compounding events, each as unpredictable as the next. Some genes and their mutant versions exhibit what are known as pleiotropic effects, meaning that they may influence multiple traits – some or all of which are unrelated.

Human history, too, has been impacted by chance events. Blaise Pascal, the French mathematician, philosopher and inventor, famously mused in his 1670 *Pensées* (Thoughts) that had Cleopatra's nose 'been shorter, the whole face of the world would have been changed'. Without the authority and elegance her distinctive nose conveyed, he argued, Julius Caesar and Mark Antony might never have fallen under her spell. The history of two critically important civilisations – ancient Rome and Egypt – might, as a result, have unfolded differently.

This deep reliance on chance makes the precise predictions typical of the exact sciences – such as physics – near impossible in biology. Despite their apparent complexity, physical systems may often be described by simple laws. For example, the behaviour of a gas can be predicted by determining the numeric values of just two parameters: temperature and pressure. The positions and velocities of individual atoms can be abstracted away. But biological systems appear to defy such simplifications, relying instead on a complex web of interconnected and often unpredictable factors.

It is perhaps this inherent unpredictability that led the nineteenth-century British astronomer Sir John Herschel to dismiss Darwin's evolutionary theory as the 'law of higgledy-piggledy'. Herschel was voicing a widely held scepticism: could a process as consequential as evolution really be attributed to a series of chance events? Charles Darwin, in a letter to Charles Lyell dated 10 December 1859, reflected drily on Herschel's critique, noting, 'What this exactly means I do not know, but it is evidently very contemptuous.'

The perceived unpredictability of evolution inspired the Harvard evolutionary biologist and palaeontologist Steven Jay Gould to further probe the role of chance in history. In *Wonderful Life* (1989), Gould examined whether evolution's outcomes were predetermined and inevitable, or whether they could have emerged in different ways. If evolution followed a deterministic script, its outcomes would be limited and repeatable – unfolding with the mechanical predictability of a cuckoo clock, producing identical, or near-identical results on each occasion.

To explore the interplay between chance and inevitability in shaping evolutionary outcomes, Gould proposed a *Gedankenexperiment* – German for 'thought experiment' – based on the idea of 'replaying life's tape'. He imagined the evolutionary history of life as a tape recording that could be stopped, rewound to any moment in life's past history, and replayed.

In Gould's view, each successive replay would likely produce a unique evolutionary outcome. While we might be tempted to believe that the emergence of our own species, *Homo sapiens* – often considered exceptional and central to life's story – was inevitable, Gould argued otherwise. The emergence of humans, he believed, was in fact highly improbable. That evolution managed to discover *Homo sapiens* at all within the vast expanses of Fred's Library was itself remarkable. 'Replay the tape a million times,' Gould argued, and 'I doubt if anything like *Homo sapiens* would ever evolve again'. Far from being the inevitable and expected outcome of evolutionary processes, humans were, on the contrary, an unpredictable, likely unrepeatable, and unplanned accident of history. We are, as such, an 'accidental species'.

From Gould's perspective, the diversity of Earth's species represents a 'subset of workable, but basically fortuitous, survivals among a much larger set that could have functioned just as well, but either never arose, or lost their opportunities, by historical happenstance'. Early evolutionary commitments – body plans, biochemistries and morphologies – locked life into specific trajectories, opening some pathways, while shutting down others. Despite the ongoing agency of Darwinian evolution natural selection, the nature of historical species carries a significant arbitrary component. Not everyone, however, shared his perspective.

The Cambridge palaeontologist Simon Conway Morris and Washington University evolutionary biologist Jonathan Losos, for example, contended that evolutionary outcomes are more predictable and deterministic than Gould imagined. This perspective was grounded in the observation that faced with comparable environmental challenges, independently evolving populations frequently converge on similar adaptive solutions.

Different species of salamander, for instance, tend to increase in size across evolutionary time, but they achieve this by invoking distinct mechanisms. While some increase the number of their vertebrae, others elongate existing ones. There are many such examples of this type of convergence in nature. More than twenty different populations of the plant *Arabidopsis thaliana* have independently evolved mutations in *Frigida*, a gene enabling them to flower early. Bats and bottlenose dolphins appear to have independently evolved the capacity for echolation – the use of ultrasound pulses for orientation and hunting – and they share genomic alterations suggestive of convergence at over 200 different positions.

Chemical and functional convergence is widespread across nature. Female Asian elephants, for example, release a pheromone into their urine prior to ovulation, to attract males. Astonishingly, this is identical to a compound used by the females of more than 126 different insect species for the exact same purpose. Despite their immense evolutionary distance, these species have independently converged on the same chemical solution to address a common biological challenge.

The South American rainforest katydid provides another compelling example. This insect has hearing organs on its legs that mimic the anatomy of human ears, complete with three human-like acoustic chambers. Meanwhile, the comb jelly – a tiny, translucent marine predator also known as the Pacific sea gooseberry – defies convention by lacking most of the ten primary neurotransmitters found in other animals. Yet it has independently evolved its own distinct nervous system, suggesting that it builds and operates its neural architecture in a fundamentally different manner to other species. This organisational diversity suggests that nervous systems have evolved independently multiple times on Earth, with each architecture achieving different types and levels of functionality. While there may be many different ways to build a brain, the eventual emergence of brain-like structures may be inevitable.

Behavioural convergence provides further confirmatory evidence for a predictable component to evolution. Aleocharine rove beetles, comprising over 61,000 different species, have evolved the ability to infiltrate the nests of army ants on more than a dozen separate occasions over the past 105 million years. These tiny beetles have managed to lose their beetle-like appearance and are now virtually indistinguishable from army ants – even adopting an ant-like gait. Indeed, they look, smell and behave like ants to such an extent that they are able to walk into the nests of these notoriously pugnacious aggressors with impunity. Once inside the foreign ant metropolis, these social parasites raid their hosts' pantries. So thoroughly do they charm their unwitting hosts that they enthusiastically transport them through their home. The intruders reward this hospitality by devouring their young.

In 2016 Joseph Parker, Shûhie Yamamoto and Munetoshi Maruyama, working at Columbia University in New York, described a 99-million-year-old fossilised ant with anatomical features suggestive of it being a social parasite. Such complex deceptions have, it seems, been in play for at least 99 million years. What once appeared to be a one-off evolutionary trick is now understood to represent a recurring evolutionary theme.

Convergence also extends to culture. In 1900, the German psychologist Carl Stumpf used the wax cylinder of his phonograph to record the Siamese Court Orchestra as they performed in the Berlin *Zoologischer Garten* (Zoological Garden). Intrigued by the parallels between traditional music derived from diverse cultures, he helped pioneer the science of ethnomusicology. He established the Berlin Phonogramm-Archiv to house his collection of recordings, as well as those of other individuals. By 1933 it contained more than 13,300 traditional world music recordings.

In 2019, a group led by Luke Glowacki and Samuel Mehr at Pennsylvania State University analysed a diverse collection of lullabies, dance songs, love songs and healing songs from across the world, searching for shared acoustic features suggestive of cultural convergence. If such patterns were present, they might imply the existence of universal principles underpinning the foundations of culture. When the songs were evaluated on the basis of formality, arousal and religiosity, several recurring acoustic patterns emerged. Dance songs, for example, were faster and more rhythmic than lullabies, while ritual healing songs were consistently melodically uniform. These cross-cultural regularities suggested the existence of conserved structural constraints that subtly influence human creativity across diverse cultures.

Not all cases of apparent convergence are, however, what they seem. In some instances, what looks like convergence actually reflects shared ancestry. The eyes of humans and fruit flies were once celebrated as prime examples of convergent evolution – as were those of about four dozen other species. But once the genetics of eye development was better understood, it became clear that the eyes of animals as diverse as flies and humans were built using the same set of evolutionarily conserved genes: *eyeless*, *atonal* and *eyes absent*. Moreover, the *PAX6* gene, central to the production of opsin – a light-detecting protein – is so conserved between species that the mouse version of the gene can be used to direct the development of an insect eye. What initially appeared to be independent inventions turned out, instead, to reflect the reuse of related components from a shared evolutionary toolkit.

To test the repeatability of evolution, Richard Lenski at Michigan State University devised an ingenious experiment, which he called the Long-Term Evolutionary Experiment (LTEE). While the tape of life's actual history cannot be directly rewound and replayed to test the veracity of Gould's hypothesis, Lenski's system allowed for the actual experimental replays of specific evolutionary histories. This made it possible to study the dynamics of evolution, and particularly its repeatability, under laboratory conditions.

Given that the bacterium *E. coli* divides around six times a day – roughly equivalent to 150 years of human history – and that bacterial cultures contain tens of millions of individual organisms, Lenski realised he could generate a host of miniature, parallel bacterial worlds. Like a divine creator and observer peering through the invisible ether of this microbial universe, he could track their evolutionary trajectories across multiple generations.

These artificial worlds were first created in 1988 when, as a young scientist at the University of California Irvine, Lenski established twelve identical bacterial cultures, each derived from the same parental *E. coli* strain. He placed the bacteria in flasks containing glucose-supplemented growth medium, and transferred them to an oxygenated incubator set to body temperature. He then followed them for more than 75,000 generations – equivalent to more than 1.5 million years of human evolution. Every 500 generations, he removed and froze samples from each flask, creating a molecular fossil record of each population's evolutionary journey. This frozen reference library allowed him to trace the genetic changes within each population across evolutionary time. When required, bacteria from the archive could be thawed and resurrected to investigate the underlying causes of any observed evolutionary phenomenon.

Given their identical ancestry, starting conditions and environments, the key question was whether the bacteria in each separate 'world' would evolve identically, or whether their genome sequences would diverge unpredictably. If the unfolding of evolution were programmed and inevitable, then the genetic changes in each flask should be similar.

While most of the populations followed predictable paths, in 2003 – around fifteen years after the experiment began – the medium in one of the flasks unexpectedly turned opaque due to extensive bacterial overgrowth. The bacteria in that flask had acquired a unique innovation: they had developed the ability to metabolise citrate, a chemical routinely added to the bacterial growth media to facilitate iron absorption. This unprecedented metabolic leap endowed them with a decisive advantage that allowed them to outcompete the other populations, which remained reliant on glucose. Here was a sudden and dramatic evolutionary breakthrough – a unicorn spontaneously emerging from a stable of horses – that exemplified evolution's unpredictability and remarkable generative power.

A molecular analysis of frozen molecular fossils from the flask that produced the unicorn, revealed that the apparently abrupt appearance of the ability to metabolise citrate had been preceded by a gradual accumulation of mutations, exactly as Darwinian theory predicted. Achieving this metabolic shift required significant rewiring of the bacteria's metabolic machinery. Lenski's analysis showed that the emergence of this new trait involved several gene duplications and modifications to gene regulatory networks. Although the bacteria in the remaining flasks never acquired the capacity to metabolise citrate, they were guided down alternative evolutionary pathways. This allowed them to accumulate their own distinct sets of mutations that helped them grow faster and metabolise glucose more efficiently.

While Lenski's experiment validated Gould's thesis – that chance events can indeed have a significant impact on evolutionary outcomes – replays using resurrected, frozen ancestral strains from representative generations revealed that some of the outcomes were repeatable, if caught at the right moment. Once certain foundational mutations had been established, particular outcomes become increasingly probable. So, although historical contingency plays a powerful role in dictating evolutionary outcomes, there appears to be a layer of – as yet poorly understood – higher-order predictability that must be factored in as well.

This may reflect the fact that the probability of a mutation occurring at a specific location is not always equal and may be influenced by the sequence's genomic environment. This generates mutational 'hot' and 'cold' spots – regions that are, respectively, more or less prone to mutation. While all positions in a genome's nucleotide sequence are mutable, some locations are more mutable than others. So, whereas any specific mutational event is random, the probability of its occurrence is influenced by its genomic context.

The characteristics of natural species are, consequently, shaped by three principal factors: chance events that generate raw mutational material for evolution, the agency of evolution by natural selection, and the historically contingent events a species encounters. These include the particular order in which mutations are introduced into genomes, and the opportunities and constraints they impose on future possibility.

One thing, however, is clear. The species alive today – including our own – represent an infinitesimally small sampling of biological possibility. Furthermore, chance events have left a deep imprint on life on Earth, resulting in vulnerabilities that might have been avoided. These have, among other things, resulted in a host of idiosyncrasies that predispose us to disease, brief healthspans and relatively short lifespans. Had history unfolded slightly differently – one viral infection more or less – the pageant of life on Earth, including human nature, might have been otherwise.

Should we wish to revisit life's history, in order to reimagine some of the haphazard structures that evolution has built into the workings of genomes, we cannot expect natural selection to help us. The unprecedented ability to design and synthesise the genomes of species, on the other hand, offers, for the first time, the possibility of restructuring the organisation of biological machines.

But as we have seen, biological machines are configured in a fundamentally different way from human-made machines and are inherently less amenable to engineering. The rational design of species from first principles will consequently require their design logic to be streamlined in ways that reduce their complexity and make them more engineerable.

5
Biological Machines

> Nature in order to carry out the marvellous operations in animals and plants has been pleased to construct their organised bodies with a very large number of machines, which are of necessity made up of extremely small parts.
>
> – MARCELLO MALPIGHI, 1697

In 1747, four years prior to his premature death at the age of forty-two – allegedly after gorging on pheasant pâté – the French physician, hedonist and materialist philosopher Julien Offray de La Mettrie published his incendiary masterpiece *L'homme machine* (Man a Machine). Its thesis was compellingly simple: humans are machines. Rather than being ethereal entities, the human mind and soul emerged, he argued, as the by-products of the body's machine-like activities. Unsurprisingly, the treatise proved controversial, and La Mettrie was forced into self-imposed exile in Berlin, where he found refuge in the court of Frederick the Great, the free-thinking Prussian king.

The notion that the human body was a machine was itself a lot to stomach. But worse still, La Mettrie's ideas raised an even more unsettling question. What if in addition to being a machine, the human body also turned out to be replete with imperfections? Excessively complicated, over-engineered and riddled with shortcomings, this flawed machine had a propensity for both malfunction and failure. This scathing reproach of human nature's constitution found a champion in the Scottish philosopher and author of *A Treatise of Human Nature* (1739), David Hume, who sensibly chose to deliver his critique from beyond the grave.

Published in 1779, three years after his death, Hume's controversial masterpiece *Dialogues Concerning Natural Religion* challenged the prevailing theological wisdom that human machines had been designed by a divine craftsman. Framed as a fictional debate between a sceptic, an orthodox Christian and a Newtonian theorist, the text critically examined the evidence for the existence of an intelligent creator through a series of reflections on the natural world. To highlight his scepticism, Hume dared to imagine the indignation and disappointment one might experience when contemplating a 'complicated, useful, and beautiful' machine like a ship, only to discover that the carpenter responsible for crafting it was a 'stupid mechanic'.

Nearly half a century later, Darwin and Wallace revealed that the 'stupid mechanic' responsible for generating life's machinery was not a deity at all, but rather the indifferent, bumbling and opportunistic process of evolution by natural selection. Constrained by its prior innovations and the machinations of chance and history, evolution has been forced to tirelessly rework its designs, rehashing them into new renditions, favouring expediency over perfection. As a result, the living 'machines' shaped by evolution – including humans – are riddled with design faults and suboptimalities that are rarely encountered in purpose-built contraptions. Such imperfections are the root cause of many of the human body's countless failings, leading to disease, degeneration and a diminished lifespan.

Fortunately, the methods of molecular biology have allowed us to zoom into the components of living machines and view their shortcomings at an atomic level of abstraction. When laid out on the metaphorical dissection table, a forensic examination of the details of our flawed structures indicates that Hume's bleak conclusion appears justified. The 'stupid mechanic' appears to have assembled its imperfect creations both blindfolded and with its hands tied. Yet this same insight also suggests a roadmap for intervention: by understanding the flawed configurations of our biological organisation, we may one day be able to repair or redesign them.

Were the human body and other biological machines built like their human-made counterparts, fixing them would be straightforward. Faulty parts could simply be removed, replaced, or modified. It turns out, however, that the evolutionary 'mechanic' has configured them using an eccentric playbook. Biological machines are, as a result, an engineer's nightmare.

While the metaphor of biological systems as machines provides a convenient way of thinking about them, they are configured very differently from human-made machines. Although machine-like in their behaviour, they incorporate several unique features. This has significant consequences for the way in which we might consider reconfiguring them to introduce new functions or modify existing ones. It also impacts how we address the issues of human disease and the attainment of healthy longevity.

One striking difference between biological machines and their human-made counterparts lies in the nature of their purpose and origin. Human-made machines are designed with intent, built according to a preconceived design and maintained by external agents such as engineers or other machines. Evolution assembles its machinery through a haphazard interplay of heredity, chance, natural selection, physical constraints and self-organisation. Biological machines are not designed intentionally according to a predetermined plan. Neither are they crafted by an actual designer, intent on realising a specific outcome. They frequently accommodate expedient outcomes that, while adequate, are far from optimal.

The purposiveness of human-made machines is extrinsic. They are designed, built and maintained by external agents, including human engineers and manufacturing systems. In the case of naturally engineered 'evolution machines', on the other hand, their purposiveness is intrinsic, with causality issuing principally from within the machine's workings. Moreover, the internal structures of biological systems change dynamically across time, in contrast to human-made machines, whose structures are static. Were evolution to operate within human-made machinery, it would invariably result in catastrophic failure. Biological machines have, on the other hand, evolved to accommodate change and to be evolvable.

Natural evolutionary engineering may overwrite historical designs, but it cannot rewrite them. Existing designs rely, instead, on the remodelling of ancient materials. Although continuously acquiring new components, biological machines regurgitate a mashed-up version of their past. As Darwin observed, nature's creations are made of 'old wheels, springs and pulleys', only slightly altered over time. As a result, natural systems cannot be optimised in any formal engineering sense.

Many features of biological systems are relics of distant chance events, with little or no relevance to the present. Some are purposeless 'free-riding' structures, architectural by-products that emerged as a result of alterations to unrelated features. As we have seen, biological machines are consequently replete with architectural flaws, imperfections and historical baggage. Dysfunction abounds; many components are functionless, suboptimal, or maladaptive. While highly optimised in some domains, biological machines also accommodate a vast catalogue of errors.

Take, for example, the human brain. While capable of remarkable feats – including love, memory and contemplation – it is also prone to distortion and dysfunction. It offers some of the most compelling examples of evolution's design botches – unavoidable consequences of its unbreakable link with history. At an anatomical level, the brain's neurons are far from ideal computing elements: they are slow, leaky and unreliable. Worse still, the genes used to build our brains are, as we have seen, similar to those used to construct the brains of other species, such as flies, worms and jellyfish. The continued use of molecular relics of an ancient past, without substantive updating, likely contributes to our various illusions, delusions, emotional disorders, inattentions and cognitive malfunctions.

We continue to use cognitive strategies that operate reflexively, based on ancient assumptions about the world that may no longer serve us. Consider, for example, the famous Invisible Gorilla Experiment performed in 1999 by Daniel Simons and Christopher Chabris at Harvard University. Subjects were asked to watch a video and count how many times three basketball players passed a ball to one another. After around thirty seconds, a person in a gorilla suit walks across the screen, faces the camera, thumps their chest and walks off. Remarkably, around half the participants failed to spot the gorilla. This phenomenon, in which we fail to notice strikingly obvious things happening right in front of us when asked to focus on something else, is known as inattentional blindness. Our cognitive apparatus, it seems, is optimised for ancient priorities, which may sometimes blind us to what's right in plain sight.

These kinds of design flaws extend beyond the human brain. The enzymes that catalyse the chemical reactions essential for living processes provide another striking example. Evolution has optimised certain kinetic parameters governing enzyme efficiency, including their 'turnover number', which represents the number of substrate molecules an enzyme can convert into products within a fixed time. While some enzymes – like triosephosphate isomerase, which helps transform one type of sugar into another – operate at near-theoretical optimality; the majority do not.

In 2011, Ron Milo and Dan Tawfik at the Weizmann Institute in Israel performed a comprehensive analysis of enzyme efficiency. Their results were striking. It turns out that most human enzymes function far below their theoretical potential for perfection. The catalytic inefficiencies of natural enzymes, for example, span five orders of magnitude, with most operating far below the theoretical limits of optimal performance displayed by 'fast' enzymes. The same holds true for the enzymes of other species. Even rubisco, the enzyme used by plants to fix atmospheric carbon dioxide, is notoriously slow and inefficient, providing yet another example of nature's tendency to favour adequacy over excellence.

In their 1995 book *Evolution and Healing*, Randolph M. Nesse and George C. Williams described the design of the human body as 'simultaneously extraordinarily precise, and unbelievably slipshod'. Some of its flaws, they argued, are so egregious, they suggest that we were 'shaped by a prankster'. Evolution, it turns out, often performs at the level of a D-grade student rather than a *cum laude*. Were evolution a person, we might at times find it exasperating. Were it a student, its unfavourable report card might describe it as 'inclined at times to be lazy, untidy, opportunistic, overly sentimental, transfixed by the past and prone to taking the path of least resistance'. Ideologically uncommitted, it is content to settle for easy-to-attain but workable solutions, rather than searching out the best ones. It's the kind of colleague that you might not be eager to share a cabin with on an overseas voyage. As Tawfik concluded in 2010, 'messiness is inherent to biology'.

The structures produced by evolution resemble the architectural layering of modern Rome: a city built atop successive archaeological strata, with remnants of classical Rome seamlessly integrated into a fabric of Renaissance, baroque and modern elements. Like evolution, Rome's city planners have had to work within the constraints of existing materials, structures and spaces. Ancient monuments like the Colosseum and Forum coexist with modern streets and buildings, each with its own chaotic beauty. But the need to accommodate such impracticalities has placed significant constraints on Rome's urban development, as planners navigate a thicket of zoning restrictions, excavation regulations and infrastructure challenges. Narrow, winding streets designed for chariots struggle to accommodate tourist buses.

The disconnect between the requirements of the past and those of the present result in a patchwork of suboptimalities that undermine biological systems at every scale. The saga of the Tappan Zee Bridge in New York State provides a fitting metaphor.

During the long, hot summer months, some New Yorkers head east in long traffic jams to vacation on Long Island, while others journey upstate to the Hudson Valley, Catskill Mountains and beyond. For the latter, there are two main crossings over the Hudson River. The closer of the two, the George Washington Bridge, is often clogged with traffic. But if one continues north, hugging the river as it snakes past Yonkers towards Tarrytown, there is a second crossing: the Tappan Zee Bridge. Its most memorable feature is that it was built in the wrong place. A glance at a map rapidly reveals that had it been constructed just a little further south, it could have been half as long.

With a span of over three miles – compared with the single-mile-long span of the George Washington Bridge – its location seems almost absurd. No rational engineer would have constructed it in its current position. Unsurprisingly, there was ongoing debate about how this peculiar placement came about. Some speculated it was to preserve the view from the Rockefeller riverside estate; others invoked geological factors. The truth, however, turned out to be much simpler. The location of the bridge was purely opportunistic, and deeply historically contingent.

The Tappan Zee was part of the New York State Thruway project, conceived in the 1940s to connect major cities across the state. The plan was to finance part of its construction through tolls. But there was a problem. The legal jurisdiction of the Port Authority of New

York extended for a twenty-five-mile radius around the Statue of Liberty. Were the bridge positioned within its range, toll revenues would flow to New York City rather than to the state. To address this issue, Governor Thomas E. Dewey lobbied against the Port Authority's proposal to build the bridge at a narrower crossing closer to the city. As a result, the bridge was constructed between Grand View-on-Hudson and Tarrytown, just outside the Port Authority's jurisdiction.

And that might have been that, were it not for the fact that Dewey's short-term expediency had significant long-term implications. For unlike the George Washington Bridge, which was anchored on solid rock, the Tappan Zee required an elaborate underwater flotation system. The cost of building this, along with the bridge's excessive dimensions, strained an already inadequate budget. A materials shortage precipitated by the onset of the Korean War served only to compound the problem, leading to shoddy construction. Completed in 1955, the bridge rapidly fell into disrepair and was demolished in 2017. This provided the perfect opportunity to correct an irrelevant, historically defined feature by relocating the bridge to a narrower part of the crossing. But by then, an extensive infrastructure had grown around its original location, making relocation impractical. The replacement bridge was, as a result, constructed alongside the position of the original, thereby perpetuating a historically contingent suboptimal design.

Biological systems, including the human body, contain their own Tappan Zees – vestiges of an irrelevant past carried forward into the present. These evolutionary relics continue to constrain our biology. The causes of many such awkward anomalies originated long before the evolution of our species, yet they reverberate across biological time, undermining our biology. One such 'Tappan Zee' is the human spine.

While well suited to the lifestyle of our quadrupedal, tree-dwelling ape ancestors, the morphology of the human spine is poorly adapted for walking on two legs. Evolution has, nevertheless, succeeded in awkwardly reconfiguring the original design to accommodate its new bipedal function. But this is akin to trying to fit a square peg into a round hole. To support an upright posture, the angle between the femur and the hip gradually widened, the pelvic structures readjusted, and the spine acquired its characteristic inward curve. This remarkable and unlikely transformation was like turning a table into a chair. That it succeeded at all is a testament to the extraordinary malleability of biology systems and the generative power of Darwinian evolution.

The resultant remodelled anatomy, however, is far from what one would design by choice, especially from a biomechanical standpoint. The human spine is, in fact, a design disaster. Were it possible to redesign it from scratch, it would not be modelled on the design of a creature that walked on four legs. Bruce Latimer, an anatomist at Case Western Reserve University, in Cleveland, Ohio, likened the unwieldy rotation of a spine by the ninety degrees necessary to make it vertical for bipedal walking to 'stacking 26 cups and saucers on top of each other', and then 'balancing a head on top'. The vertebrae are the cups; the discs between them, the saucers. In this new configuration, the weight of the head places constant, excessive pressure on these fragile structures.

But this 'jerry-rigged' redesign of the spine had to operate within very tight parameters. Chief among these was the need to accommodate the pelvic birth canal. Humans give birth to babies with much larger bodies and brains than primate babies, necessitating a large pelvis. Unfortunately, these two features – bipedal locomotion and childbirth – have conflicting design needs. While delivering a baby requires a wide pelvis, walking upright favours a narrow one. The evolutionary compromise was an inward-curving spine with its characteristic 'S' shape, accommodating both functions, albeit imperfectly. This compromise has left us prone to back pain, vertebral fractures and obstetric complications.

Our feet, too, are poorly designed, representing another anatomical compromise. Anthropologist Jeremy DeSilva at Boston University has used the human foot as an example of how evolution 'works with duct tape and paper clips' to patch solutions together quickly. Had you asked an engineer to design a human foot, they would have been unlikely to make it with so many moving parts. Its exuberant deployment of bones is especially notable when one realises that the twenty-six bones in the foot account for a quarter of all the bones in the human body.

While this configuration is well suited for tree-dwelling apes that need to grasp branches, it is both unnecessary and cumbersome for a bipedal species. Though evolution has attempted to bolster its rigidity, the human foot remains susceptible to inward and outward twisting, and its arches prone to collapse. These vulnerabilities lead to a host of ailments and afflictions including tendonitis, plantar fasciitis and broken ankles. Had we been consulted, we might have fashioned our feet more like those of ostriches, whose ankle and lower-leg bones are fused into a single, powerful structure optimised for walking and running.

One of the most striking biological examples of an anatomical 'Tappan Zee' is the eccentric distribution of the left recurrent laryngeal nerve. The anatomy of this nerve, which facilitates sound production in the larynx, has a highly irrational feature. While the right recurrent laryngeal nerve branches off from the vagus nerve as it exits the brain at the base of the skull and proceeds directly to the larynx, the left eschews this direct course, embarking instead on a bizarre detour. It passes down in front of the aorta, loops underneath and behind it, performing a U-turn before ascending back up to the larynx, ending up essentially where it started. It is hard to imagine any advantage to this circuitous path, and a rational engineer would never have designed it that way

The left recurrent laryngeal nerve's eccentric route is, however, more than just a curiosity. It comes with jeopardy. Compression, ischaemia or injury to the left recurrent laryngeal nerve is more common than on the right, and may result in left-sided vocal cord paralysis. While relatively rare in humans, in long-necked species like horses, this anatomical quirk can cause significant issues. Horses, for example, are prone to develop a condition called equine recurrent laryngeal neuropathy. This is predominantly left-sided, and may cause respiratory problems. It is especially pronounced in racehorses, their size and elongated necks making them particularly susceptible. In the long-necked African giraffe, the nerve's convoluted route can extend to a staggering fifteen feet. Had the nerve been routed directly to the larynx, a length of one foot would have sufficed. The left recurrent laryngeal nerve of the gargantuan, long-necked, sauropod dinosaur *Supersaurus* – living between 230 and 66 million years ago, and with a neck six times the length of a giraffe's – would have been subject to a detour of up to ninety-two feet. It provides a striking example of evolution's tomfoolery.

The explanation, as always, lies in its evolutionary history. In our fish-like ancestors, the nerve followed a short, direct path to its target, as is still the case in modern fish. But as the mammalian neck lengthened and the position of the heart progressively lowered, the nerve lengthened too – its development constrained, as ever, by its original genetic blueprint.

Evolution's nostalgia for the past isn't just confined to anatomy. It may also occur at the molecular level. While most animals synthesise their own vitamin C, humans – along with certain primates, fruit bats and guinea pigs – have acquired mutations in an enzyme called GLO, critical for vitamin C synthesis. As a result, we can no longer produce our own vitamin C and must obtain it from dietary sources. There is no obvious reason why it should be beneficial to lose the ability to make vitamin C. The loss is most likely a whimsy of evolution.

A lack of adequate dietary vitamin C may result in scurvy, a potentially fatal disease. Scurvy was common in sailors, who on long sea voyages lacked access to fresh fruit and vegetables. It results in weakness, bleeding gums, joint pains, poor wound healing and infections. While sailing around the Cape of Good Hope in 1498, Vasco da Gama lost 100 of his 160 crew to the disease. In *A Treatise of the Scurvy* (1753), the naval physician James Lind described how sailors suffering from scurvy aboard his ship, the *Salisbury*, were treated with a diet of oranges and lemons. The beneficial effects were, he wrote, 'sudden and visible', one sailor 'being at the end of six days fit for duty'.

While evolutionary oddities are common in physical structures, they may also occur in psychological ones. What we might refer to as behavioural 'Tappan Zees' are mental patterns rooted in behaviours that evolved to keep us safe in the past but that in the context of modern life may prove problematic, or even maladaptive. They may, for example, trigger emotional responses that undermine our well-being. Often, we are unable help ourselves, as our wiring isn't configured to do so. Yet such apparent imperfections in human nature may sometimes be components of its most essential features.

The German composer Richard Wagner's cynical contemplation of the human condition in *Die Meistersinger von Nürnberg* (The Mastersingers of Nuremberg), written between 1862 and 1867, brilliantly captures this tension. In Act III, the poet-musician and cobbler Hans Sachs laments the penchant humans have for misery, wondering why we find it so hard to be good and happy, proclaiming: 'Madness! Madness! Everywhere madness!' In a letter drafted to the king of Bavaria, Wagner described this famous 'madness monologue' as a 'cry from hidden depths'. Yet Sachs also observes that beauty and order can, paradoxically, emerge from such chaos. As inconvenient and unpleasant as it may appear, the coexistence and juxtaposition of happiness with strife is an intrinsic part of the human condition, and most likely an existential necessity.

The philosopher Jerry Fodor echoed this sentiment in his 2007 essay 'Why Pigs Don't Have Wings', stating bluntly – while alluding to Hans Sachs's monologue – that 'there isn't yet a science of the human condition that can tell us what is wrong with the way we are'. At least some of the blame, however, lies with our evolutionary origins. Our minds did not evolve to navigate the intricacies of the modern world. While culture evolves exponentially, historical, linearly and incrementally progressing genetics continuously lags behind, leaving it poorly equipped to keep pace with our rapidly changing world. The success of human culture serves only to exacerbate the inadequacy of genomes to track high-frequency environmental change. Some aspects of our dysfunctional behaviour may result from behavioural Tappan Zees that, while well suited to the lifestyles of our primate ancestors, serve us poorly in the present. Fodor concluded that evolution has left us with a brain that 'doesn't work very well in third millennium Lower Manhattan'.

A common misconception about natural selection is that every feature of a biological structure has been optimised. Were this the case then structures would bear the hallmarks of adaptation at every scale. This approach of 'atomising' the individual features of organisms, known as the 'adaptationist programme', assumes that the traits of organisms can be broken down into their components, each of which has the potential to be independently optimised. Every 'trait' is then provided with its own uniquely tailored adaptive narrative.

There is a tendency to routinely attribute adaptive utility to traits. But to do so is to ignore the constraints and trade-offs that natural evolutionary engineering invariably contends with. It also risks confounding the reason for a structure's origin with its current utility. A human foot, for example, has twenty-six bones – not for adaptive reasons, but because it was *once* adaptive in a different setting. Modern feet function *despite* having twenty-six bones, not because of them. We would likely benefit from having fewer bones in our feet, but it would be easy to construct an adaptive narrative supporting their utility that would be difficult, if not impossible, to falsify.

These kinds of speculative adaptive narratives resemble Rudyard Kipling's *Just So Stories*, first published in 1902. These fables – designed to delight young children – offer playful examples of how various animals acquired their distinctive features. The camel, for example, in 'How the Camel Got His Hump', acquires its hump as a punishment from a djinn or 'genie' for its idleness. Similarly, in 'How the Rhino Got His Skin', Kipling recounts how the 'great folds' and creases in the hide of a rhinoceros resulted from the actions of a vengeful Red Sea Parsee intent on teaching an uncompromising rhinoceros a lesson after it devoured his plum cake. While speculative and fantastical, *Just so Stories*, and their adaptive evolutionary narrative equivalents, are irresistibly digestible and have a veneer of plausibility, despite being erroneous and unfalsifiable.

In their 1979 essay 'The Spandrels of San Marco and the Panglossian Paradigm: A Critique of the Adaptationist Programme', the Harvard biologists Stephen Jay Gould and Richard Lewontin used an architectural detail from the Basilica of San Marco in St Mark's Square, Venice, to challenge this pervasive type of baseless, adaptive thinking. They argued that it had become 'deeply engrained' in evolutionary biology, while overlooking the reality of natural selection's messy workings. They emphasised that many biological structures are not subject to direct selection, and are neither optimal nor adaptive. Sometimes, features emerge not because they were selected for, but because they were unavoidable by-products of other processes. Such epiphonema clutter biological machines with irrelevancies and suboptimalities. Even Darwin, in the final edition of *The Origin of Species* (1872), adopted a pluralistic view of evolution, writing: 'I am convinced that natural selection has been the main, but not the exclusive means of modification.' A more holistic and integrated view of evolution was required.

To make their point, Gould and Lowentin turned to the spandrels of San Marco. The central dome of the basilica of San Marco is supported by four massive arches. Four triangular tapered three-dimensional spaces are generated at the points of intersection of adjacent arches, known as 'spandrels' or 'pendentives'. These distribute the weight of the dome, and are necessary 'architectural byproducts of mounting a dome on rounded arches'. In the basilica, each is adorned with mosaics depicting Christian iconography. They dominate the visual aesthetic of the dome, incorporating overpowering images of Christ and angels, accommodating the four evangelists and four biblical rivers. One may be forgiven for thinking that the imagery should be the natural starting point for any architectural analysis.

But to put the mosaics at the centre of the spandrels' purpose would be to invert causality. The artwork is an opportunistic adaptation, a creative afterthought, which leverages an architectural constraint. It was the architectural constraint that drove the arrangement, not the iconography. The decorative artwork simply capitalised on the non-adaptive leftover empty spaces created by the construction of the spandrels.

Applying this metaphor to biological traits, Gould and Lewontin urged caution against the 'reliance on plausibility alone as a criterion for accepting speculative tales'. They challenged the reluctance of many evolutionary biologists to consider plausible alternatives to adaptive stories. Evolution by natural selection, they argued, did not possess an unbounded ability to fashion the 'best of all possible worlds'. They compared the reductionist 'adaptationist' mind-set to the optimistic rantings of Dr Pangloss in Voltaire's satirical 1759 novel *Candide, ou l'Optimisme* (Candide, or Optimism), who famously insisted that 'everything is made for the best purpose', and that things 'cannot be other than they are'. Indeed, 'our noses were made to carry spectacles' and 'legs were clearly intended for breeches'. Voltaire used satire to highlight the folly of perceiving every outcome as the signature of intentional design.

The idea of an optimised, perfectly designed world was first mooted by the polymathic German mathematician and philosopher Gottfried Wilhelm Leibniz. Sometimes working in felt socks and a fur-lined gown, Leibniz, the inventor of mathematical calculus, produced just a single philosophical text in his lifetime. Published in 1710, his *Essais de Théodicée sur la bonté de Dieu, la liberté de l'homme et l'origine du mal* (Essays of Theodicity on the Goodness of God, the Freedom of Man, and the Origin of Evil) detailed his vision of an infinite collection of possible worlds, each differing from the next in some subtle detail.

In this metaphysical scheme, all these worlds existed simultaneously in God's mind. At the moment of Creation, God – having peered into each possibility – actualised just a single version, which contained the most optimal balance of good and evil. This, Leibniz argued, represented the best of all possible worlds. Imperfections were to be expected, being integral to the optimality of the system. But as Voltaire realised in *Candide*, – his lampoon of optimistic philosophy – to adopt such a position is untenable, as it is not falsifiable. The evolutionary biologist George C. Williams echoed this scepticism when, in 1966, he cautioned that the term adaptation should not be applied to a characteristic 'unless it is clearly produced by design, and not chance'.

There are many examples of apparent non-adaptive 'spandrels' in the designs of species. The tiny bones of the human inner ear – the malleus, incus and stapes – critical for transmitting sound, provide one example. These structures evolved opportunistically from the jawbones of ancient synapsids, a group of vertebrates that includes mammals and their extinct relatives. At one point in evolutionary history, these bones formed part of the jaw. This decoupling of chewing and hearing evolved independently in several different mammalian species.

Other cases abound. The human retina appears to be wired back to front. As a result, the optic nerve passes through the photoreceptors, creating a 'blind spot' in our vision. The pelvic bones of whales are vestigial remnants of the hind limbs of their land-dwelling ancestors. Ostriches, emus and other flightless birds have wings but are unable to fly; flightless ants and beetles, similarly, have wings that are permanently encased within protective wing covers. The human coccyx – a remnant of our ancestors' tails – lacks a function. The palmaris longus, a slender muscle in the human forearm, is absent in up to a quarter of individuals, but produces no obvious functional deficit. Human males have a vestigial structure known as the prostatic utricle, which is associated with the urethra, but serves no known function.

These features provide compelling evidence against the simplistic notion that every biological feature has an adaptive purpose. What may appear superficially to be a suboptimal design often reflects, instead, a compromise between conflicting evolutionary constraints. Considerable care must therefore be taken before concluding that any particular structure or feature is suboptimal. For example, while the back-to-front configuration of the retina is often cited as a flaw, it has been suggested that this arrangement may actually enhance colour vision – a compromise solution that satisfies conflicting constraints. Similarly, there is evidence that pelvic bone size in whales may correlate with mating success, indicating how evolutionary remnants may, over time, acquire novel functions.

Evolutionary spandrels may also occur at the molecular level. In 2015, biologist Alexandre Morozov proposed that many protein-binding interactions exist not because they are inherently functional, but because they enhance protein folding. The interactions themselves, he suggested, may be largely irrelevant for the function of the protein.

Consider the example of insulin, the protein essential for regulating blood sugar. A single leucine amino acid at position sixteen of insulin's primary amino acid sequence makes no apparent contribution to the insulin receptor. Instead, it acts as a molecular 'buttress', reinforcing the shape of the receptor-binding surface. The leucine appears to be an obligatory architectural feature, persisting because it supports the protein's overall stability.

Having observed the occasional irrationality of natural selection's handiwork, it is instructive to explore some of the other differences between human-engineered machines and their evolutionary counterparts. One critical distinction lies in their modularity – or, more precisely, their lack of it. Unlike human-designed systems, which are characterised by orthogonal components that function independently, biological machines are composed of interdependent, multifunctional parts.

Consider the Boeing 777, a marvel of modern engineering. This long-range commercial airliner, with its approximately 3 million parts (including 133 miles of wiring), is meticulously designed so that each component performs a distinct, optimised function. Each part is precisely assembled and wired into an elaborate network of control systems featuring 150,000 subsystems and 1,000 onboard computers, all integrated into an electronic fly-by-wire interface that reduces the need for pilot intervention. There is a sense in which a 777 is an elaborate network of control systems that happens, almost coincidentally, to fly. Despite its near-biological degree of complexity, the systems-level description of a 777 relies on the functional independence of its components and subsystems.

Biological machines, in contrast, have a fundamentally different mode of assembly and functionality. The equivalent parts list in biological machines is detailed in their genomes, but the way in which the parts of biological machines assemble to generate functionality and the nature of each part, diverges from that of the components of designed machines like aeroplanes. When a part of a biological machine breaks, it cannot simply be replaced or repaired without disrupting multiple other processes. This is because many of the components of biological machines are involved in other processes and functions. This pervasive interconnectivity makes biological machines difficult to engineer, as optimisation in one domain often entails trade-offs in several others – a feature alien to human-made systems.

This lack of orthogonality in biological machines is both a challenge and a necessity. It endows living things with a remarkable robustness and ability to tolerate faults. Unlike engineered devices – typically designed to operate just above their failure thresholds, and discarded or replaced when they break – 'evolution machines' must absorb damage, preserve function and persist in order to replicate and ensure their continuity.

While there is some degree of orthogonality and modularity in biological systems, their components are generally more interconnected and interdependent. The parts of living machines are, in effect, non-orthogonal. Although proteins such as enzymes are often considered exquisitely specific – catalysing individual chemical reactions with a single substrate – they often display a surprising promiscuity, interacting with multiple substrates with varying degrees of efficiency.

This metabolic 'inaccuracy', reflecting what the biologist Max Delbrück in 1969 called 'limited sloppiness', creates what is known as an underground metabolism. These off-target catalytic activities, while existing beyond an enzyme's usual biochemistry, result in metabolic cross-wiring that can serve as new starting points for evolutionary innovation. Much like the mosaics harboured within the spandrels of San Marco's basilica, these 'accidental functions' initially emerge as molecular spandrels – direct by-products of promiscuous interactions that provide metabolic networks with robustness and plasticity.

This molecular promiscuity is not, however, confined to enzymes. Unlike the parts of human-made machines, natural components often multitask, either simultaneously or at distinct points in development. In Arthur Miller's *Death of a Salesman* (1949), the protagonist Willy Loman supplements his primary work as a travelling salesman by taking on additional odd jobs, often at night. This kind of 'moonlighting' is common in many different protein types.

For example, the enzyme DNA-dependent protein kinase, critical for DNA repair, also helps assemble the machinery responsible for protein synthesis. Another multitasking protein, signal transducer and activator of transcription 3 regulates both gene expression and energy production in mitochondria. The immune system protein C-C chemokine receptor type 5 (CCR5) moonlights to assist memory formation in the brain. Even the histone protein H3, known primarily for its DNA packaging role in the nucleus, leads a double life as an enzyme with copper reductase activity.

While many proteins are multifunctional, some are especially prone to versatility. Intrinsically disordered proteins, or IDPs, provide a compelling example of this. Their structural plasticity, which allows them to adopt multiple conformations, enables them to switch effortlessly between unrelated functions. This challenges the classic notion of proteins as predictable entities, each folding into a specific shape to perform a single, discrete function. Instead, IDPs embody an unprecedented versatility – a kind of biological thrift – unlike anything seen in the fixed, purpose-built components of engineered machines. Acting as low-affinity molecular interfaces, they facilitate decision-making processes and execute multiple context-dependent roles.

Yet the flexibility and economy of multitasking confer a risk of dysfunction. Moonlighting proteins are more than three times as likely to be implicated in disease processes compared with their more specialised counterparts.

Moonlighting is not, furthermore, confined to proteins. Long non-coding RNAs (lncRNAs) – key players in gene regulation – also multitask. Even cellular structures, and entire cells, may have dual roles. Red blood cells, for example, are known for transporting oxygen and carbon dioxide, but they also perform an immune function, surveilling the blood for signs of injury or infection.

However, the differences between human-made and naturally engineered machines extend far beyond moonlighting. Engineered systems thrive on precision, with interchangeable, structurally stable, context-independent parts designed for stability and ease of repair. Biological parts, in contrast, lack uniformity, are often imprecise, and behave differently between individuals, species, and across time. Proteins, such as haemoglobin, behave like invariant miniature machines, but many others, like IDPs, lack predictable behaviour. Some exist in multiple activation states and conformations, while others lack fixed structures, toggling between different conformations.

Much of the information in biological systems remains enigmatic. While a protein's amino acid sequence is encoded in the genome, its conformational details are not formally specified. In fact, much of the cellular information required for life is not coded at all. Genomes limit their encoded information, and as a result, the structures of living things are under-specified. But coded information amplifies its impact by manipulating and entraining the natural order of self-organising systems. While a typical genome contains between 10 million and 100 billion bits of information, the biochemistry generated by genomes is substantially more information-rich. It has been estimated that a cell contains up to a nonillion (a 1 followed by 54 zeros) bits of information – nineteen orders of magnitude more than that of the largest genome.

This informational disconnect between genomes and the biochemistry they orchestrate sets limits on the precision and efficiency of biological systems. The complex network of interactions between proteins is largely unencoded, generating countless evolutionary spandrels. Indeed, the genome's ability to control cellular biochemistry is assisted by the fact that much of its own information has not been selected. Motoo Kimura's neutral theory of molecular evolution, first proposed in 1968, suggests that most genomic changes confer neither a significant advantage nor a disadvantage. They are, as a result, fixed in genomes by random drift, rather than by natural selection.

Like human-made machines, biological machines receive a host of inputs and transform these into corresponding outputs. But whereas human-made machines operate according to mechanical principles, the events within biological machines are governed by statistical principles, which describe the behaviour of large numbers of chemical interactions. This fundamental difference between evolved machines and engineered ones, rooted in differences of scale, was explored in detail by the British-Indian biologist J. B. S. Haldane.

In his essay 'On Being the Right Size' (1926), written while at the University of Cambridge, Haldane observed that physical forces vary dramatically with scale. Tiny insects, for example, rely on simple diffusion for oxygen transport, while larger organisms require sophisticated circulatory systems to address their metabolic needs. Gravity, too, operates very differently across scales. While a 'nuisance' to a mouse, it can be catastrophic for a man, or a horse. As Haldane noted: 'You can drop a mouse down a thousand-yard mine shaft; and, on arriving at the bottom it gets a slight shock and walks away, provided that the ground is fairly soft.' The situation would be very different for a horse.

These differences are the predictable result of physics: 'the resistance presented to movement by the air is proportional to the surface of the moving object'. Due to their minuscule size, cells and molecules experience forces very differently from the organisms built from them. In this sense, microscopic entities like cells inhabit a different world from whole organisms. In contrast to the gravity-dominated world we inhabit, at the microscopic level, where cells reside, it is Brownian motion – the random jostling of molecules – that reigns supreme. Were human-made machines the size of cells, they would be unable to function the way they do at the human scale. The engineering metaphor of human-made machines works well in the macroscopic world, but becomes distorted in the microscopic world of biology.

In human-made machines, all critical functions are programmed into their operational logic. Biological systems, in contrast, generate order not just through encoded instructions, but also through the non-programmed self-organisation of their components into higher-order patterns. The development of a multicellular organism from a single fertilised egg, for example, relies extensively on self-organising processes that are not explicitly programmed in the genome. Genomes simply guide, channel and exploit these self-organising processes.

Viruses provide another striking example of self-assembly in biology. Their protein coats, known as capsids, self-assemble into elegant icosahedral geometries with twenty triangular faces. Others self-organise into helical structures, producing rod-shaped particles. The hierarchical ordering of individual coat proteins into these 'supramolecular' assemblies is not coded. It emerges spontaneously, governed by as-yet poorly understood laws of physics and chemistry. During viral assembly, disorganised collections of capsid proteins spontaneously transition into ordered, multi-protein states. These are textbook examples of 'order for free'. Viruses have learned to masterfully manipulate this generous feature of nature. Such structures are no more programmed than the vortex that spontaneously forms when you pull the plug out from a bathtub.

Termites in Africa, Australia and South America similarly exploit self-organisation to build towering mud mounds – insect skyscrapers that may be up to two metres high. These intricate architectural masterpieces of the insect world are constructed without centralised oversight. Each termite follows a simple set of behavioural rules that coordinate the cooperative construction of their insect metropolis. Together, their collective action generates a mound design that is highly optimised for structural stability and incorporates a ventilation system that regulates internal temperature. It emerges as a result of the termites' collective behaviour, without a master architect. Such emergent phenomena have no equivalent in human-made machines. In devices such as smartphones, washing machines, automobiles and lawnmowers, every function is rigidly programmed. There is no opportunity for spontaneity, self-assembly, or the exploitation of free order.

Biological machines are deeply historical, and indelibly imprinted by the agency of chance. Almost any aspect of an organism's design could, in principle, be improved, whether at a component or higher-order structural level. But the random selection of particular evolutionary pathways closes down alternative opportunities, making some designs impossible to realise. Once evolution selects a given pathway, certain routes to potentially optimal solutions become inaccessible. Evolution proceeds in a ratchet-like manner, locking in changes that are often irreversible. While some features can be repeatedly lost and regained, the dependence of mutations on their genetic background makes the reversal of some design commitments difficult or impossible to achieve.

A significant constraint on the design of biological machines is that they must be optimised for evolvability. If they could not change, they would be unable to adapt to new challenges. Yet evolvability comes with trade-offs. Over-optimisation in one domain can create vulnerabilities in others, leaving organisms trapped on evolutionary 'peaks' and making future innovation difficult.

Several additional features of naturally engineered machines make them difficult to engineer. Biological machines are, for example, gratuitously intricate. This complexity is reminiscent of Gaudí's Sagrada Família cathedral in Barcelona – a fantastical vision rendered into an elaborate, concrete reality. While equally dazzling, the complexity of natural genomes has consequences for those seeking to modify them. Natural biological systems are difficult to understand and hard to engineer. Like the promiscuity of their molecular components, much of this reflects the overlapping nature of the control systems and regulatory feedback loops that contribute to their robustness.

Organisms are built to function, not to fail. Examples of this are the compensatory mechanisms that allow the functions of damaged genes to be subserved by surrogates. A common strategy used to determine the function of a gene is to knock it out. Such 'knock out' experiments delete or incapacitate a gene, allowing the 'lost' function to be determined. Except, sometimes, no loss of function is observed. Didier Stainier at the Max Planck Institute in Germany showed that when a premature stop signal – known as a termination codon or 'nonsense' codon – is introduced into a gene, it can trigger a compensatory mechanism called nonsense-induced transcriptional compensation (NITC). This mechanism switches on ancestrally related genes, which in some cases are able to fill in for the missing functions. This remarkable resilience helps preserve function in the face of genetic damage, serving to buffer organisms against disease and dysfunction.

Some of the complexity of biological machines issues from the haphazard, historical nature of their construction. The interconnected network-like organisation of biological systems further compounds this complexity, generating complex relationships between genes organised into distributed networks. These architectures often give rise to emergent non-linear behaviours, where small changes can produce outcomes that are difficult to predict from the behaviour of individual components. Such inherent complexity poses major challenges for the future rational design and modification of complex species.

While biochemical pathways tend to follow modular, linear and deterministic routes, gene and protein networks are more versatile, and less constrained. As a result, their outputs are less predictable, and their dynamic states can generate a diversity of behaviours. This introduces redundancy into biological systems, which allows similar outcomes to be achieved using alternative strategies. This provides biological machines with back-up systems. If multiple parts perform overlapping functions, then if one breaks, another can fill in. Biological machines are also degenerate, meaning that individual components may sometimes perform overlapping functions, once again providing a mechanism to buffer design change or damage to biological systems, thereby allowing for greater flexibility. The connectivity of biological systems is best studied in simple species. It is also discernable in non-biological networks.

A 2023 investigation by Matthias Mann at the Max Planck Institute examined the global organisation of the baker's yeast protein 'interactome' – the complete map of all its protein-to-protein interactions. Remarkably, the data showed that most yeast proteins interact with at least sixteen others. The highly organised yeast interactome includes 3,927 proteins linked by 31,004 interactions. These proteins form a single interconnected system that may be subdivided into forty-one sub-compartments, each comprising around eighty-eight proteins. Much like human social networks (such as Facebook), the average shortest path between any two proteins in yeast involves just four interactions. So while most protein 'nodes' are not directly connected, there is on average just four degrees of separation between them. This organisation, characterised by local clustering and relatively short average path lengths between nodes, is known as a small-world network.

The concept of 'small-worldness' is not unique to biology. It's a hallmark of many complex natural and artificial networks, from the internet and electrical power grids to road systems, social networks, ecosystems, metabolic pathways and supply chains. The term 'small-world network' was first introduced by the psychologist Stanley Milgram in 1967 in his famous experiment that determined how likely any two individuals are to be connected. Letters were sent to random individuals in the American Midwest, asking them to forward a letter to a specified individual (if they knew them), or to someone they thought might know them. Surprisingly, many letters reached the intended target recipient in just six steps, giving rise to the now-iconic phrase: 'six degrees of separation'.

Another defining feature of biological machines, which distinguishes them from engineered systems, is the remarkable diversity of their operating systems. In computer science, an operating system (OS) manages the interaction between hardware and software, directing the flow of information. In biology, the equivalent is the regulatory network, which determines when, where and how protein-coding and non-coding genes are switched on and off. These regulatory networks have the ability to manipulate the behaviour of small-world network components. The estimated 10 million different 'biological operating systems' on Earth (reflecting the estimated number of living species) greatly exceeds the number of engineered systems designed by humans.

Human-made machines usually run on digital 'Boolean' logic – simple yes-or-no rules that link input and output. For example: if Input 1 or Input 2 is active, then Output 3 turns off. If both inputs A and B are active, then Output C turns on. Biological systems, in contrast, seem to use something closer to 'fuzzy' logic, where values can fall anywhere between 0 and 1. While introducing more flexibility, this also generates unpredictability.

Given their networked organisation, it's hard to see how multicellular biological machines – as currently configured – could be substantially engineered in a piecemeal manner. Natural biological systems are far more intricate than engineered machines, with structures and interconnections that resist straightforward analysis or manipulation. In their current form, biological machines are barely engineerable, let alone programmable. While some encoded components behave like the interchangeable parts of engineered systems, and can be manipulated and repaired, much of their behaviour emerges from dense, overlapping networks involving vast numbers of independent elements. This densely connected architecture makes precise interventions challenging. It also has profound consequences for how we treat and cure complex diseases, and may impose fundamental limits on what conventional component-based genomic medicine can realistically achieve.

6
Rewriting Genomes

> Whole-genome engineering could one day create cells unbound by biochemistry as we know it.
>
> – GEORGE CHURCH, 2011

In his masterpiece *Don Juan*, published in instalments between 1819 and 1824, Lord Byron, the bohemian hearthrob, literary sensation, Romantic poet and one-time keeper of a tame bear, reimagined the legend of the fictional Andalusian libertine in a new and radical way. This reformulation was conveniently sympathetic to his own philandering inclinations. The poem, almost code-like in its structure, was written in the rhyming iambic pentamer of the *ottava rima* stanza. Byron, however, died on 19 April 1824 before completing it.

At the time of his passing, his estranged and only legitimate child, Augusta Ada – later the Countess of Lovelace – was just eight years old. The scandal surrounding Byron's *ménage à trois* with his wife and his half-sister had forced him into self-imposed exile in Italy and Greece in 1816, barely one month after Ada was born. He would never return to England.

While lacking her father's gift for crafting verse with code-like precision, the brilliance of Ada's imagination and her skill in symbol manipulation would later prove instrumental in the creation of a new type of codified notation – computer programming. This field, which she helped pioneer, would ultimately lay the foundations for the kind of artificial intelligence that today promises to reveal the generative rules of biology and enable humankind to design the genomes of living things.

Encouraged by her mother to study mathematics, astronomy and music as an antidote to her father's questionable morality, Ada was tutored by Augustus De Morgan, the famous logician and proponent of Boolean algebra. Routinely accompanied by her pet cat, Mrs Puff, she had, by the age of twelve, already designed a flying machine.

On 5 June 1833 at the age of seventeen, Ada attended one of Charles Babbage's soirées – frequented by the likes of Charles Darwin, the Duke of Wellington and astronomer Sir John Herschel – at his home in Dorset Street, London. The 41-year-old Babbage, a polymath and the Lucasian Professor of Mathematics at Cambridge University, was known for his diverse writings – ranging from chess and lighthouses to cryptography, lock-picking and the causes of broken plate-glass windows. But it was as the inventor of a mechanical calculator known as the Difference Engine that he was best known. This device was, among other things, designed to produce mathematical tables, and, in particular, the type used in naval navigation.

Their mutual fascination with machines led Lovelace and Babbage to become regular correspondents. In a letter dated 9 September 1843 to the legendary discoverer of electromagnetic induction, Michael Faraday, Babbage described Lovelace as 'a youthful fairy' and an 'Enchantress who has thrown her magical spell around the most abstract of sciences'.

When Babbage outlined his plans for a second and more advanced machine, the Analytical Engine, he believed it would provide society with 'greater minds'. The military engineer and future prime minister of Italy, Luigi Federico Menabrea, summarised a lecture that Babbage gave on the topic at the University of Turin in an 1842 monograph titled 'Sketch of the analytical machine invented by Charles Babbage Esq'. It was the first published description of a computer programme.

Babbage enlisted Lovelace to translate the monograph, which she completed between 1842 and 1843, annotating it with an extensive set of accompanying notes. These 'Notes by the Translator', published in 1843 alongside the treatise *Elements of Charles Babbage's Analytical Machine*, contained a prescient analysis of the engine's capabilities. While it could 'weave algebraic patterns just as the Jacquard-loom weaves flowers and leaves', its ability to manipulate symbols transcended the humdrum computation of arithmetical problems.

This differentiated it from 'mere calculating machines' and extended its potential utility to such activities as playing games like solitaire and noughts and crosses, generating graphics, and composing musical works 'of any degree of complexity or extent'. Lovelace concluded that Babbage's new machine held a 'position wholly its own', and was the harbinger of 'a new, a vast, and a powerful language'. This unique language went beyond all existing ones, enabling humans to talk to machines, and empowering them to accomplish complex tasks. Babbage's Analytical Engine promised to be far more than 'the mind of a mathematician embodied in metallic wheels and levers'.

In her remarkable synthesis, Lovelace articulated the earliest known vision of universal, general-purpose computation – the idea that a single machine could be programmed to perform any computation or logical operation. This realisation was a defining moment in the future of human progress. Although never constructed, this fully programmable, general-purpose device became the prototype for modern computers.

Casting herself as 'the high-priestess of Babbage's engine', Lovelace began to sketch out the grammar of this new type of language based on logical principles. She outlined a step-by-step method for computing Bernoulli numbers – an important sequence of rational numbers that are of general significance in mathematics, but immensely tedious to calculate by hand – using Babbage's machine, which she effectively referred to as 'code'. She had unwittingly devised the first-ever source code for a computer programme – a machine-readable language containing codified instructions that could be sequentially followed by a computing device to execute specific functions. Although the algorithm was never tested in her lifetime, it was eventually transcribed into the modern programming language 'C' by computer programmer Sinclair Target. Despite its visionary nature, the code proved unusable due to a likely typesetting error, producing the first known software 'bug'.

Prior to her premature death in 1852 at the age of thirty-six, Lovelace was visited by Charles Dickens, who read to her from his book, *Dombey and Son* (1848). Sadly, her enthusiasm for Babbage's engine was not shared by the establishment, who failed to provide the funds necessary for the machine's construction. In a scathing assessment, the prime minister, Robert Peel, enquired: 'What shall we do to get rid of Mr Babbage and his calculating machine? Surely if completed it would be worthless as far as science is concerned?'

Today, computers are ubiquitous. The average European or American home may contain up to fifty microprocessors – embedded in devices such as mobile phones, refrigerators, dishwashers, hair-dryers, microwaves and televisions. Microprocessors house the integrated circuits that serve as the computer's central processing units (CPUs). CPUs process digital inputs and convert them into corresponding outputs. They are the 'mini-brains' of a computer, functioning as control centres where processing, decision-making and communication occur. Their operation relies on two types of code structures: operating systems, which manage the software and orchestrate the processes necessary for the machine's basic function, and application software written to perform specific tasks. The OS provides the essential context that enables programmes to run and forms the interface between the user and the computer hardware.

In his 2019 book *Coders*, Clive Thompson described computer programmers as 'amongst the most quietly influential people on the planet'. Indeed, their creations have shaped much of modern life. Any yet, paradoxically, coders spend far less time writing programmes than they do fixing them. Debugging – the painstaking and often frustrating process of identifying and correcting software coding errors – dominates their working lives. Without such efforts, bugs proliferate, systems glitch and failures occur. Another significant part of their craft involves refactoring code – restructuring and streamlining it without changing its function.

The term refactoring was first introduced by the American computer scientist William F. Opdyke in his 1990 paper 'Refactoring: An Aid in Designing Application Frameworks and Evolving Object-Oriented Systems'. If factoring is the process of breaking down complex code into manageable modules, then refactoring is the painstaking refinement of those modules, which improves the structure of the code. It can transform tangled and outdated programmes into more streamlined and comprehensible systems. Refactoring is especially important for 'legacy' software – older programmes that have been patched and repurposed so many times that they have become convoluted, chaotic and barely manageable.

The process of refactoring typically unfolds in microsteps, governed by a specific set of design rules. While each individual line edit may appear trivial, together these code rewrites help eliminate unnecessary complexity, fix bugs and address vulnerabilities. Refactoring is also part of the routine maintenance of newly authored software; neglect it, and code has a tendency to degrade into spaghetti code. Developers call this accumulating code burden a 'technical debt' – the future cost of reworking code that has been neglected and allowed to become disordered.

Biological machines are replete with dysfunctional code and are a software programmer's nightmare. Whereas computer programmers can rework their convoluted programming, evolution lacks an external programmer to refactor its handiwork. Natural selection can delete faulty and imperfect code, but it cannot refactor genomes in a formal engineering sense, or redesign them from scratch. The genomic code scripts of living things have not, in fact, ever been refactored. Over the approximate 4 billion years of life's history, they have, as a result, amassed a staggering quantity of technical debt. No wonder they are so complex, difficult to interpret and riddled with bugs, glitches, redundancies and inconsistencies.

The pervasive incoherency woven into genetic code scripts is reminiscent of the spaghetti code still entangling the computer programmes used by Wall Street, the US federal government and at least a dozen US states. These and many other institutions continue to rely on a now-antiquated but stubbornly functional software programming language known as COBOL – Common Business-Oriented Language – first developed in 1959. Legacy COBOL code is poorly documented, inefficient, hard to modify, difficult to understand, sluggish, expensive to repair and highly inflexible. And yet, despite having long been superseded by newer and more adaptable languages like Python and Java, software written in vintage COBOL continues to run vast swathes of the global economy. Today, companies scramble to find programmers who understand its quirks, and can navigate its arcane architecture. They rely on a dwindling, superannuated group of programmers – many of whom are now in semi-retirement – to fix their problems and keep the old systems running.

Working from his home in northern Texas, Bill Hinshaw – known in some circles as the 'COBOL Cowboy' – manages a seasoned team of COBOL engineers. Many of these were first active in the 1960s and 1970s, at a time when computers were programmed by punch cards and took up entire rooms. The youngest among them are in their fifties. And yet, the global shortage of programmers familiar with COBOL means that their expertise is always in demand. The situation is, in fact, so dire that IBM has run free training seminars in this vintage, pre-internet programming language in an attempt to entice young programmers to learn COBOL. As Hinshaw himself observed, 'Some people like to say that COBOL is going away soon. But it isn't going anywhere. It's running the world.'

There are many parallels between programmes written in COBOL and the code of genomes, including that of humans. Both are highly functional, yet frustratingly opaque – the products of iterative tinkering, rather than deliberate design. As such, the code of the human genome faces many of the same issues as computer software. Programmed not by a sentient author, but by billions of years of evolutionary improvisation, its code is tangled and hard to understand. Like COBOL software, the human genome is difficult to overhaul. Darwinian evolution, the author of natural software, can amend the structure of its code through selective revision, but cannot comprehensively refactor it.

No natural process can rewrite part or all of a genome from scratch. Evolution has no 'COBOL cowboys' to address the insufficiencies in the genomic code it authors. It is like a negligent programmer, content to sit back and watch the mess accumulate – indifferent to the jeopardy that disorganised human genome source code wreaks on individual lives. It has no concern for the child afflicted with leukaemia, for the arthritic patient losing their mobility, for the relentless march of dementia, or for the slow, ignominious process of ageing and bodily decline.

Yet there is reason for hope. Where evolution falters, human ingenuity has the potential to intervene. The shared digital structures of computer programmes and genetic code suggest that the methods of computer science could, in theory, be applied to biology.

If we could manipulate the source code of living organisms in much the same way that computer programmers refactor software, the architecture of living things could be made more amenable to engineering. This would require a degree of genome deconvolution and removal of some of the dependencies and interconnectivities. Doing so would allow individual components to be locally engineered without introducing unintended consequences.

The prospect of refactoring the genomes of species has the potential to revolutionise our understanding of life itself. But achieving this will require more than just the ability to edit genes, which offers only a limited capacity to modify genetic source code. Genomic refactoring requires the ability to rewrite whole sections of genetic code. To achieve this, we need technologies that allow genomes to be synthesised in their entirety, and from scratch. The capacity to synthesise genomes is, however, insufficient on its own. To refactor existing species – and eventually design new ones – we must learn the natural programming language of genomes. This will enable us to introduce alterations into the genomic code in a rational and predictive manner. For the time being, however, our efforts to streamline the code script of living organisms are necessarily limited to the more tractable genomes of the simplest biological entities.

Drew Endy, a synthetic biologist at Stanford University, is one of the pioneers of this field. In the 1990s, he set out to build a computer model simulation of a biological entity: the virus bacteriophage T7, which infects bacteria. This early attempt at creating a 'digital twin' capable of simulating T7 at the molecular scale was both an intellectual challenge and an interdisciplinary adventure. With a background in civil engineering, Endy's leap into biology would prove anything but straightforward.

Viruses have the advantage of being one of the simplest known biological entities. They are not technically alive, as they must infect bacteria to replicate. After infecting a host bacterium, the virus – a raw piece of DNA wrapped inside a self-assembling icosahedral protein coat – instructs the bacterium to make around

250 copies of its tiny genome. In a show of ingratitude, it then destroys its host, breaking it open and releasing its newly synthesised particles.

The T7 genome is a linear sequence of about 40,000 nucleotides, encoding fifty-seven genes and fifty-one regulatory elements. As Endy worked to model the system, he found himself wondering whether its genetic architecture was truly optimal, or merely adequate. The T7 virus had, after all, evolved over millennia and, like all other evolved biological entities, was the product of both natural selection and unpredictable factors such as chance and historical contingency. Could its genome be re-engineered to make it easier to modify – literally more designable?

As at the time he lacked the tools to physically manipulate DNA, Endy decided to turn to computer simulations. He built a model system using all available experimental data, then rewrote his code to simulate the behaviour of artificially redesigned genomes *in silico*, rearranging genes and regulatory elements. One prediction stood out: repositioning the gene for RNA polymerase could enhance the virus's replication efficiency.

When Endy finally tested this and other predictions by engineering actual T7 phage mutants, the results were sobering. Most of his hypotheses failed. The RNA polymerase tweak, in fact, had the exact opposite effect of his predictions – the mutant virus grew more slowly. The lesson was clear: even the simplest biological systems are far more complex than their modest genomes would suggest.

Accustomed to analysing engineering failures, such as why bridges collapse or walls crack, Endy approached the results of his experiments with a similar root-cause failure analysis. Unsurprisingly, he found that his mathematical models were overly simplistic. His computer simulations assumed complete knowledge of the system's behaviour, yet critical gaps remained. The function of many of the genes and potential regulatory elements were, for example, unknown. It wasn't even clear if the parts list for the virus was correct.

One design feature of the natural virus, however, was particularly perplexing – some of the viral genes overlapped. The virus's tangled natural design, coupled with an incomplete understanding of how its components functioned, together thwarted Endy's attempts to produce an accurate model. Evolution had, it turned out, not configured its creations to be either engineerable or comprehensible.

By 1998, Endy realised that he needed to chart a new course. While the T7 phage had been designed by nature to be evolvable and offered a wealth of insights into how each part of the system contributed to the whole, it was not systematically modifiable in the manner required by a human engineer. Endy concluded that if nature couldn't provide him with a biological entity suitable for his experiments, then he would have to create one himself. He realised, in short, that to model biological systems efficiently from an engineering perspective, he would need to 'rebuild the living world'. Like a trainer breaking in a wild horse, he planned to transform the awkward design of the T7 genome into a more tractable, domesticated version.

The genetic code of T7 phage would, much like an indigestible piece of COBOL spaghetti code, need to be refactored. The natural, overlapping state of its genes would have to be untangled to produce a more rational, non-overlapping and coherent set. In achieving this extensive genomic reorganisation, Endy would become the first person to redesign a genome's logical structure. He had defined an entirely new way to engage with natural genomes. While cells and genomes are not computers in the strict engineering sense, the genome-as-computer metaphor provided a useful heuristic – forming the basis of a framework that made a complex problem more tractable.

Viewing the T7 phage genome as a kind of biological USB stick encoding a computer programme and the bacterium it infects as computing hardware, Endy began to map out his next steps. The code of T7's overlapping genes would need to be disentangled, much like optimising legacy computer code.

In digital computing, source code is written in binary sequences – strings of 0s and 1s – corresponding to physical states in the machine's hardware, such as the presence or absence of electrical charge. Transistors, the semiconductor building blocks of digital circuits, act as switches that perform logical operations, transforming binary inputs into outputs. Billions of transistors are combined to form integrated circuit architectures capable of executing complex computational tasks. These microprocessors, or central processing units, comprise arrays of transistors organised into wafer-thin pieces of semiconductor material. They are typically made from silicon and are sometimes referred to as 'silicon chips'.

The coordinated actions of the multiple transistors within CPUs execute instructions and perform programmatically defined tasks. While the term 'code' refers generically to any set of instructions written by a computer programmer, there are several different subtypes. Source code refers to the human-readable instructions a programmer writes. This higher-level language is then translated into a series of lower-level languages, including assembly language and machine language. Machine language is the lowest level programming language, directly interfacing with the computer hardware.

DNA, similarly, can be viewed as a kind of source code – analogous to computer software – but written in combinations of the four nucleotides: A, T, C and G. The code of DNA is quaternary, as it comprises combinations of four possible states, rather than two. Unlike the binary code of computers, quaternary code allows for the encoding of greater informational complexity. While binary notation is well suited for use in computer programming, as it corresponds to the on–off states of transistors, biological code is not similarly restricted.

The similarities between computer software and genomes led Endy to consider whether biological systems could be programmed, using DNA as their programming language. In this sense, living things might be thought of as a distinct class of biological computing machines, executing programmes written in the four-letter code of DNA rather than in the binary code of conventional software.

Programming biological systems comes with unique challenges that greatly surpass those faced by computer engineers. While writing computer software involves the manipulation of binary machine code, biological systems can be programmed through the design and synthesis of artificial nucleotide sequences. Yet biological programming is subject to many of the same issues encountered by software engineers, including bugs, inefficiencies and unpredictability.

Given there is no general theory of software programming, most computer code is written by trial and error. Bug-filled and error-prone, the code is run, tested, retested and patched up in real time until it works. This strategy, however, is not well suited to biological programming, as biological systems are far less forgiving. Furthermore, the lack of rapid, accurate and low-cost methods for constructing genomes at scale continues to be a major limiting factor.

The physical reality of the genome-as-a-programme metaphor was most convincingly demonstrated by Craig Venter's genome transplantation experiments. By replacing the natural genome of one cell with an artificial one derived from another species, Venter's team showed that the host cell's molecular machinery could be reprogrammed. The foreign genetic programme 'ran' on the cellular machinery of the host cell much like a piece of computer software, subverting its processes and transforming it into the species specified by the transplanted genetic source code. It was like retooling a car assembly line to manufacture aeroplanes.

The experiment underscored a fundamental principle: genomes contain an internal representation of the organism they encode and the minimal description necessary to build and operate it. Sydney Brenner once stated that this 'is a fundamental feature of the living world, and must form the kernel of biological theory'. The code script, he argued, contains 'a description of the executive function, not the function itself'. As far as Brenner was concerned, the notion of computation was 'the only valid approach' to understanding biological complexity. The cell, quite literally, behaves like a computer, and the genome like source code. Genomes are both necessary and sufficient to define the nature of a species. They are codified symbolic representations of organisms.

One key distinction between natural biological computers and their human-made counterparts is the extent to which their information is specified. Cells – and non-living biological entities like viruses – are pervasively under-specified from an informational perspective. As we've seen, the self-assembly of a virus's coat into a precise icosahedral geometry isn't specified by genetically encoded instructions. The components come together of their own accord, guided by the laws of physics and chemistry. If a process doesn't need to be coded, it won't be.

The same principle governs protein folding – the process by which linear sequences of amino acids assume the complex three-dimensional shapes that endow them with one or more functions. Intriguingly, the genome doesn't specify the fine details. It 'assumes' that these processes play out as they always have. In this way, the laws of physics and chemistry are anticipated, and are indirectly represented within biological structures and processes, hard-wired into the invisible rules of self-assembly. A viral genome doesn't 'know' how to assemble the icosahedral structure of its protein coat; it only 'knows' that if it produces the right protein parts they will assemble themselves correctly. This contrasts with digital computers, where all the information necessary for functionality must be explicitly specified.

Another critical difference between biological and human-made computing is how computation in cells sidesteps the constraints of time. In cells, vast numbers of regulatory elements – which may be viewed as the biological analogues of microprocessors – operate simultaneously. This enables an unparalleled degree of flexibility and complexity. Digital computers, in contrast, run on synchronised microprocessors embedded in rigid architectures, limiting how many can run at once. Moreover, while the separation of the operating system from the programme is a fundamental feature of digital computer architecture, genomes appear to blur this distinction. The genome of a species seems to function simultaneously as both the software and operating system, integrated into a single dynamic entity.

When Endy arrived at MIT in 2002, he set to work planning how he might deconstruct the bacteriophage T7 genome. To make it more predictable and controllable, he would need to engineer out its byzantine entanglements. His vision of streamlining the genetic structure of an organism and refactoring it like a piece of computer code was conceptually different from all prior forms of genetic manipulation. During the process of remaking T7, he planned to systematically hunt down every bit of biological information hidden within its genome, producing an inventory and complete description of its encoded instructions. The refactored version of phage T7, called T7.1, would be a completely new beast.

Endy was set to become the 'T7 Cowboy', wrangling the viral genome's unruly, wild and tangled code and coaxing it into a domesticated, engineered form. His goal was nothing short of revolutionary: to clean up the code, disconnect it from its evolutionary history and create 'an engineered surrogate, which, if viable, would be easier to study, and extend'. Reflecting on Endy's achievement, after its completion, synthetic biologist Thomas Knight declared, 'biology will never be the same again'. Knight invoked the words of the Hungarian-American aerospace engineer Theodore von Kármán, who famously remarked: 'A scientist discovers that which exists. An engineer creates that which never was.'

Endy's efforts to refactor the T7 phage, rebuilding its genome and dividing it into discrete 'parts' that could be independently manipulated, marked the dawn of a new era in biology: one where the mechanical principles of engineering could be systematically applied to the genomes of living species. His work helped establish the conceptual foundations for what later became known variously as synthetic biology, synthetic genomics, generative biology, generative genomics, and engineering biology.

At the time, the tools available for reprogramming DNA's digital sequences were rudimentary, and the computing power far from adequate. But Endy figured he would find a way forward and resolved to engineer the genome of T7 manually, rather than synthesising it chemically from scratch. Constructing the refactored T7 virtual genome 'by hand' proved a Herculean feat, consuming three years and his entire research budget.

In computer science, there are two different approaches to software design. The first, known as 'genetic programming', is an evolutionary process that mimics natural selection. Software emerges from a pool of candidate algorithmic solutions, which are then mutated, reshuffled, tested, retested and selected in an iterative trial-and-error process. Structured design, on the other hand, proceeds in a more orderly, deliberate and stepwise manner. While evolved computer software programmes typically match or even exceed the performance of human-designed systems, they frequently lack 'human-readable descriptions', making them nearly impossible to debug or modify. The evolved genomes of natural species are highly functional, but they resemble the messy, opaque and disordered products of evolved computer software.

Endy resolved to change this. He patiently set about replacing 11,515 nucleotides of T7's natural 39,937 nucleotide genome with 12,179 synthetic nucleotides, increasing its total size by 664 nucleotides. One of the key innovations of the synthetic virus was the removal of overlapping gene sequences – a common feature in compact viral genomes where different genes share parts of the same DNA segments. By eliminating these overlaps, he managed to functionally separate each gene, thereby simplifying the architecture of the synthetic viral genome.

The synthetic T7 virus, T7.1, was shown to retain its essential functionality despite the extensive modifications. However, it was less 'fit' than its natural counterpart, displaying growth defects – likely caused by some of the engineered changes. As with software programming, Endy's modifications had inadvertently introduced 'bugs' into the code. He had, nevertheless, demonstrated that living organisms could be redesigned through systematic refactoring. The reconfiguration and streamlining of T7's genetic structure served as a prototype for the potential redesign of all species. The results were published in 2005 in a paper titled 'Refactoring bacteriophage T7'.

Suspecting that T7.1's defects resulted from 'scars' left by the unsophisticated methods he had used to assemble its genome, Endy turned his attention to a new challenge: the 5,386-nucleotide genome of bacteriophage phiX174. Unlike T7, which had a relatively straightforward genomic organisation, phiX174 was more complex, with eleven extensively overlapping gene sequences. One region of the phiX174 genome was particularly entangled, containing three overlapping genes.

Shifting from his manual genome assembly method to chemical genome synthesis, by 2021 Endy had crafted a 'scarless' synthetic phiX174 genome that had been meticulously redesigned on a computer. The redesigned genome, expanded to 6,302 nucleotides, managed to 'unpack' the overlapping sequences, eliminating the regions of overlap. The 'decompressed' synthetic phiX174 genome encoded a fully functional virus which, unlike the growth-impaired T7.1, grew vigorously.

Endy had demonstrated an important principle: while some viruses have evolved compressed genomes – likely for reasons of efficiency – compactness wasn't necessary for functionality. When carefully designed, decompressed genomes can perform just as well as natural ones. The densely packed nature of viral genomes may simply reflect a historical accident – their compressed genome structure likely reflecting the imprint of chance, rather than design or necessity.

While Endy's refactoring approach expanded viral genomes by 'unpacking' overlapping sequences, György Pósfai and Frederick Blattner, at the Institute of Biochemistry in Szeged, Hungary, and the University of Wisconsin, Madison, respectively, set out to do the exact opposite. They wondered whether they could shrink genomes, stripping them down to their bare essentials. The behaviour of a cell with a minimal streamlined genome was likely to be more predictable and programmable.

Their test organism was the bacterium *E. coli*. Known as the 'workhorse' of the biotechnology industry, it is widely used to synthesise proteins for industrial and therapeutic purposes, including enzymes, antibodies and hormones. The minimisation and streamlining of its genome offered an opportunity to enhance the organism's utility.

Through sequencing the genomes of three different strains of *E. coli*, Pósfai and Blattner found that each had a unique set of gene 'islands' inserted into a nearly identical genetic 'backbone'. While the backbone was 3.7 million nucleotides long, each strain's gene islands added another 0.9 million nucleotides, comprising around 20 per cent of the genome. They noticed that the backbone housed most of the core metabolic functions necessary for survival. The genes in the islands, on the other hand, appeared more specialised and potentially dispensable, coding for toxins, virulence factors and specialised metabolic capabilities relevant to survival in specific environmental niches. The islands also housed genomic detritus – remnants of inactivated genes, transposable elements and viral sequences. This suggested that the islands were non-essential, and the fact that they were clustered together – making them relatively easy to excise – made them prime targets for deletion.

They figured that the genomic islands could be stripped away, leaving just the essential 'backbone' of the genome, to generate a 'backbone-only' strain. They began cautiously, removing just 8.1 per cent of the bacterial genome, equivalent to 376,180 nucleotides. This eliminated 409 genes, about 9.3 per cent of the bacteria's total count. Remarkably, the pared-down *E. coli* thrived under standard culture conditions. The results, published in 2002, provided a definitive proof of concept for the genome minimisation engineering approach.

Encouraged by their success, the team decided to see whether they could shrink the bacterium's genome even further. By broadening their comparative analysis to include additional *E. coli* strains, they identified additional redundant regions that were potentially suitable for excision. This enabled them to achieve a 15.27 per cent reduction in genome size, involving the deletion of a total of 708,267 nucleotides and 743 genes. This was implemented using a natural engineering process called homologous recombination, which enables 'scarless' genomic excisions. The genome-reduced strain not only functioned normally, but had acquired some 'emergent' benefits. It was significantly more stable and exhibited a tenfold improvement in its ability to absorb foreign DNA through electroporation, a process that uses an electric current to make tiny holes in cell membranes through which DNA passes. This streamlined genome, now a simplified 'chassis', formed the basis of a versatile platform for industrial and commercial applications.

The concept of genome minimisation – reducing organisms to their most essential genetic components – began to take shape in the 1990s. This was driven by the falling costs – and therefore the increased accessibility – of whole-genome sequencing. The pioneers of the field, Craig Venter and Clyde Hutchison, posed a provocative question: was it possible to identify an irreducible genome for living organisms? This could, in principle, be achieved by systematically removing all non-essential genes to reveal the minimal set compatible with life. It was not just a quest to tidy up genomes, but an attempt to understand their origins.

Specialised cell types likely evolved through the embellishment of hypothetical ancestral genomes, present in the earliest cells. In 1995, Venter's team at the Institute for Genomic Research sequenced the first-ever complete genome of a free-living organism: the 1,830,137-nucleotide genome of the bacterium *Haemophilus influenzae*. This was swiftly followed, in the same year, by the determination of the full genome sequence of *Mycoplama genitalium*, whose 580,070-nucleotide genome was, at the time, the smallest known for a self-replicating organism. These bacteria can survive with smaller genomes because they are parasites that rely on the metabolic activities of their host.

Once these two genome sequences had been decoded, it was possible to compare them to determine whether a shared group of essential genes, representing a putative ancestral 'minimal set', could be defined. This minimal gene kit would have been present in the ancestral cells from which complex life evolved, and was subsequently elaborated to generate new species. Comparisons of the genome sequences from different organisms would allow for 'a more precise definition of the fundamental gene complement for a self-replicating organism, and a more comprehensive understanding of the diversity of life'.

In 1996, Eugene Koonin and Arcady Mushegian at the National Institute of Health in Bethesda, Maryland, made a direct comparison of the genomes of *Haemophilus influenzae* and *Mycoplama genitalium*. They identified twelve core processes that any minimal cell would need to perform, including DNA repair, replication, protein synthesis and the transport of metabolites. These basic functions likely represented minimal metabolic requirements for all cellular life. After examining the functions of the 468 predicted protein-encoding genes in *Mycoplama genitalium* and the 1,703 in *Haemophilus influenzae*, they identified 256 genes with functions that were conserved across both species. These were the candidates for inclusion in the minimal gene set, the rest being non-essential and dispensable.

At the J. Craig Venter Institute in La Jolla, California, Craig Venter and Clyde Hutchison were already contemplating the next step forward. Producing a design for a minimal genome was one thing – building it from scratch, quite another. They planned to design and construct the refactored genome of a minimal cellular organism, excising all non-essential genes and retaining only those necessary for free-living life. The starting point was the genome of *Mycoplasma mycoides* – the bacterium used in their 2010 genome transplantation experiment. They had already made a fully synthetic version of its genome, *Syn 1.0*, which served as a convenient point of embarkation.

The first attempt, called *Syn 2.0*, was a minimal, gene-deleted synthetic genome designed on a computer. The refactored genome was then chemically synthesised and transplanted into *Mycoplasma capricolum*. The newly written genetic 'software' could now be 'run' on the hardware of a biological computer. But the experiment failed. As Venter himself commented, it seemed as if the understanding of biology at that time was 'not sufficient to sit down, and design a living organism and build it'. Acknowledging that they had been blindsided by biology's complexity and their incomplete understanding, the team adopted a new approach.

They divided the *Mycoplasma mycoides* genome into eight sections and began to independently modify each fragment through trial and error. They then mixed and matched the engineered segments, testing the essentiality of genes by knocking them out one by one. Through this painstaking, iterative approach, they generated a miniaturised genome called *Syn 3.0* – a leaner version that was approximately half the size of *Syn 1.0*. It was a triumph of bioengineering. Functioning with just 473 genes, *Syn 3.0* was a synthetic cell with the smallest genome known to science.

Once they had isolated this minimal gene set, the compression was further extended by reordering the genes and clustering them according to their function – much in the way that the digital data in computer files is compressed and reorganised to conserve disk space. The slimmed-down genome of the compressed and refactored *Mycoplasma mycoides* organism was 531,000 nucleotides long, as compared to the 600,000-nucleotide-long parental genome. They might have reduced it even further, but there was a trade-off between genome size and the organism's growth efficiency. A set of so-called 'quasi-essential' genes, while not strictly necessary for survival, proved important for maintaining efficient growth. The final genome was a compromise that provided the organism with both near-minimal artificial genetic software and an acceptable growth rate.

Remarkably, the functions of at least one-third of the genes were unknown. While *Syn 1.0* was an example of pure genetic plagiarism, in which the wild-type genome sequence of *Mycoplasma mycoides* was essentially copied, the Lilliputian genome of *Syn 3.0* had been extensively redesigned and rewritten to meet human-defined objectives. Though built upon a natural foundation, Venter and his colleagues had succeeded in going beyond nature's design to create the prototype for the first truly artificial species.

Intriguingly, the synthetic organism they built was constructed without any fundamental insights into, or comprehension of, the generative rules governing genome creation. While miniaturised and refactored, the genome was built according to an existing plan detailed in the 'software' of the parental organism's genome. Nonetheless, the work marked a defining moment in the history of species and would have significant implications for future organisms. They had shown – albeit in an indirect and rudimentary manner – that it was possible to design and synthesise artificial genomes from scratch. Natural and artificial had begun to converge.

The combined work of Endy, Blattner and Venter had revealed the astonishing power of the genome-as-computer-software metaphor, and the extent to which the source code of genomes could be modified through a blend of rational design and trial-and-error refactoring. But their work raised another provocative question. Was it possible to refactor the organisation of genomes more substantially, by altering the structures of chromosomes – the storage units in which genetic information if compiled and housed?

Unlike microorganisms, which lack a cell nucleus and whose DNA floats freely as a single closed loop, the genomes of complex organisms are partitioned into distinct code files known as chromosomes. The human genome comprises twenty-three pairs of such files, one of each chromosome pair inherited from each parent. The genome of the fruit fly has just four pairs of chromosomes, while that of the marbled lungfish genome has a sumptuous 132. Each cell of the tiny fern *Ophioglossum reticulatum* contains a remarkable 630 pairs. Possessing a specific number of chromosomes does not appear to confer an organism with any particular advantage, as witnessed by the fact that closely related organisms have very different chromosome numbers. And the rules determining how a particular genome is partitioned into a specific number and arrangement of chromosomes and how these evolve and are maintained, are only partially understood. Were chromosome number a product of historical contingency rather than selection, then it should, in theory, be possible to alter it without disrupting overall cellular function.

To test whether the sixteen linear chromosomes in the yeast *Saccharomyces cerevisiae* genome represented an accident of history or an essential design feature, and whether the organism could tolerate a change in chromosome number, in 2018 Jef Boeke and his team at the Institute for Systems Genetics in New York used gene editing to fuse yeast's sixteen chromosomes together into two giant chromosomes.

Surprisingly, the yeast genome hardly seemed to notice. Its growth was mildly impaired, but it otherwise appeared to function normally. As Boeke himself put it, the yeast just seemed to 'shrug its shoulders' and continue as before. Because it was now reproductively isolated from the parental strain which contained sixteen chromosomes, Boeke had technically created a new species. Yet the appearance and behaviour of the natural and engineered organisms remained almost identical, owing to the similarity of their source code.

That same year, Zhongjun Qin and his colleagues at the Chinese Academy of Sciences in Shanghai went even further, fusing all sixteen chromosomes into a single gargantuan chromosome. The synthetic organism grew more slowly than both its natural parent and the two-chromosome organism made by Boeke. It also showed reduced versatility across different environments, reduced viability, and produced spores less efficiently. But otherwise it appeared to function normally. While the artificial yeast did not compete effectively with the natural version, it was remarkably robust and tolerated the extreme restructuring surprisingly well. This suggested that chromosome number in yeast may, at least to some extent, reflect historical contingency rather than the imprint of an adaptive process of natural selection.

Together, these studies provided the basis for a new engineering paradigm for biology, in which the code script of living things could be broken down and rationalised to create streamlined, programmable organisms. For the first time ever, life could be detached from the idiosyncrasies of its evolutionary history and the whimsy of natural design processes. It suggested that species generation might one day be transformed into a formal engineering discipline.

But it is one thing to reconfigure, refactor, and rewrite genomes using existing biological source code – crafted by evolution over billions of years – to tame the baroque complexity of natural biological systems, and quite another to script entirely new biological source code, unconstrained by anything that has been written before.

While the ultimate goal of generative biology is authorship, we don't yet understand genome design well enough to move beyond the plagiarism involved in making alternative versions of existing life. We are like monks in a scriptorium – copying and rewriting ancient texts – rather than true authors. Like pre-literate children, we are learning to copy genomes before we can contemplate creating them. Drew Endy succinctly summarised the situation: 'can write DNA, little to say'.

To progress towards the authorship of artificial species and transform biology into an unbounded engineering technology, we must first discern the generative rules of life. These hypothetical principles would be analogous to the physical laws that enable us to launch spacecraft to other planets, or harness the electrons that generate electrical phenomena.

The construction of new genomes will play a central role in uncovering the generative principles of biology, which will eventually allow organisms to be programmed. Mastery of these rules could enable the eventual compilation of a user's guide to life and transform biology into a predictive engineering material, limited only by our imagination, the laws of physics and chemistry, ethics and societal notions of acceptability. There is a profound difference between engineering biology and merely tweaking it through genome editing.

Harvard Medical School synthetic biologist George Church has stated that the ability to design and write synthetic genomes – transcending nature's designs – represents an opportunity 'bigger than the space revolution or the computer revolution'. Realising this potential will require the ability to chemically synthesise genomes of any sequence complexity, rapidly, efficiently, accurately and at low cost. But it's only a matter of time before technology catches up. As Church has noted: 'It's going to get easier and easier with time to build large genomes.'

Like the generative grammar of natural languages, we will – once fluent in the generative rules of biology and equipped with next-generation genome-writing technology – be able to make infinite use of finite means. In other words, just as a finite written alphabet can generate an infinite library of books, genome design – the at will, specific, and deliberate ordering of nucleotides into DNA sequences – combined with powerful construction technologies, will make it possible to generate any possible species. Eventually, we will use our mastery of natural genetic grammar to construct the first truly authored genome – conceived with the same creative flair as James Joyce's *Dubliners* (1914) or Fyodor Dostoevsky's *Crime and Punishment* (1866). Life will, at last, have become an art.

As synthetic biologist Christopher Voigt observed, 'We're almost taunted by what exists in nature.' Yet the creative potential of synthetic biology is overwhelming. Beyond merely imitating nature's biological languages, we will one day transcend them, and invent entirely new biological tongues. These foreign biological languages will have their own distinct grammars and vocabularies, enabling the generation of species that go beyond what is possible using natural rules alone. They may even expand the natural palette of biological building materials, surpassing that used by all organisms throughout life's history.

To define the generative rules of life necessary to author the genomes of new species, we must first study the sequences of a vast array of existing genomes to uncover the grammatical principles that define their nature. This will involve interrogating their source code and examining how it is choreographed in time and space to reveal how genome architecture influences the execution of genetic programmes and shapes the unique characteristics of each species.

In so doing, it will be possible to identify patterns across the genomes of different species that offer insights into the rules for generating living things. Artificial intelligence will be indispensable for this task. Its capacity to detect structure in complex datasets far exceeds our own and promises to reveal the elusive generative rules of biology. Mastering these rules will one day allow us to author life, by designing and writing the genomes of new species.

7
The Generative Grammar of Life

> Just as mathematics turned out to be the right description language for physics, biology may turn out to be the perfect type of regime for the application of AI.
>
> – DEMIS HASSABIS, 2021

On 11 May 1997, on the 35th floor of the Equitable Center in Midtown Manhattan – amid television cameras and the buzz of reporters – Garry Kasparov, the Russian world chess champion and creative genius, finally met his nemesis. It came in the familiar form of the Caro-Kann Defence, of which he was the undisputed master. But the opponent on this occasion wasn't his arch-rival Anatoly Karpov, nor indeed any other grandmaster. In fact, it wasn't even human. It was a 1.4-ton electronic machine called Deep Blue. A hulking colossus of a supercomputer built by IBM over a twenty-two-year duration, Deep Blue was capable of calculating over 200 million chess moves per second, or about 12 billion positions per minute. In this contest, heralded as humankind's last chance to assert its authority over machines, the stakes could not have been higher. The match attracted global media attention.

Deep Blue's dramatic arrival on the world stage was a poignant moment in the history of life on Earth. It marked the occasion on which the extraordinary and viscerally palpable power of machine intelligence first flagrantly challenged the dominance of natural biological intelligence. It also signalled a turning point – the twilight of an antiquated, lumbering type of rigid and pre-programmed artificial intelligence that was realised through the agency of brute-force computation. In its place, a new kind of machine intelligence was emerging based on deep learning, adaptation and flexibility.

This new type of AI, enabled by distinctive neural network architectures, would go on to play a pivotal role in establishing the emerging science of generative biology. It suggested, for the first time ever, the tantalising possibility of achieving fluency in the generative grammar of life's language. Unlike the grammatical rules of human languages, which are relatively simple, explicit and fully comprehensible, the generative rules of biology are vastly more complex. While AI might ultimately help master them, its inner workings are, for now, barely understood.

The 5 May 1997 edition of *Newsweek* magazine billed the game as 'The Brain's Last Stand'. On the seventh move of the deciding game, Black, played by Kasparov, made what turned out to be a fatal error. White responded unhesitatingly and uncompromisingly, launching a new attack through the sacrifice of a knight. Just eleven moves later, Kasparov – considered by some to be the greatest chess player in history – had been ruthlessly defeated. Of the six games played over nine days, he won the first, lost the second and drew the next three. His defeat in the final, decisive match meant that Deep Blue was the overall winner. The machine had achieved an unprecedented victory. While games of chess at this level typically lasted at least four hours, Deep Blue had dispatched its biological opponent in under sixty minutes.

While it was expected to play by brute force, Deep Blue ended up dazzling the international chess community with its subtle creativity and outlandish virtuosity. Stunned by the outcome, a disgruntled Kasparov grudgingly admitted that, at times, the machine had 'played like a god' and 'made moves beyond anyone's mind'.

But while it superficially appeared to be a tremendous victory for AI, the match paradoxically marked the death knoll for Deep Blue's old-fashioned brand of machine intelligence. Demis Hassabis, joint winner of the 2024 Nobel Prize in Chemistry and founder of the AI company DeepMind, noted that, unlike his soulless electronic opponent, which had been 'hard-coded with a set of specialised rules distilled from chess grandmasters and empowered with a brute-force algorithm' to perform a single narrow task, Kasparov could perform the full and varied repertoire of activities commensurate with being human. This included walking down the street, having lunch, playing tennis, experiencing happiness, enjoying a movie, writing a letter and reading a novel.

While this 'computational leviathan' was an undisputed chess savant, beneath its skin-deep veneer of sophistication, Deep Blue was more like the Tin Man in L. Frank Baum's *The Wonderful Wizard of Oz* (1900) than an intellectual heavyweight. Rather than emerging from its own deliberations, its moves reflected the preferences and prejudices of its programmers. Its 'mind' comprised nothing more than hundreds of thousands of preprogrammed, rigid and inflexible lines of computer code.

Deep Blue was coded to perform just a single task. It was unable, for example, to play simpler games like checkers without being reprogrammed. As Kasparov himself later stated, 'Much as airplanes don't flap their wings like birds, machines don't generate chess moves like humans do.' Its success was based on a formidable ability to crunch numbers rather than on true intelligence. This competence was itself a direct result of Moore's Law. Gordon E. Moore, the co-founder of the Californian semiconductor chip company Intel, famously predicted on 19 April 1965 in his paper 'Cramming More Components onto Integrated Circuits', that the number of transistors that could be printed onto a microchip would double roughly every two years, while the cost would halve.

This has largely turned out to be the case over the last few decades and has resulted in an exponential increase in computing power. While chips in 1971 could accommodate just 200 transistors per square millimetre, today's chips can squeeze up to 130 million transistors into the same space – each operating tens of thousands of times faster. Without this type of technological development, the brute-force approach of Deep Blue would not have been possible. But Moore's eponymous law was not intended to last for ever. There is a physical limit to how small a transistor can become. Transistors can now be made at the three-nanometre scale, and while they could, in principle, be made even smaller, the cost of doing so is currently prohibitive. Moore's Law can no longer be sustained.

The cold, reptilian logic of Deep Blue lacked the visceral features of human intelligence, including the capacity to learn, adapt and generalise across different domains. Its superficial veneer of intelligence also concealed hidden vulnerabilities. While behaving as if it were able to think, it was in fact incapable of doing so. It certainly did not possess the type of human-like intelligence that the founding fathers of AI aspired to. And despite its brief moment of glory, Deep Blue's triumph over Kasparov turned out to be a pyrrhic victory, precipitating its own demise.

Deep Blue was what the American mathematician Claude E. Shannon, in his 1950 paper 'Programming a Computer for Playing Chess', described as a general-purpose machine deploying a 'type A strategy'. This was a machine that used brute force and number-crunching intelligence to compute its solutions, and whose responses to inputs were rigid, inflexible and incapable of learning from prior experience. This hulking, posturing electronic giant would, as a result, soon end up being as relevant to AI's future as the dinosaurs were to life's future. But unlike the dinosaurs, which were obliterated by a chance event, Deep Blue was deleted from AI's future because it was unfit for purpose.

Deep Blue's ignominious reward for its unprecedented victory over natural intelligence was to be split in two and dismantled. One of its two racks was sent to the Smithsonian National Museum of American History in Washington, while the other was deposited 3,000 miles away at the Computer History Museum in Mountain View, California. Despite its trailblazing performance, it had ultimately failed because its engineers had made a fundamental philosophical error. They had tried to model the complexity of human thought and actions as a logical process. As a result, its moment of glory was also its epitaph.

Deep Blue was destined to be eclipsed by a new and higher-performing class of type B machines. These emerging systems harnessed biologically inspired architectures known as artificial neural networks, which were capable of adaptive learning. Unlike their predecessors, the outputs of these machines were not fixed but evolved based on the history of their inputs. These AI models employed 'deep learning', relying on iterative training and reinforcement to acquire the abstract structures underlying their decision-making capabilities.

Although these systems embodied foundational generative rules, the nature of those rules remained inscrutable to users. They could not be readily dissected or abstracted in the way scientists might unravel the principles governing a physical phenomenon such as a planetary orbit, or the acceleration of a falling object. They were the direct opposite of the rigid, fully comprehensible 'expert systems' drilled into Deep Blue and coded into the 480 specialised chips welded into its circuitry.

The 'hand-coded' systems of machines like Deep Blue were brittle, deterministic, and written line by line by humans. They were incapable of making inferences or adapting to scenarios lying outside their pre-programmed knowledge base. Despite their domain-specific expertise and the fact that they had been built on a foundation of extraordinary computing power, they were unable to do anything beyond the task they were intended for. Their intelligence was painfully non-generalised, and more reminiscent of artificial 'unintelligence' than a step towards authentic artificial 'intelligence'.

In hindsight, the demise of the type of inflexible, rule-based AI logic powering Deep Blue's now defunct electronic circuits was both predictable and inevitable. What few could have anticipated, however, was that its successor – and cause of its premature superannuation – could be traced back to the dingy arcades and cafés of the late 1970s and early 1980s, where teenagers and young adults gathered to play video games. The origins of this technological revolution may be pinpointed even more precisely to the launch of *Space Invaders*, during the golden age of coin-operated arcade gaming. This blockbuster game not only inspired a massive global video game industry but also prompted innovations in computer graphics that would prove pivotal to the evolution of AI.

Driven by the public's insatiable desire for increasingly realistic and visually compelling experiences on their gaming consoles, the burgeoning video game industry would ultimately lead to a technological arms race focused on the quality of computer graphics, and to the development of a new type of computing architecture called graphics processing units (GPUs). These specialised chips were designed to execute large numbers of parallel operations simultaneously, and greatly accelerated the processing of images and rendering of graphics. GPUs serendipitously turned out to be perfectly suited for enabling the evolution of next-generation AI. Unlike the deterministic algorithms powering Deep Blue, these new systems were fluid, adaptive and capable of learning. They were not constrained by the rigid behaviour of machines that computed their solutions using predetermined sets of rules. This new embodiment of AI will be of critical importance in defining the generative laws of biology that will enable the seamless refactoring and creative authorship of complex genomes.

The journey to this transformation began in Tokyo, with an unassuming electronics engineer called Tomohiro Nishikado. Inspired by H. G. Wells's 1898 science-fiction novel *The War of the Worlds* depicting the chaos of a Martian invasion, he envisaged a game where players defended Earth from an advancing phalanx of extraterrestrial invaders. In the final version of *Space Invaders*, a grid of fifty-five luminescent, pixelated aliens – arrayed in eleven rows and five columns – marched menacingly across the screen. Players, armed with a single cannon, had to shoot down and destroy as many of these antagonistic combatants as possible, while dodging enemy fire and sheltering behind protective shields. Occasionally a flying saucer would whirr across the top of the screen, offering bonus points to sharpshooters. Launched in 1978, *Space Invaders* was an instant success and became a worldwide sensation.

The game was highly addictive, as witnessed when a twelve-year-old boy held up a bank in Japan with a shotgun, took an employee hostage, and demanded 5,000 yen in coins to operate a *Space Invaders* machine. In England, the Labour Member of Parliament George Foulkes warned of the 'glazed eyes' of 'crazed' *Space Invaders*-addicted teenagers, and on 20 May 1981 introduced – unsuccessfully – a 'Control of Space Invaders and Other Electronics Games' bill. In 1982 Martin Amis wrote an anthropological account of the now lost world of New York's arcade scene, titled *Invasion of the Space Invaders: An Addict's Guide to Battle Tactics, Big Scores and the Best Machines*. He speculated that despite the moral ambiguity of arcade gaming, there might be something more noble underlying this apparently pathological teenage obsession. The escapades of these 'foul-mouthed arcade youths' may, he conjectured, represent a search for 'the meaning of life'. In his foreword to the book, the film director Steven

Spielberg thanked the author for undertaking his 'horrific odyssey around the world's arcades' and warned readers of the risks of becoming 'video junkies'.

A key part of *Space Invaders*' appeal lay in its remarkable graphics innovations, which pushed the boundaries of what was possible at that time. The pioneering use of microprocessors in a video game provided the computational power required for the aliens to move independently across the screen. This illusion of movement was achieved using an innovative grid mechanism. Another feature – the ability to record a player's highest score – fostered the emergence of a competitive gaming culture. With its distinctive graphics, engaging mechanics and innovative use of 'bleeps' and other sound effects, *Space Invaders* set a new precedent for video games.

Historically, the fixation on logic as the cornerstone of human intelligence had led AI research to focus on games like checkers and chess as its model systems. These games had, for decades, functioned as AI's equivalent of the fruit fly *Drosophila melanogaster*, a widely used workhorse species for genetic research. But while chess had long been viewed as the epitome of human intelligence, solving it proved to be a relatively trivial task for machines with sufficient computational power.

What proved far harder were many of the other tasks that humans were able to perform, such as writing prose, composing a song, driving a car or navigating city streets. The advent of deep learning and neural network architecture heralded a fundamental shift, allowing machines to tackle such challenges for the first time. Unlike their rigid predecessors, these systems could learn organically, adapting their behaviour through training on huge datasets. They did not need to be programmed with expert knowledge like Deep Blue. Although inspired by biological neural networks, their internal processes were incomprehensible to their creators – black boxes that discerned patterns and inferred rules, but withheld the secrets of their operations. AI engineers had relinquished control of their creations, allowing the machines to chart learning paths themselves.

The concept of neural networks was not new. It had been a core component of AI since the field's formal inception at the 1956 Dartmouth Summer Research Project on Artificial Intelligence, in

New Hampshire, convened by John McCarthy, Marvin Minsky and Claude Shannon. However, these early efforts faced an important limitation. They lacked the appropriate hardware necessary to support these new computing architectures. Researchers also continued to debate AI's key issues, namely whether the field's future and ability to simulate human-like Artificial General Intelligence (AGI) – an aspirational type of artificial intelligence with the potential to equal or even surpass human intelligence – resided in the narrow logic-driven approach of expert systems, or within neural network architectures inspired by the architecture of the human brain. As a result, work on neural networks languished, while the narrow intelligence of its logic-based cousin flourished. But a few devoted individuals kept the flickering light of this, at the time obscure, branch of AI alive. Its first big win had occurred a decade before the Dartmouth meeting as a result of the pioneering work of Arthur Samuel, 'the father of machine learning'.

Following a spell at the Bell Telephone Laboratories, in 1946 Samuel became a professor of electrical engineering at the University of Illinois, before joining IBM in 1949. There, he developed a checkers programme, which ran on an IBM 701. By 1962, his machine had defeated the self-proclaimed checkers expert Robert Nealey.

The neural network architecture of Samuel's programme learned by trial and error, much like the human brain. Samuel trained it on a large database of checkers games, and then allowed it to improve further by playing against both him and itself. He had no choice other than to built it that way, as the computing power required by Deep Blue wasn't available at that time. This type of adaptive artificial intelligence was very different from the algorithmic intelligence of Deep Blue. As the founder of NVIDIA, Jensen Huang stated, 'deep learning is not an algorithm', but rather 'a new way of developing software'.

Samuel was forced to adopt a neural network architecture by necessity. The available computing power at that time was wholly inadequate for his purpose. In the absence of neural networks, he would need to find a computing capacity capable of programming in the region of 500 quintillion (10 to the power of 18) possible outcomes. He described machine learning as a 'computational method for achieving artificial intelligence by enabling a machine to solve problems without being given problem-specific programming'. It subsequently became 'the most important way most parts of AI are done'. The approach was revolutionary as it allowed machines to learn without being externally programmed. They programmed themselves. But even neural networks had formidable computing requirements, and the development of his prototype of adaptive machine intelligence eventually stalled due to a lack of this. It soon became overshadowed by the successes of the monolithic algorithmic approach championed by Deep Blue. For the time being, machine learning would remain an interesting, but largely peripheral academic pursuit.

But after languishing for almost four decades, a remarkable thing happened. The global intercompany arms race to develop increasingly realistic video game graphics – initiated by *Space Invaders* in 1978 – culminated in 1999 with the invention of GPUs, which where designed and manufactured by the technology company NVIDIA. A GPU (graphics processing unit) is a circuit board built around a powerful microchip containing billions of electronic transistors. Its literally game-changing chip architecture allowed the processing of high-resolution images to be greatly accelerated.

GPUs were unique. In contrast to CPUs (central processing units), which process data sequentially, they broke complex mathematical tasks down into multiple subtasks and associated subcalculations. They then processed these simultaneously, using an approach known as parallel computing. The parallel architecture of GPUs provided them with transformative properties. The gaming community had never seen anything like it, and NVIDIA would go on to become one of the most valuable companies in the world. It was as if someone had turned on the light in a dark room. The quality and speed of the graphics in the showcase video game *Quake* – which deployed parallel computing for the first time to render monsters targeted by players with grenade launchers – was extraordinary. It also enabled multiplayer combat. NVIDIA had begun the process of making supercomputing mainstream.

The GPU's architecture, designed for rendering high-resolution images, had serendipitously turned out to be perfectly suited for deep learning in artificial neural networks. What began as a technological arms race in video game graphics had catalysed an AI renaissance. Prior to the GPU revolution, neural net researchers like Samuel had been described as 'prophets in the wilderness'. Operating at the edges of acceptability – pariahs cast out from the mainstream – they had been relegated to irrelevancy.

The computational power that GPUs enabled breathed life into the virtual non-playable characters (NPCs) of a new generation of video games, making them more compelling and lifelike, and generating vibrant, immersive digital environments. Although developed for the purpose of enhancing graphics, GPUs opened up a whole new universe of additional opportunities, one of which was in deep learning AI.

The GPU-driven AI revolution was initiated by the computer scientists Alex Krizhevsky Ilya Sutskever, and their academic mentor and 2024 Nobel Prize winner in Physics, Geoffrey E. Hinton. Working at the University of Toronto, in 2012 they connected two NVIDIA GeForce 256 graphics cards – originally designed for gaming – together to power a visual image recognition and classification neural network, which they called AlexNet. Each card contained a GPU, video memory, a cooling system, circuitry to power the chip and an interface to connect it to the motherboard.

Krizhevsky began training the device at his parents' house. Prior to this seminal moment in AI's history, all neural networks had been built using CPUs. This was the first time that anyone had tried to construct one using GPUs. GPU technology provided the calculating speeds and power necessary to process the massively parallelised large-scale computations that occur in artificial neural networks. The theoretical basis for using GPUs to implement artificial neural network computations had been established three years earlier by a group at the Stanford Computer Science Department led by Andrew Ng. They first explored the potential for applying GPUs to AI in their influential and visionary 2009 paper, 'Large-scale Deep Unsupervised Learning using Graphics Processors'.

The results were astonishing. With just two GPUs, AlexNet performed tasks that had previously required up to 16,000 CPUs. Moreover, GPUs enabled neural networks to be trained up to 100 times faster. AlexNet rapidly excelled at complex visual discrimination tasks, for example, accurately distinguishing a leopard from a scooter. If deep learning was able to solve a problem as challenging as computer vision, the world was its oyster.

The meaning of the word 'supercomputer' suddenly became obvious. Without GPUs, neural nets were slow, clumsy, and untenable. Implemented using GPUs, however, they took on a superlative new life. GPUs opened the door to an avalanche of computing complexity. They empowered artificial neural networks and deep learning in ways never previously imaginable. The use of GPUs in AI marked a new beginning, and resulted in the rapid thaw of a deep-set AI winter fuelled by the dead-end, logic-based computational approaches of machines like Deep Blue.

Hinton described the unleashing of GPUs' genie-like powers as AI's 'Big Bang' moment. The age of GPU-driven supercomputing had begun. It was poised to impact almost every aspect of human existence, and began to demonstrate its formidable capabilities swiflty and definitively by securing some landmark achievements.

While the invention of AlexNet marked a turning point for AI, its true potential emerged in 2016 during an epochal moment in the world of board games. At 1 p.m. local Korean time on 9 March 2016, 33-year-old Lee Sedol – one of the world's highest-ranking Go players and holder of multiple international titles – sat down in the Grand Ballroom of the Four Seasons hotel in Seoul. He was optimistic, as were most of the tens of thousands of spectators watching in real time as the game was live-streamed on YouTube. As in Kasparov's 1997 chess match against Deep Blue in Manhattan, his opponent was not made of flesh and bones. It was instead a computer programme called AlphaGo that had been developed by the company DeepMind, which had been acquired by Google two years earlier.

Wei Qi, known as Go in the West and Baduk in Korea, has played a critical role in Chinese culture since the game's inception more than 2,500 years ago. During China's Tang Dynasty, it was revered as one of the four necessary arts that cultivated elites should master, along with calligraphy, painting and lute playing. It is a strategic game of territorial control played out on a 19 x 19 grid with 361 black and white disc-shaped pieces, known as stones. The American diplomat and former United States Secretary of State Henry Kissinger, in an analysis of the Taiwan Strait crisis of the 1950s in his 2011 book *On China*, argued that both the People's Republic of China and Taiwan behaved in a manner consistent with the gaming strategy of Go. The dynamics of the Cold War with the Soviet Union, on the other hand, were more aligned with the strategic imperatives of chess, the board game favoured by the Russians. Go's apparent simplicity belies its staggering complexity. Unlike chess, the pieces in Go are physically identical and lack piece-specific rules and dynamics.

The consequence of this is that a player's power emanates not from the intrinsic capabilities of the pieces, but from their acquired importance arising from their relative positions on the board. Whereas in chess the imperative is to attack, the objective of Go is to outmanoeuvre one's opponent and win through territorial gain in a way that avoids direct conflict. The position of each stone defines its sphere of influence. The emphasis is on building and consolidating dominant structures, and only secondarily on undermining enemy ones. As a consequence of this conflict avoidance, unresolved tensions gradually accumulate until, at a certain point, conflict can no longer be avoided.

While the rules of Go are simpler than those of chess, the game itself is far more complex. A player must typically choose between around 200 possible moves at each turn, compared with just 20 in chess. In an average 150-move game, the number of possible board configurations is an astronomical 10 to the power of 170. An algorithmic expert system approach would be incapable of exploring even the tiniest corner of this enormous search space to define the next best move. The number-crunching approach of Deep Blue would fail catastrophically. The game of Go embodies many of the key challenges facing AI. These include a demanding decision-making task, an intractable search space, and an optimal solution so complex that it is barely comprehensible. Compared to Go, chess is child's play. Go was the game computers couldn't crack. But that was about to change.

Unlike the neolithic, number-crunching colossus Deep Blue, AlphaGo had a very different type of computational intelligence. It was not pre-programmed to play Go. It was equipped instead with nothing more than the capacity for deep learning. Its inventors, Demis Hassabis, David Silver and their colleagues at Google's DeepMind, described its structure in a 2016 paper.

The first component, a 'policy network', employed a 'supervised learning' process in which around 30 million Go games were hoovered up from digital archives. Using machine learning, this model was then used to build a predictive model, and subsequently optimised through reinforcement learning that generated an experiential database by playing against itself over a million times a day. This enabled it to predict an opponent's moves with uncanny accuracy. The second component, the 'value network', assessed board positions to calculate the probability of winning from any given configuration. These computations were powered by Google's tensor processing units (TPUs), which, like graphics processing units, were specifically engineered to accelerate machine learning in deep neural networks. AlphaGo also utilised 1,202 central processing units. Together these helped propel AlphaGo to meteoric success.

At a certain point in the game, panic set in. Sedol stood up and smoked a cigarette while gazing sullenly out at the rooftops of Seoul. Three and a half hours after commencing the game, he was was mercilessly defeated. Sedol told reporters that he wanted 'to apologise for being so powerless' and conceded that AlphaGo had played a near-perfect game.

The next spectacular advance in AI transcended the abstract realm of board games and addressed something far closer to human existence – the process of language generation. This extraordinary development arrived in the form of ChatGPT, a conversational AI programme capable of generating remarkably human-like narrative text by analysing language patterns. The programmatic analysis of alphabetic letters employed by ChatGPT is reminiscent, in some ways, of the manner in which nature programmes the order of the chemical letters in genomes through evolution by natural selection. Yet, while Darwinian evolution serves as the passive author of undirected biological narratives, the authorship of AI is deliberate. The philosophical roots of this technology trace back to linguistic debates in the late nineteenth century.

The Swiss wunderkind Ferdinand de Saussure was an aristocrat, scholar, horserace enthusiast, poker player and intellectual giant. Of his two major works, only one – that addressed the use of vowels in Indo-European languages – was published during his lifetime, in 1878, when he was just twenty-one. The second, and more consequential work, *Cours de linguistique générale* (Course in general linguistics), was the magnum opus that he never wrote. It was published in 1916 as the result of a remarkable act of devotion by two of his former students, Charles Bally and Albert Sechehaye, three years after his death. They painstakingly constructed the book from notes jotted down by the few students who attended the lectures that he gave at the University of Geneva between 1907 and 1911, along with Saussure's own scribblings.

Widely considered the father of modern linguistics, Saussure argued that the grammatical structures of natural languages, both spoken and written, were derived from underlying generative patterns. He termed these invisible structures *langue* and distinguished them from *parole* – the actual speech and written words used by individuals. According to Saussure, the coherence of language emerges from 'a system of signs that expresses ideas'.

Saussure's pattern-based approach to language generation contrasts sharply with the rules-based framework espoused by Noam Chomsky. A towering figure in twentieth-century linguistics, Chomsky, a political scientist and philosopher at MIT, introduced his theory of generative grammar in his 1957 book *Syntactic Structures*. He argued that language arises from a finite set of grammatical rules capable of generating an infinite variety of sentences.

Language, in his view, had its roots in deep biological structures, defined by their capacity for infinite creativity. In *Aspects of the Theory of Syntax* (1965), Chomsky suggested that all languages are generated by the same underlying 'universal grammar'. A finite set of grammatical rules is deployed to generate all possible verbal and written constructions, thereby making 'infinite use of finite means'. Universal grammar was the engine empowering the articulation of the coherent regions of Borgesian libraries.

Using an approach to language generation more akin to Saussure's pattern-based conception than the rules-based generative models of Chomsky, in December 2022 the California-based company OpenAI, led by Sam Altman, launched ChatGPT – a language model powered by cutting-edge artificial neural network technology. This AI chatbot captivated the world's interest by achieving an unprecedented mastery of the English language. It was able to effortlessly compose stories and poems, generate narratives, answer complex questions on a wide range of topics and offer advice in flawless grammar.

ChatGPT acquired its formidable generative prowess by training on vast datasets, uncovering patterns in text and refining its ability to predict statistically probable outputs. It uses an advanced AI deep learning architecture called a large language model (LLM), built around a revolutionary type of artificial neural net called a transformer. These specialise in processing language by identifying relationships between words in a sentence. Their key feature and innovation is that they focus on language syntax, which defines the relationships between the words in a sentence. Transformers are able to capture complex long-range dependencies and contextual relationships in sentences. This ability to weigh the relative importance of words in a sentence is known as an 'attention mechanism'.

But despite their linguistic competence, LLMs lack any true understanding. ChatGPT doesn't 'know' anything. It is simply a statistical engine for pattern detection on a huge scale, and is devoid of comprehension. Although giving the impression that it has human-like thought processes, ChatGPT is in fact a 'stochastic parrot'. This lack of understanding has profound implications.

Noam Chomsky has pointed out that ChatGPT is incapable of making authentic moral judgements – its outputs are unconstrained by the ethical principles governing human creativity. Its brilliance is marred by what Chomsky terms 'the moral indifference born of unintelligence'. It is, furthermore, prone to 'hallucinations', resulting in factually incorrect or nonsensical assertions, due to the fact that it lacks a real-time verification mechanism. Early versions, for instance, referenced a fictitious Agatha Christie book titled *The Mystery of the Missing Manuscript*.

Despite such limitations, the parallels between human languages and the nucleotide sequences of DNA – first appreciated by Friedrich Miescher – indicate that ChatGPT's deep learning-based LLM approach could help illuminate the hidden grammar of life. Just as written texts comprise constructions of letters, words and sentences, so genomes are built from sequences of four chemical 'letters'. The source code of genomes must be assembled according to its own unique type of universal grammar, with the genome of each species being akin to a distinct work of literature.

The proof that genetic sequences are physically structured like words has its origins in the work of the geneticist Seymour Benzer. Born in New York City in 1921 to Polish immigrants, he received a microscope for his thirteenth birthday, an event that 'opened up' his 'whole world'. It led him to become fascinated with the nature of genes, and in particular the question of whether they were indivisible or built from smaller parts. In the 1950s, this was still unknown. Using the fact that when two viruses infect the same cell they produced daughter viruses that blend the genetic material of both types together, he showed in 1955 that genes – previously thought to be indivisible – were composed of chemical letters reminiscent of the alphabetic letters of written languages. Genes could, as a result, be viewed as chemical words, capturing units of meaning.

But words alone are insufficient to generate the complex meaning embodied by language. To construct nuanced and meaningful utterances, sequences of words must be organised into the higher-order structures of sentences, paragraphs, chapters, volumes and sets of volumes. In 1971 the pioneering molecular biologist Erwin Chargaff likened the genetic code to a cryptographic language governed by a grammar. Sequences of the genetic alphabet's four letters are transcribed into RNA sequences, and then translated into the language of proteins, which is articulated in combinations of the amino acid alphabet. This shared chemistry forms what Arthur Kornberg called the 'universal language' of all life on Earth.

As Chargaff once lamented, despite decades of molecular genetic advances, we remain 'far from defining an actual grammar of a living cell, not to speak of that of an organ, or organism, or, even more, a thinking organism'. Chargaff went on to identify what he saw as the central problem at the heart of life: 'We lack a scientific knowledge of the whole.' Providing an example of this shortcoming, he observed: 'I look out of the window. There is a dog; he wags his tail. What is his molecular biology?' Chargaff concluded by referencing the German-language author Franz Kafka who used a similar phrase to describe his own inadequacy in the domain of human relationships, stating 'somehow I cannot rid myself of the feeling that we still lack an entire dimension, that is necessary for the understanding of the living cell'.

Chargaff was not implying that metaphysical life forces dictated the generative rules of life. It was just that we appeared to be missing an as-yet-to-be-defined critical piece of understanding that had eluded scientific investigation. The rules governing biology's generative grammar – how genetic sequences coded for the structures and operation of different species – remained an enigma. The full comprehension and mastery of these elusive rules, and the ability to execute them through genome design and construction, is the currently obscured North Star of biology that I have called Artificial Biological Intelligence.

One possibility is that this missing principle arises from the way in which nucleotide sequences are typically viewed as one-dimensional texts, akin to written language or computer code. In reality, DNA is a three-dimensional molecule that exhibits dynamic, complex behaviours. These additional dimensions layer further tiers of meaning onto the linear code script of life, rendering it unlike any written language. It is as if each page of biology's book of life jumps out in three dimensions – twisting, turning, and contorting as it is read. The genome is a 'living' text, animated in ways that printed books can never be.

We are at the babbling stage of our ability to speak the language of biology. We babble a little when we edit or engineer genes, and when we successfully copy the smaller genomes of existing species into synthetic sequences, recapitulating nature's blueprints. But our knowledge of life's generative grammar of biology is rudimentary and lacks fluency. We are currently able to glimpse just the smallest sections of its rule book, and lack a comprehensive grammar of biology. To author entirely new organisms, we will need to master this language, embodied in the concept of ABI (Artificial Biological Intelligence). We will need to learn how to speak the language of life.

Here, large language models like ChatGPT offer a promising way forward. The digital, four-letter nucleotide alphabet of DNA suggests that, like human languages, biology's language is computable. As such, LLMs of the type used by ChatGPT may help identify hidden patterns in the code script of the genomes of living organisms, much in the way that they are able to navigate natural language.

Demis Hassabis, the co-founder of DeepMind, has argued that 'biology can be thought of as an information processing system' akin to a living computer. 'Just as mathematics turned out to be the right language for physics,' he conjectured, so 'biology may turn out to be the perfect type of regime for the application of AI.'

With sufficient data, it might be possible to discern the generative grammar of biology – a linguistic framework that defines the possible 'utterances' of genetics – and holds the key to understanding the rules by which DNA encodes structure and function. These putative generative rules of life comprise 'the language of biology', which describes how biological source code computes the form and function of species. It will enable the source code of synthetic genomes to be transformed into new biological structures and generate new behaviours. In conjunction with emerging technologies for genome synthesis, the ability to define the generative rules of biology will unlock remarkable possibilities and provide the basis for a new science of generative biology – not only unlocking biology's secrets, but making it fully programmable.

The vast biological databases assembled over decades, are well suited for such explorations. One such database, the Worldwide Protein Data Bank (wwPDB), distributed across locations in the US, Europe and Japan, forms the repository for all the three-dimensional protein structures determined by experimentally using X-ray crystallography. When initially established in 1971, it contained just seven structures. By 2025, it contained over 194,000. While it seemed self-evident that the process by which a protein 'folds' into its three-dimensional structure should be computable, the inability to discern the underlying rules of protein folding meant that X-ray crystallography offered the only means by which protein structures could be determined.

On 11 December 1962, the pioneering protein crystallographers John Kendrew and Max Perutz presented their Nobel Prize lectures in Stockholm. Working together at the Cavendish laboratory in Cambridge, they had determined the first ever three-dimensional structures of two closely related proteins. The structure of a protein is a molecular map detailing the positions of all its atoms. Proteins are the principal molecular components of biological machines and carry out many of life's tasks. While some form the structural components of cells or have regulatory roles like hormones, others, such as enzymes, catalyse the chemical reactions underlying the metabolic processes of life.

Proteins are made from one-dimensional sequences of twenty different amino acids joined end to end into chains. In order for proteins to perform their functions, the one-dimensional amino acid sequence must adopt a specific structure in which the position of every atom attains a precise position relative to all the others in three-dimensional space. Although the raw information of life is stored in one-dimensional amino acid sequences, biology operates in three dimensions.

While the amino acid sequences of proteins are encoded by genes, the information required for a protein to fold into its correct structure is neither specified nor encoded. As discussed, proteins fold into their three-dimensional structures spontaneously as a result of the laws of physics and chemistry. Like the young ladies in Victorian England who required chaperones to accompany them to society events, some proteins require assistance from molecular chaperones that guide amino acid chains into their correctly folded three-dimensional forms. Others lack structure, remaining completely or partially disorganised. Since the functions of proteins are determined by their three-dimensional folded structures – or in some cases lack of structure – the challenge for these pioneering molecular cartographers was to track down the position of every atom in a protein, in order to define a corresponding atomic map.

Kendrew concluded his Nobel Prize lecture, 'Myoglobin and the structure of proteins', by reflecting that 'In determining the structures of only two proteins we have reached, not an end, but a beginning; we have merely sighted the shore of a vast continent, waiting to be explored.' He realised that they had opened the door to a limitless protein universe containing the amino acid sequences and structures of all possible proteins. In the same way that Fred's Library contained the sequences of all possible genomes, 'Max's Library' contained the complete protein sequence and accompanying structural universe. Looking to the future, Kendrew stated that 'Though the detailed principles of construction do not yet emerge, we may hope that they will do so at a later stage of the analysis.' In other words, long before protein structural prediction using artificial intelligence had become a mature science, he speculated that it would one day be possible to predict the three-dimensional structure of a protein from its primary amino acid sequence.

While X-ray crystallography is an effective method for mapping the shapes of proteins at an atomic level of resolution, it is a time-consuming process that yields unpredictable results, and often takes many years to complete. It involves synthesising and purifying proteins, getting them to crystallise, taking high-resolution X-ray photographs, interpreting the data, modelling and validation. More recent methods include cryogenic electron microscopy (cryo-EM) – pioneered by Richard Henderson at the LMB, who built on the work of Jacques Dubochet and Joachim Frank – which takes two-dimensional photographs of frozen proteins at different angles, producing lower-resolution structures.

Like Kendrew, the Columbia University physicist Cyrus Levinthal had in 1969 wondered whether there might be a faster and more direct way to determine protein structures. He speculated that it might be possible to compute a protein's structure from its primary amino acid sequence. These ideas were detailed in a paper 'How to Fold Graciously' presented at a conference at the University of Illinois. But there was a significant barrier to doing this, as even relatively short amino acid chains had an astronomical number of potential configurations. Systematically sampling each of them would take longer than the age of the universe. Yet proteins fold in milliseconds – a conundrum that became known as Levinthal's paradox. From this, Levinthal inferred that proteins must navigate preferential folding pathways generated as a result of local interactions influencing the folding landscape. It suggested that there must be an underlying 'protein-folding code'. The next half-century in the field of protein-folding was spent in an ultimately futile attempt to crack this hypothetical code, through an analysis of physical forces, biochemistry and evolutionary comparisons between related proteins.

In 1994, the computational biologist John Moult at the University of Maryland in College Park co-founded a computer-based protein-folding competition. The Critical Assessment Structure Prediction (CASP) was a global, community-based initiative that allowed researchers to test the performance of their prediction algorithms – designed to determine the three-dimensional structure of proteins from their primary amino acid sequences – against competing methods. But despite some progress, the available methods fell far short of atomic accuracy. Protein folding appeared to be an elusive and computationally intractable problem. The idea that the hidden universe containing the structures of all the natural and artificial proteins could be illuminated using computational methods, seemed to be fanciful.

This situation changed dramatically, however, in 2020 with the release of AlphaFold 2, an AI programme developed by DeepMind. Using the same deep learning and neural network technologies that power programmes like ChatGPT, overnight AlphaFold revolutionised protein structure prediction. It utilised a deep learning algorithm to consider both local and long-range interactions within proteins, and iteratively refined its structural predictions. In essence, it was the ChatGPT of proteins. But instead of training AlphaFold2 with text, it was trained using the three-dimensional X-ray crystallographic structures of proteins housed within the PDB database.

This enabled the model to establish an intuitive understanding of protein sequences, and of how the atoms of each amino acid map onto the shapes of proteins. It was able to do this with an accuracy of around the width of one atom, and illuminated the structures of about 350,000 proteins from organisms as diverse as soybeans, yeast and flies. After observing its performance, John Moult suggested that 'in some sense the problem had been solved'. Despite being unable to articulate the underlying mechanics of AlphaFold2's insights and processes, DeepMind had functionally cracked the protein-folding code.

AlphaFold2 was trained on the approximately 180,000 three-dimensional structures in the PDB at that time, and their corresponding amino acid sequences. It quickly learned to identify hidden patterns in the data. The results were published by John Jumper and Demis Hassabis in 2021 in a paper titled 'Highly Accurate Protein Structure Prediction with AlphaFold'. Although not fully understanding the inner workings of their programme, or why proteins fold up the way they do, they had nevertheless succeeded in developing a computational approach for predicting many natural protein structures to near-experimental atomic accuracy.

In August 2022, DeepMind unveiled a spectacular treasure trove of information containing the predicted protein structures of nearly every protein in the known natural protein universe. This database, containing more than 200 million protein structures from organisms spanning bacteria to humans, was heralded by Hassabis as 'the beginning of a new era of digital biology'.

Meanwhile, as AlphaFold2 was focused on solving the structures of individual proteins, in July 2021 David Baker and Minkyung Baek at the University of Washington, Seattle, released an AI-based structure-prediction model called RoseTTAFold, which was able to predict the structures of complexes of interacting proteins. AlphaFold 3, released by DeepMind in May 2024, further expanded its capabilities by being able to predict the organisation of complex biomolecular assemblies, as well as the impact of chemical modifications to proteins, and the interactions of proteins with DNA, RNA, ligands and ions.

The mapping of most of the known protein universe by AlphaFold and related AI models promises to revolutionise drug development and the redesign of living systems. Yet its implications stretch much further, outlining a path towards the invention of novel proteins that nature has overlooked. Evolution has sampled only a tiny fraction of the vast landscape of possible proteins contained within the amino acid sequence space of Max's Library. By going beyond nature, we gain the ability to construct proteins that are precisely tailored to meet challenges that evolution was unable to anticipate.

But while the ability to predict protein structure allows us to pencil in some of the early chapters of life's grammar book, it does not, by itself, illuminate the generative grammar of genomes – the hidden logic through which genomes compute the unique behaviours and identities of living species. To author new biological systems – and design entirely new forms of life – we must first decode these generative rules.

The most fundamental of these rules are likely encrypted within the genome's non-coding 'dark matter'. Rather than coding for proteins, these regions generate RNA molecules that regulate gene activity. This genomic dark matter – and indeed the entire genome – is folded into a complex, dynamic three-dimensional structure. It is within these folds – the twists and turns of the non-coding regions of genomes – that the missing chapters of life's grammar book are likely to be located.

8
The Design of Species

> The story of the field so far is can write DNA, nothing to say.
>
> – DREW ENDY, 2011

One of the most striking features of biology is its astonishing versatility. From the same basic starting material – sequences of just four different nucleotides – genomes generate an extraordinary diversity of life forms: from spiders and kangaroos to panda bears, bacteria, flowers and pythons. All life on Earth emerges from the same simple genetic alphabet.

If we wish to send a person to the Moon, physicists can sit down and, by referencing well-established laws of physics, confidently plot a course to the spacecraft's destination. Biology, in contrast, is fundamentally different. To date, we have not been able to establish predictive laws that reliably determine particular biological outcomes. There is, for example, no Newton's law or Kepler's law of biology. If you wanted to create a new species – say, a cross between a giraffe and a bison – biologists would be unable to define a genome sequence that could make this a reality. Nor can we look at a genome and infer that it computes a dandelion, an oak tree, or a zebra. Biology is not, at present, a predictive science.

We currently have only the most rudimentary understanding of the generative laws of biology – the hypothetical principles that map genome sequences to the characteristics of species. While we can read the language of DNA, we cannot yet speak it, nor fully comprehend it. The processes by which genomes compute the identities of different species remain elusive. Yet, such predictive generative laws must exist. Understanding the rules of biology's generative grammar – how genome sequences specify the unique features of species – is the most important outstanding question in biology. Unlocking those rules would not only allow us to create new and highly modified species, but also offer profound insights into the origins and mechanisms of human diseases. Moreover, sustainable, biologically inspired technologies hold the potential to transform nearly every aspect of human existence. They may help us to preserve Earth's fragile ecosystems and to protect the diversity of its natural species.

Were we to unlock the code of this biological grammar we would gain unprecedented control over living systems. Combined with the ability to synthesise genomes, this knowledge would transform biology into a predictive engineering material. The creation of new organisms from synthetic genomes would, at that point, become an algorithmic, rules-based process. We would have transitioned from being passive observers of life to becoming its authors.

While the exact nature of these biological rules remains unknown, they must be encoded in DNA – represented not only in the sequences of nucleotides, but also in the way that the genome folds and packages itself. DNA is not just a linear repository of genetic information, it is also the engine that translates genetic texts into biological meaning. It is simultaneously both the pages of the manuscript and the reader – a self-referential grammar book that interprets itself. The relationship between the grammar hard-wired into DNA and its three-dimensional folding is reminiscent of advances in origami, where many of the folded structures derivable from a single sheet of paper can now be predicted mathematically. In a similar way, the folding of genomes – the spatial arrangement of the nucleotide sequences of DNA molecules – may eventually be understood as a mathematically governed process.

In 1992, Jun Maekawa, a member of the Tokyo-based Japanese Origami Tanteisan Convention, sat down with a sheet of paper. With the precision of a structural engineer, he began to meticulously fold it. Before long, he produced something spectacular. His creation emerged lifelike, in its full animated glory. He had made a bug. A folded paper species. But it wasn't just any bug. It was a magnificent, multi-horned rhinoceros beetle, its flamboyant wings outstretched in flight. Maekawa's virtuosity made him an overnight sensation within the origami community, establishing him as its Michelangelo. And this audacious offering had an unexpected effect. It ignited what would later become known as the 'Bug Wars' – a surge of creative rivalry that pushed the boundaries of origami complexity and realism to astonishing new levels, ushering in a period of unprecedented paper insect evolution.

While creating a paper beetle from a single sheet of paper without making any cuts may not seem like a big deal, prior to Maekawa's breakthrough, insect-like origami was considered an impossibility. The existing repertoire of paper 'bases' forming the folding foundations and starting points for origami innovations were simply inadequate. They struggled to generate the fat-bodied abdomens and spindle-like appendages required to mimic insect morphologies. To transcend these limitations, entirely new bases were required. Having thrown down the gauntlet, the origami community sprang into action.

What followed next was a proliferation of weird, wonderful and increasingly more complex paper insect forms – a creative explosion of paper forms that transformed the once-constrained world of origami creations into one of seemingly endless possibility. A deluge of folding one-upmanship conjured from the mathematical depths of the 'Maekawa Library', the conceptual archive of all possible origami insect forms. These included long-legged wasps, praying mantises, dragonflies and grasshoppers. Maekawa had instigated an origami arms race, in which appendages proliferated, spots flourished, antennae sprouted, and spiders transmogrified into scorpions.

As is often the case in the context of competition, the need to secure a technological advantage fostered innovation. While earlier designs could be implemented with just a few dozen folds, it soon became routine for new designs to require several hundred. The relationship between paper creases and their resulting shapes could now be defined mathematically. As a result, the number of possible designs multiplied exponentially.

Robert J. Lang, a former laser physicist turned origami master, was one of the first to apply mathematical models to origami design. Known as the 'renaissance man of origami', and working at NASA's Jet Propulsion Laboratory in Pasadena, California, Lang sought to understand the generative principles underlying paper folding – aiming to go beyond the formulation of step-by-step instructions.

His goal was to develop algorithms that could map the anatomical designs in his imagination to specific, tangible outcomes – such as producing an insect with six or eight legs, rather than four. If successful, these algorithms would form the basis of an origami generative grammar, capable of generating a virtually unlimited range of innovative paper species. Ideally, such generative folding rules would act as mathematical instructions for creating a folded paper organism of any type. In 2003, Lang published *Origami Design Secrets: Mathematical Methods for an Ancient Art*, and the science of computational origami, which turned flat paper sheets into prespecified three-dimensional forms, was born.

The intellectual roots of this algorithmic approach to paper folding can be traced to the work of T. Sundara Row, an Indian tax collector and amateur mathematician, whose book *Geometrical Exercises in Paper Folding* (1893) explored the mathematical principles behind origami constructions for regular polygons, symmetrical structures and algebraic curves. Aiming to use 'math like glue', Lang developed two computer programmes, *Tree Maker* and *Reference Finder*, which translated computer-generated images into human-readable folding instructions, known as 'crease patterns'. These allowed him to create origami designs of unprecedented complexity, including an origami cuckoo clock, a dun-coloured centipede, a turtle with a patterned shell, a rattlesnake with 1,000 scales and a moose.

Like the origami organisms constructed from paper using computer algorithms, the structures of living species should also be amenable to a predictive, mathematical analysis. In the case of biology, however, rather than being translated into crease patterns, computer-generated predictions would be converted into synthetic genome sequences. These sequences would compute not only an organism's structure and function, but also its biochemistry and behaviour. The outputs of such predictive algorithms would be made from flesh and blood, rather than paper. To realise this vision of computer-generated species, we would need to develop a computational science of species design. The grammatical rules governing this would need to accommodate both the shared features of species and their idiosyncrasies.

Yet, as Lang's work in origami demonstrated, predictive algorithms have their limitations. For a start, they are rarely fully predictive. Lang was never able to compile the definitive grammar book of origami forms he had aspired to create. While opening up significant new possibilities, his algorithms were incomplete. Many of the design subtleties required offline experimentation and hands-on folding expertise. The grammar of origami life, in its full magnificence, complexity, and intricacy, had ultimately eluded him. Lang himself acknowledged that despite the spectacular insights that his algorithmic approach had revealed, we were 'still nowhere near the limits of what's possible'.

Lang's work inspired a new generation of origami artists to go beyond conventional representations of natural species and explore the untapped corners of origami's mathematical possibility. Eric Vigier, known for his crafting of hybrid origami organisms, created a grenoceros – a blend of a frog and a rhinoceros – while the Belgian origami artist Michael G. LaFosse folded a unicorn.

Origami design principles have proved to have a general utility beyond the creation of folded paper forms, notably inspiring the design of material structures that mimic biological systems. In 2017, Daniela Rus of MIT's Computer Science and Artificial Intelligence Laboratory used origami-inspired designs to develop artificial muscles capable of lifting up to 1,000 times their own weight and producing up to six times more force than human muscles. Another intriguing application emerging from the mathematisation of origami lies in its application to space travel, particularly in the design of deployable structures such as folding telescopes.

NASA's James Webb Space Telescope (JWST) is the largest and most powerful infrared telescope ever placed in space by humankind. It has allowed us to look back in time at distant worlds and to explore some of the earliest moments in the universe's history. But building the telescope presented a formidable and seemingly intractable design problem. The JWST's mirror and heat shield were too large to fit into the chassis of the rocket designed to launch it into orbit. The telescope's primary mirror alone was 21.5 feet across, while its heat shield spanned an enormous 46 by 69 feet. The engineering team realised that in order to fit these structures into the Ariane 5 rocket that would transport them, both components would have to be folded up – much like a picnic table packed into the boot of a car. Once in space, the mirror and heat shield would, at the right moment, then need to be meticulously unfolded like a delicate desert flower opening in the sun, without generating marks, creases or misalignments.

That was when the NASA team contacted Robert Lang. He swiftly produced an origami-inspired solution to address the problem, showing that the mirror could be assembled from eighteen identical, gold-coated, hexagonally shaped segments, each just 4.3 feet in diameter. On day 29 after the launch, following a close to million-mile journey through space, the mirror segments and heat shield unfolded with astonishing precision, aligning with an accuracy of around 1/10,000 the width of a human hair. The telescope deployed successfully, and on 8 January 2022, it began transmitting images back to Earth.

Around half a century earlier, the American biochemist Roger Kornberg, the son of Arthur Kornberg, encountered a different type of folding problem. Arriving at Cambridge's MRC Laboratory of Molecular Biology in 1972 on a one-year sabbatical, he learned that his supervisor – the brilliant South African-British crystallographer and laboratory head Aaron Klug – was on holiday. Kornberg soon realised that the ongoing projects in Klug's group were too advanced for him to make a meaningful contribution. He needed a project of his own, and was inspired by an eloquent, though as it turned out technically incorrect, paper published by Francis Crick in 1971. Among other things, Crick speculated that chromatin – the nuclear substance comprising proteins and DNA – organised itself into higher-order structures stabilised by proteins called histones. Kornberg had finally identified a topic for his studies. He would, like Lang, focus on a folding problem – but in his case, it related to the way the human genome folded itself rather than to paper insects, artificial muscles or space telescopes. He would study the origami of DNA.

Like the Ariane 5 rocket accommodating the JWST telescope, the nucleus in a human cell – a small membrane-bound organelle in the cells of higher organisms that houses the hereditary information – must also solve a challenging storage problem. It must somehow, Houdini-like, fit approximately two metres of DNA, the entire human genome, into a space of just five to ten microns – five-millionth to ten-millionth of a metre – in diameter. This 'miracle of packaging' had been likened to cramming twenty-four miles of thread into a tennis ball. Without the benefit of an origami master like Lang to assist them, cells had to evolve their own strategy for solving this extraordinary folding problem.

The term 'chromatin' was coined in 1882 by Walther Flemming, the German biologist, pioneer of cell division and advocate of 'microscopic anatomy'. In his *Zellsubstanz, Kern und Zelltheilung* (Cell Substance, Nucleus, and Cell Division), Flemming described chromatin as 'that substance in the nucleus, which upon treatment with dyes known as nuclear stains does absorb the dye'. Chromatin would soon become central to our understanding of how DNA works. Its ability to fold and package itself within the nucleus plays a central role in regulating gene expression, by determining which genes are switched on and off. Chromatin was known to have properties that caused it to be fluid, suggesting that DNA was dynamic rather than static.

Far from being a passive repository of linear information, DNA located inside the cell nucleus operates as a dynamic machine – undulating, folding, and refolding into the intricate three-dimensional structures that govern gene activity with remarkable precision. These movements of the genome represent a distinctive kind of genetic programme – one which is written not only in sequence, but in time and motion.

Chromatin exists in two principal forms. When DNA is tightly packed, it 'condenses' into a state known as heterochromatin. In this compact, folded form – much like books locked in a library – genes are inaccessible. When unfolded, however, and adopting a loose and extended state called euchromatin, the library door is unlocked and genes become accessible for transcription. The precise mechanism governing the folding and unfolding of DNA, and the conversion between these two states, remains unknown. Moreover, the fact that chromatin appeared to lack a consistent, static structure, suggested that its elucidation might not be possible using classic X-ray crystallographic methods.

This perspective changed, however, when Aaron Klug returned from his holiday and appeared with a research paper in his hand that he had managed to extract from the towering piles of documents in his office. This showed that chromatin produced 'a striking, simple, X-ray diffraction pattern', suggesting that it did, in fact, have a regular structure. Klug recognised that the problem of chromatin structure was therefore likely soluble, and encouraged Kornberg to pursue it. What began as a one-year sabbatical evolved into a four-year secondment, during which Kornberg discovered the fundamental building block of chromatin structure and the atomic unit of DNA packaging known as the 'nucleosome'. This discovery, along with his more general work on the molecular basis of eukaryotic transcription – for which he was awarded the 2006 Nobel Prize in Chemistry – transformed our understanding of DNA's structure and function.

Kornberg likened nucleosomes to 'beads on a string', each bead comprising DNA wrapped nearly twice around a spool of eight histone proteins, like yarn on a bobbin. Nucleosomes are analogous to origami bases – foundational units from which more complex DNA-folding patterns are derived. The histones anchor DNA in place much like Velcro, attaching, detaching, and reattaching, enabling the genome to form stable structures or unfold as required. Each nucleosome has a small 'tail' that functions like a bookmark, helping to orchestrate the choreography of DNA folding. Kornberg described strings of nucleosomes as a 'flexibly jointed chain of repeating units'. Comprising roughly 25 million nucleosomes connected by short DNA linkers, this structure forms a scaffold that helps organise the architecture of the human genome.

At Stanford, Kornberg and his colleague Yahli Lorch discovered that nucleosomes were not merely passive units of DNA packaging. Instead, they function more like switchboards, regulating gene expression and forming part of the genome's operating system. When DNA is wrapped around histones, the nucleosome functions as an 'off' switch, but when a gene is activated, it is pulled off the histone proteins in the nucleosome, allowing it to be transcribed into RNA.

The enzyme that initiates this process, RNA polymerase II, operates in concert with around sixty other proteins to coordinate gene expression by communicating through interactions with regulatory sequences known as enhancers. This collection of proteins is known as the mediator complex. It recruits and stabilises RNA polymerase II, forming a bridge between the enzyme and transcription factors – proteins that bind DNA sequences and regulate gene expression activity. If nucleosomes are the storage units of DNA, and RNA polymerase molecules the copying machinery, then mediators are the control panels that determine which genes are read and when.

Beyond nucleosomes, DNA organises itself into increasingly complex structures. In 1976, Aaron Klug and John Finch showed that strings of nucleosomes fold into even more tightly folded structures, known as 30-nanometre fibres, which condense DNA a further sixfold. The compacted genome then attains an even higher level of folded organisation through the formation of topologically associated domains (TADs).

These structures emerge when stretches of nucleosome-packed DNA balloon out into loops, bringing distant regions of DNA into contact with one another. A protein called CTCF acts like a molecular clasp, drawing together two distant points on DNA, much like the ends of a shoelace. CTCF marks specific sites on the genome, enabling another protein, cohesin, to extrude the intervening DNA into a loop. These loops – which can span up to 2 million nucleotides and encompass thousands of nucleosomes – are essential for the genome's organisation. While TADs dominate the genomes of complex species like humans, simpler organisms like yeast lack them, suggesting that the evolution of TADs created new regulatory opportunities that were critical to the emergence of complex life.

In 2014, Suhas Rao and Erez Lieberman Aiden at Baylor College of Medicine in Texas used artificial intelligence to map the estimated 10,000 loops in the human genome, which they called the 'loopome'. Each loop creates a distinct zone, which insulates genes from one another and ensures their activity is precisely controlled. The loops regulate gene expression, forming and dissolving to fine-tune gene function. They demarcate distinct neighbourhoods where genes may be expressed without affecting one another, and introduce complex patterns into genome folding. Together they define a 'loop code', which forms a key part of the genome's regulatory grammar.

The universality of these loops across multicellular species underscores their central role in the genome's regulatory machinery. Further exploration of genome folding – and of the mechanisms underlying loop formation – are likely to provide insights into the generative rules of life. Just as a city map reveals the structure and logic of urban planning, so too AI-augmented mapping of the genome's three-dimensional architecture will illuminate the regulatory programmes that underpin biological complexity. This increasing fluency in life's language will open up new frontiers for synthetic biology and enable the programmatic design of life.

The genome's organisation, however, doesn't stop at TADs. These domains fold into even finer structures, called sub-TADs, which connect distant regions of DNA and contact domains where interactions between DNA segments are more frequent. Insulated neighbourhoods, on the other hand, minimise their contact with other regions of the genome, forming self-contained regulatory microenvironments.

These various layers of genome folding complexity help position enhancers – the master regulators of gene expression – in close proximity to the genes they control, even if positioned millions of nucleotides away. The human genome contains at least 810,000 enhancers. This complete repertoire, along with all other known regulatory elements, has been catalogued by the ENCODE project (Encyclopedia Of DNA Elements) – an initiative aimed at uncovering how DNA regulation works, and elucidating the function of the approximate 98.5 per cent of the genome that does not code for proteins. The instructions written into these non-coding regions of the genome play a key role in gene regulation.

While the Human Genome Project initially focused on protein-encoding genes, we now know that at least 75 per cent of the human genome – once dismissed as 'junk' – is transcribed into non-coding RNAs. The number of RNA-producing genes greatly exceeds those coding for proteins. Their discovery has stimulated a new interest in the 'RNA-ome' – the complete repertoire of RNAs. It is likely that many of these play a pivotal role in gene regulation and cellular function.

This 'RNA renaissance' has reframed our understanding of genome functionality, refocusing attention from protein-coding genes to the human genome's non-coding 'dark matter'. The new emphasis on RNA, whose full significance is still emerging, is reshaping our understanding of genome function. Protein-coding genes are no longer perceived as the sole regions of importance.

Yet another layer of DNA-folding complexity involves lamina-associated domains (LADs), which are regions of the genome tethered to the nuclear lamina – a dense protein network lining the inner surface nucleus. These anchor points help organise the overall shape of the genome by attaching it to the edge of the nucleus. LADs, which are often gene-poor and inactive, generate silent genomic landscapes that are densely packed with nucleosomes.

At the top level of organisation, the genome divides into two higher-level compartments. The gene-rich 'A' compartment, located towards the centre of the nucleus, actively transcribes RNA. The gene-poor, transcriptionally silent 'B' compartment resides close to the nuclear periphery. During cell division, the genome adopts a radically different structure – condensing into rod-like chromosomes to ensure that each new cell receives an uncorrupted copy of the genetic material.

If life has a 'grammar' – an underlying system of rules that governs gene activity and transforms it into the unique forms and functions of living species, from pelicans and flamingos to ants and humans – then it must reside here, within the uncharted and fluid world of DNA-folding dynamics. Comprehending this grammar of life is biology's final frontier. To confidently design and build new forms of life, we will need to decipher the genome-folding code and compile its rules into a coherent grammar book of life.

This hidden generative grammar extends far beyond the deterministic DNA-to-protein code. It spans every scale of chromatin organisation, and encodes the logic by which folding patterns are decoded. It is the rule book that describes how remodelling of the chromatin landscape finely maps onto biological structure and function. Cracking this elusive code will allow us to programme genomes with the same mastery that origami artists apply to paper species – enabling life to be built from the ground up.

What we do know is that the rules of this genomic grammar book are highly distributed, operating through multiple layers of regulation at every level of biological organisation. For example, DNA undergoes methylation – a chemical modification that typically represses gene activity – in a process known as epigenetic regulation. The histone proteins within nucleosomes can also be modified, forming part of what is known as the 'histone code'. In addition, specialised histone variants fine-tune chromatin structure. A complex interplay of regulatory RNA molecules, transcription factors, and other proteins adds still further complexity to gene regulation.

As we have seen, RNA now appears to be far more important than originally imagined. In fact, most of the genome is dedicated to making RNA. RNA uses a language nearly identical to that of DNA, but, unlike DNA, it is single-stranded. This allows it to adopt a wide variety of different three-dimensional shapes. While DNA contains a deoxyribose sugar, RNA uses ribose, which includes an extra oxygen atom. This subtle difference makes RNA less stable and shorter-lived, which is ideal for its regulatory functions. RNA also differs from DNA in that one of its four nucleotides replaces thymine (T) with a short-lived nucleotide called uracil (U).

Many RNAs – especially those longer than 200 nucleotides, known as long non-coding RNAs (lncRNAs) – play essential roles in genome regulation and influence chromatin structure. Estimates suggest that the human genome makes between 16,000 and 100,000 different lncRNAs. Some of these have been shown to persist for years and regulate crucial biological processes. One called Maenli plays a key role in limb development. Mutations in Maenli, inherited in a Mendelian fashion, cause severe congenital limb malformations. LncRNAs also control the features of non-human species, including the colourful wing patterns of butterflies, such as the painted lady butterfly. Genes that encode RNA are clearly every bit as important as classic protein-encoding genes.

Currently, researchers can only infer the three-dimensional structure of the human genome through indirect measurements. A true understanding of the genome-folding code will, however, likely require direct visualisation of entire genomes within single cells. We will need to produce a high-resolution map of the dynamics of the entire folded human genome at the nanometre scale. Much like how DeepMind's AlphaFold2 used AI to revolutionise protein structure prediction, an AI model trained on large-scale datasets of three-dimensional genome structures derived from single cells may help decode the grammatical principles of genome folding.

By integrating genome-folding data with RNA and protein-expression data from cells of different types and species, it may be possible to develop predictive, rules-based models of genome authorship. Such an approach could usher in a new era of algorithmic biological design – akin to the rules-based creation of origami species pioneered by Robert J. Lang. This represents the most plausible route to achieving the design component of Artificial Biological Intelligence.

Researchers at the University of Washington have developed a technology called GAGE-seq, which indirectly maps the three-dimensional architecture of the genome and links these structural insights to gene expression within the same cell. Although it does not directly visualise chromatin structure, it generates contact maps that reveal how different genomic regions interact. By correlating chromatin organisation with gene expression, GAGE-seq makes it possible to explore how genome folding influences RNA production within individual cells.

To fully define the genome's generative grammar – the rules governing how genomes compute organismal structure and function – it will be necessary to build a comprehensive atlas of genome structures across diverse cell types, development stages, and physiological states, including both health and disease. Such a database could be used to train artificial intelligence models capable of recognising specific genome folding patterns and predicting their influence on morphology and function. This deep structural understanding may lead to new therapeutic strategies for treating human diseases based on the correction of misfolded genomes. Furthermore, since the genome's three-dimensional structure changes throughout life, its study could illuminate the biological basis of ageing, and potentially offer new insights into the attainment of healthy longevity.

The definitive elucidation of the genome's grammar will require going beyond folding patterns and its source code to address its mechanical properties. The physical attributes of DNA are known to vary depending on their sequence position. Regions like nucleosomes, for example, where DNA undergoes sharp bending, exhibit great flexibility, whereas the linker regions between nucleosomes tend to be stiffer, helping to anchor them in place. These variations in 'bendability' across the genome suggest the existence of an underlying 'mechanical code' superimposed onto the genetic code. The need to preserve the integrity of this code may impose constraints on how genomes can evolve and may also influence the way in which genetic instructions are read and executed.

Beyond housing genetic information, the genome also functions as a major physical structure within the cell, serving as a scaffold that transmits mechanical forces from the nucleus to the cytoplasm. It also contributes to the assembly of the nuclear membrane and pore complexes, and anchors regulatory proteins.

The application of deep learning and large language models to three-dimensional genome structures will be vital in elucidating the grammar of human biology. Such approaches will help reveal how the human genome functions as a generative machine, and to uncover the design rules that will help guide future genome synthesis efforts and the predictive engineering of biological systems.

Remarkably, the three-dimensional structure of genomes is highly robust and may persist in the cells of extinct species. In 2024, Erez Lieberman Aiden and Marcela Sandoval-Velasco demonstrated that the genome structure of a 52,000-year-old woolly mammoth remained intact in skin cells preserved in the Siberian permafrost. These molecular 'subfossils' offer a rare glimpse into the enduring nature of DNA's three-dimensional architecture.

One-dimensional sequence data – derived from DNA, RNA, and proteins – can also be harnessed to train machine learning models that search for patterns in biological languages. In 2024, Brian L. Hie and Eric Nguyen, working together at the Arc Institute and Stanford University, introduced Evo, an AI DNA-language model designed to generalise across these three components of biological languages. Evo is capable of both prediction and generative design. Its goal is to build a model capable of designing viral and bacterial genomes from scratch, for use in a variety of practical applications – from the development of cancer treatments to carbon dioxide capture and biofuel production.

Evo treats DNA as a language and uses a variant of a model called StripedHyena – optimised for extremely long sequences – which improves on the architectures used in AI systems like ChatGPT. While ChatGPT identifies patterns in human languages, DNA is far more complex and requires tools capable of making predictions at the single-nucleotide level of resolution while maintaining coherence across much longer sequences.

Evo's architecture was designed specifically for this purpose, integrating information across long stretches of DNA while maintaining single-nucleotide-level resolution. That enables it to make predictions at both the molecular (single nucleotide) and genome-wide scales. Trained on a data set comprising 2.7 million viral and bacterial genomes – totalling approximately 300 billion nucleotides – Evo demonstrated its ability to identify essential genes, predict protein functions, and design artificial gene-editing systems using proteins and lncRNAs. Although its ability to design whole genomes was rudimentary, Evo was the first multimodal model capable of working across all three of biology's core languages.

Yet the complexity of the human genome, with its highly intricate grammatical rules, far exceeds that of bacteria and viruses, posing a formidable challenge to decoding its functional logic. To go beyond the limitations of Evo, a far larger and more diverse training dataset would be required, incorporating sequences from complex multicellular organisms.

Released on 19 February 2025, Evo 2, the updated successor to Evo (now known as Evo 1) – developed at the Arc Institute, Stanford, UC Berkeley, UC San Francisco and NVIDIA by a team led by Brian Hie and Garyk Brixi – was trained on thirty times more data than Evo. This represented the largest integrated dataset of its kind, encompassing over 9.3 trillion DNA base pairs of sequence across 128,000 different genomes, as well as metagenomic data (recovered directly from environmental samples) spanning all the domains of life, including plants, animals (including humans), yeast and bacteria. Viral genomes were deliberately excluded from the training set to prevent the open source Evo 2 model from being used to generate dangerous pathogens. At the time of its release, Evo 2 was the largest AI model ever constructed in the field of biology.

As a result of the development of a next-generation version of the StripedHyena architecture, called StripedHyena 2, Evo 2 can process genetic sequences of up to one million nucleotides long at a time.

This expansive 'window' enables the model to detect connections and relationships between distant regions of an organism's genome, as well as the underlying mechanics of cell function, gene expression, and disease. This broader 'context' window greatly enhances its ability to read and write the language of DNA. Evo 2 learns the statistical properties of DNA, transforming what is effectively a foreign biological language into something comprehensible. Just as ChatGPT works on sentences or chunks of human language, Evo 2 operates on DNA to generate biologically plausible propositions.

Increasing the size of the context window would typically require hugely more compute than was used in Evo 1. This is because as sequence length increases, the compute required increases quadratically. A DNA sequence that is one million nucleotides long therefore requires 100 times more compute power than one that is one-tenth its size. The StripedHyena architecture is able to surmount this problem by providing a wider context window at a significantly lower computational cost.

Patrick Hsu, co-founder of the Arc Institute, summarised the significance of Evo 1 and Evo 2, by stating that they 'represent a key moment in the emerging field of generative biology, as the models have enabled machines to read, write, and think in the language of nucleotides'. Unlike Evo 1, which was trained exclusively on viral and bacterial sequences, Evo 2 has 'a generalist understanding of the tree of life'. This broader scope makes it useful for a multitude of applications, ranging from 'predicting disease-causing mutations to designing potential code for artificial life'.

Evo 1 and Evo 2 are reminiscent of the large language models used in natural language processing models, such as ChatGPT. They function as a kind of ChatGPT for the genome, trained not on words, but on DNA sequences. As Brian Hie explained, 'Just as the world has left its imprint on the language of the Internet used to train large language models, evolution has left its imprint on biological sequences.' These patterns 'refined over millions of years, contain signals about how molecules work and interact'.

After identifying new patterns in the DNA nucleotide sequences database it is trained on, Evo 2 uses this knowledge to predict what the next nucleotide in a new DNA sequence is likely to be. This enables it to generate DNA scripts that have never previously existed in nature.

While the full scope of Evo 2's generative potential remains hard to fully comprehend, the model may serve as a foundation for building specialist applications. One immediate use of Evo 2 has been in analysing mutations in *BRCA1*, a gene that is frequently mutated in breast cancer. Evo 2 has demonstrated 90 per cent accuracy in predicting mutant versions of this gene that are likely to be capable of driving tumour development. The model also shows great promise in identifying the genetic causes of a wide range of diseases and in accelerating the discovery of new therapeutics. Other uses include the design of non-coding regulatory elements that could help ensure that the activity of gene therapies is restricted to targeted cell types.

Beyond healthcare, Evo 2 holds immense promise in the environmental sciences, for example, by helping to guide the design of the genomes of organisms capable of degrading plastic or cleaning up oil spills. It is also able to design sequences at the scale of whole genomes, and used its understanding of genomic language to generate 250 unique human mitochondrial genomes, 10 minimal bacterial genomes, and 20 artificial yeast chromosomes. But these designed genomes were neither constructed nor tested and given that they lacked various critical structural elements, it seemed unlikely that they would function. There was, in short, no clear evidence that any of these virtual genomes would be viable if actually constructed.

Evo 2, and its anticipated future iterations promise to help rewrite regions of the genomes of cash crops and to design biological systems that have never previously existed in nature to address a broad range of applications. As a result of Evo 2, the possibility of designing and exploring complex biological systems is, for the first time, on the cusp of becoming a reality. In natural evolution, mutations occur randomly and are subsequently filtered by natural selection. In contrast, Evo 2 is able to rapidly identify optimal mutations that achieve defined tasks, dramatically accelerating the incremental process of evolution by natural selection, and generating new, and unexpected pathways within DNA sequence space that can be tested and explored.

One of Evo 2's limitations is its tendency to 'hallucinate', generating outputs that do not correspond to biological reality. Supplementing its training set with a database of three-dimensional genome structures in addition to one-dimensional sequence data is likely to help mitigate this issue. However, there are significant differences between the gene regulatory systems in the three-dimensional genomes of simple organisms and those of humans. For this reason, it may not be possible to merge three-dimensional genome data across the different kingdoms of life. In simple organisms like yeast, for example, the genome's three-dimensional structure can be radically reconfigured – even turned inside out – without significantly altering its function. In animals, however, chromatin structure is essential to gene regulation.

It is easy to see how the quality of Evos 2's predictions and design outputs could be further improved. The incorporation of population-scale human genomic variation is one of the most promising ways to enhance its ability to comprehend and refactor regions of the human genome.

A number of other AI models have been developed to address specific biological tasks. Oracle, created by Aviv Regev and Eeshit Vaishnav at MIT in 2022, was trained to predict how strongly a gene will express under the control of natural and synthetic promoters. By analysing data from 30 million promoters, Oracle enables the design of synthetic gene circuits that can precisely modify gene expression. Another model, GPN, trained on data from the plant *Arabidopsis thaliana*, predicts the effects of introducing changes into non-coding regions of DNA. Although developed using plant data, it could be adapted for other species. ChatNT is similar to Evo, but more user-friendly, allowing individuals who lack coding expertise to achieve design goals in a conversational interface modelled on ChatGPT.

However, while adept at identifying relationships within data, AI models like Evo 2 and AlphaFold 2 have no true understanding of the patterns they uncover. Nor are we, the human operators of these systems, currently able to fully grasp how they function at a granular level, or to discern the mathematical principles underlying their extraordinary capabilities. This disconnect, between the ability to identify patterns and our understanding of how they originated, underscores a fundamental limit of AI. These systems lack any intrinsic comprehension of the biological rules they appear to navigate. Even expert users struggle to explain their internal mechanics – the invisible 'calculus' behind their predictions. In their current form, they are highly performing black boxes that conceal the basis of their operations. The rules they execute can only be inferred indirectly, through the solutions they produce.

In response to this, an emerging field of research called neural network interpretability is focused on peering 'under the hood' of AI, to understand how neural networks process data, make decisions, and arrive at their solutions. This should, eventually, make AI's decision-making more transparent and intelligible to humans. It may also help ensure that the outputs of these models are fair, equitable, and free from hidden biases and discrimination.

In order to determine whether Evo 2 was a 'stochastic parrot,' using probability to generate plausible DNA sequences as opposed to utilising true understanding, the Evo 2 team trained a specialised model called a sparse autoencoder. This was able to look into Evo 2's inner workings. Surprisingly, it showed that rather than just memorising examples from the training data, Evo 2 appeared to have developed representations of some key biological concepts. These enabled it, for example, to discern structural features of genes within a wooly mammoth genome, despite the fact that it had not been exposed to its sequences during training. This suggested that Evo 2's representationts were able to 'capture a broad spectrum of biologically relevant signals'.

While providing compelling evidence that their genome language model could design small genomes at scale – including human mitochondrial genomes and minimal bacterial genomes – the Evo 2 team had stopped short of demonstrating that the designed, artificial genomes they had generated were actually viable. To do so, they would need to physically synthesise and test them. This situation changed, however, with the online publication of a landmark paper in September 2025, in which Hie and his collaborators demonstrated that they could design and build functional bacteriophage genome sequences that had never previously existed in nature. These became the first biological entities in history to be designed computationally using artificial intelligence at the genome-scale, synthesised, and then shown to be viable.

While AI had already been used to script and construct new genes and collections of genes, designing whole genomes was a far more complex task. This is because genomes are characterised by 'emergent complexity,' resulting from regulatory interactions between regions that are often located at considerable distances from one another. Genomes are also exquisitely sensitive to alterations to their sequence composition, with even minor changes having the potential to produce deleterious effects.

To script the sequences of new viruses they needed a design template to 'seed' the AI. They chose the genome sequence of the 5,386-nucleotide-long phi-X174 virus. First discovered in a Paris sewer in the 1930s, its modestly-sized genome contained just 11 genes, many of which were overlapping. This virus was of historical significance, as it was the first biological entity to have its genome sequenced (by Fred Sanger), as well as the first to be chemically synthesised (by Craig Venter and Hamilton Smith). The Evo 2 team would introduce another first into phiX174's portfolio – it would become the first ever genome to be designed using AI and shown to be viable through construction and testing.

Although the Evo genome language models had already been trained on more than two million bacteriophage genomes, in order to fine-tune them to manage this new task, the team used a method called 'supervised learning' to further train the models on a database of natural Φ-X174 genomes. They then went on to create thousands of AI-generated designs, finally narrowing these down to 302 candidate sequences. While most of the artificial sequences shared around 40 per cent DNA sequence identity with the natural virus, some were completely different. After successfully chemically synthesizing 285 of the artificial AI-generated viruses, they identified 16 that could be 'booted up' to form viable viruses. They then showed that these were able to kill antibiotic-resistant *E. coli* bacteria. One of these, Evo-Φ2147, had just 93 per cent sequence identity to the natural parental virus, meeting some taxonomic thresholds for a new species.

A virus is not itself a living organism. But in developing a computational framework for the AI-augmented generative design of complete viral genomes – some of which were distinct from anything previously seen in nature – Hie and his colleagues had successfully established the foundations of the new field of 'generative genomics'. This emerging science promises to leverage AI to redesign the genomes of existing species, and to design the genomes of entirely new species from scratch. Rather than building genetic circuits piecemeal in the slow, incremental, and plodding manner of evolution by natural selection and human genetic engineering, AI had – with the masterful brilliance reminiscent of a computational Leonardo da Vinci – grasped the design features of these diminutive viral genomes in their entirety.

Rather than treating genomes as cogs and gadgets, this 'supervised' incarnation of Evo had adopted a holistic view of genome function, generating new and highly innovative renditions on a viral genome theme, in a single masterstroke. While a remarkable and foundational achievement, to go beyond the design of simple biological entities like viruses will require significant future innovation, far more compute capability, huge biological datasets, and the ability to construct DNA at scale. The next level of biological complexity after a virus, namely a bacterium, has a genome size that is around three orders of magnitude greater than that of a bacteriophage. But, as Hie correctly stated, this achievement 'suggests a future' in which 'genome design could become a core technology' with the potential to enable 'the generation of complete living organisms'. They had shown, for the first time, that it was possible for genome authorship to go 'off script' to create viable genomes that humans could never have achieved using rational design.

In his 1620 *Novum Organum* (New Instrument), Francis Bacon proposed that by assembling a comprehensive collection of nature's anomalies and errors in one location, it would 'be an easy matter to bring nature by art to the point reached by chance'. This reverse-engineering approach – inferring function from failure – to understanding the nature of living species was revolutionary. It established a new paradigm for the scientific method, one that would define all future biological enquiry. Bacon had established the philosophical basis for a method to unravel the mysteries of the human machine. By examining how biological systems break down and fail, it would be possible to deduce their usual mode of operation. It would then be possible to mimic them.

Fortunately, nature provides an abundance of such clues, with every member of a species displaying several naturally occurring construction errors, known as mutations. These mutations may also be introduced into organisms artificially to observe their effects on function. Many mutations cause disease or malfunction, thereby offering powerful insights into the healthy operation of biological systems.

Bacon believed that this empirical, observational approach to studying biological machines, grounded in natural phenomena, would replace Aristotle's deductive system, which was based on logic rather than experimentation. He understood that the essence of life lies in the invisible rules – the genomic generative grammar – that guide the nature of living organisms.

Today, with our emerging ability to design and synthesise genomes, it is possible to envision a new paradigm for biology based on the construction of living things rather than on their degradation – a pivot to forward engineering, defined by design principles, rather than by invoking conventional reverse engineering that throws a spanner into the works of machines to document how they fail. Within this build-to-learn paradigm, genomes can be iteratively designed, tested, and redesigned. This elevates biology from the level of components to the level of systems, offering a deeper understanding of life.

Four years after the sequencing of the human genome had been completed, and nearly 400 years after Bacon's visionary insights had defined the basis for the modern scientific method, the geneticist Ronald (Ron) Davis issued a bold challenge to attendees of the 2004 Yeast Genetics and Molecular Biology Meeting in Seattle. He urged the community to consider constructing a synthetic yeast genome. If you truly wanted to understand an organism, he argued, the best way to do so was to build one.

At the time, the only fully synthesised genome was that of the poliovirus. As we've seen, this tiny genome, synthesised in August 2002 by Eckhard Wimmer using mail-order parts, was just 7,741 nucleotides long. In comparison, the genome of budding yeast *Saccharomyces cerevisiae* was a tremendous 12,157,105 nucleotides in length, around 1,500 times larger, and far more complex due to its organisation into sixteen separate chromosomes.

Sitting in the audience was a yeast geneticist called Jef Boeke, then at the Johns Hopkins University School of Medicine in Baltimore. Boeke recalled thinking, 'Why on earth would you want to do that?' He saw little point in copying nature's designs, and designing and synthesising a genome of that magnitude appeared to be beyond the limits of what was possible. The idea faded into obscurity, and no attempts were made to follow up on Davis's audacious suggestion.

Two years later, in 2006, Boeke was having coffee at Johns Hopkins with his colleague Srinivasan Chandrasegaran, who was exploring ways of using gene editing to introduce multiple, parallel modifications into the yeast genome. It was during this conversation that Boeke had an epiphany. Maybe Davis had been right all along. He realised that beyond a certain point, genome editing became so complex and unwieldy that it was more efficient to perform a complete rewrite. He joked that synthesising a yeast genome might not be such a bad idea after all – especially if the design could be customised and remade in novel ways. To his surprise Chandrasegaran agreed. And so they decided that they would set about reimagining the genome of this unsuspecting organism in a new and artificial form.

The next step was to find a computer scientist who could help them design a synthetic yeast genome on a computer screen. They were fortunate to find Joel Bader, a bioinformatics expert, also at Johns Hopkins, who set about designing a yeast genome in silico.

Synthesising a yeast genome offered many potential benefits. Yeast, like humans, is a eukaryote, but as a unicellular organism, it is far easier to study. Its cells share many features with those of humans, including nuclei and specialised organelles such as mitochondria and lysozymes. Moreover, it is one of the most extensively studied and best understood eukaryotic species, having been scrutinised for decades by a tightly knit and highly collaborative research community.

Yeast also holds a unique cultural and historical significance. It has played a seminal role in human civilisation for at least 9,000 years, being widely used in the winemaking, brewing and baking industries, and serving as an ingredient in a range of foodstuffs. It was also, in 1996, the first eukaryotic species to have its genome sequenced. Many yeast genes have corresponding versions in the human genome, and they share numerous cellular processes. As a result, insights gained from yeast often have direct relevance to human biology. Perhaps most importantly, the yeast genome is much larger than that of the poliovirus genome, yet still 260 times smaller than the human genome, making its synthesis a far more tractable problem.

While rewriting the code of the yeast genome would not constitute true authorship – *Saccharomyces cerevisiae* already exists as a natural species – it would nonetheless demonstrate that genomes relatively similar to our own can be redesigned and built from scratch. It would pioneer genome redesign and synthesis as a general method for reconfiguring the features of living systems.

If successful, the project would signal the beginning of a new era of biological design, a moment reminiscent of when the origami master Jun Maekawa first transformed a sheet of paper into a multi-horned rhinoceros beetle. Like the origami bug wars, it might trigger an analogous wave of genome synthesis rivalry that fostered innovation in genome design. Most significantly, it would represent a giant leap forward for biology, marking the point at which humankind moved beyond the revision of nature and began, instead, to reimagine it.

Success in creating a non-natural 2.0 version of the naturally evolved 1.0 version of the yeast genome – designed to meet human specifications rather than those of evolution – would constitute a bold demonstration of the feasibility of synthesising complex genomes. It would also test the extent to which genomes can be reconfigured without compromising the behaviours wired into their natural genomic 'software', which was compiled over billions of years.

The Synthetic Yeast Genome Project, or Sc2.0, was conceived as an international consortium, involving teams from the United States, United Kingdom, China, Singapore, and Australia. Each team was tasked with synthesising one or more of yeast's sixteen chromosomes. Collectively, their work showed that the yeast genome is remarkably resilient and tolerant of extensive redesign. As Boeke later remarked, 'You can torture the genome in a multitude of different ways, and the yeast just laughs.' Of the approximately 250,000 changes made to its genome, only around 0.1 per cent resulted in fitness deficits, which the team referred to as 'bugs'. Reflecting on this robustness, Peter Enyeart and Andrew Ellington at the University of Texas at

Austin concluded that 'God's fingerprints' on genomes were 'fainter than creationists would have you believe'.

Biology has, to date, relied on a Baconian reverse-engineering approach – breaking down the components of living systems to determine how they function; The ability to synthesise genomes from scratch provides the basis of a new forward-engineering paradigm for decoding life, one founded on its construction. This new kind of 'building-to-understand' methodology for the natural sciences invokes a note found on the blackboard of the Caltech physicist Richard Feynman at the time of his death: *What I cannot create, I do not understand*. This new approach is particularly well suited to understanding biology at the systems level. Manufacturing a species from first principles is the most compelling way to understand how coded genetic instructions cooperate to produce complex biological systems.

Boeke never concerned himself with the issue of whether it was feasible to synthesise an entire yeast genome. What interested him was how its sixteen chromosomes could be made 'different from a normal chromosome', and how novel features could be introduced into the synthetic genome to make the entire enterprise 'worthwhile'. A simple replica of the natural sequence would not be scientifically or philosophically interesting. Though lacking the full generative grammar necessary to invent a new species, the synthetic genome should, at a minimum, transcend the nature of the existing one. From the outset, Boeke emphasised the importance of designing a genome that produced 'something that was very heavily modified from nature's design'.

Building upon nature's design and engineering a genome from the ground up required ingenuity, imagination, and patience. Ultimately, Boeke and the Sc2.0 team formulated a set of rational design rules that could be systematically applied across all of yeast's sixteen natural chromosomes to construct corresponding synthetic versions. One team, led by Patrick Cai in the UK, went even further, synthesising an artificial seventeenth 'neochromosome', which was unlike any sequence existing in nature. This was one of the first true acts of creative genome authorship.

The design rules defined by the Sc2.0 team were governed by three key principles. First, the modifications should 'do no harm'.

Alterations to the yeast genome should not disrupt the yeast's natural appearance or behaviour. The overall 'fitness' of the yeast would be assessed by measuring colony size and morphology under a variety of conditions. Second, the genome should be streamlined to remove and disentangle as much complexity from the natural design as possible. It would also be engineered for greater stability. It was not inevitable, however, that this 'death by a thousand cuts', as Boeke called it, would result in a viable organism.

But it was the third principle that was the most ambitious. Boeke didn't just want to construct a stripped-down version of the yeast genome. He wanted to control it in real time and restructure its code at will through the use of inducible accelerated artificial evolution. This, he believed, would help reveal the yeast genome's inner workings and create novel versions by merging artificial design with natural evolution.

To enable this, the team introduced an inducible evolution system called SCRaMbLE (Synthetic Chromosome Rearrangement and Modification by LoxP-mediated Evolution) into the synthetic genome. When activated it results in random deletions, relocations, duplications, and inversions of the sequences located between the LoxP sites. As a result, the genome is reshuffled, producing combinatorial diversity. Boeke called this 'evolution on hyperspeed'. Variants could then be isolated that were optimised for specific tasks, such as identifying disease targets, producing therapeutic proteins, and manufacturing biofuels.

The genome was recoded to allow for the incorporation of non-natural amino acids. New sequences were also introduced into the synthetic chromosomes, including watermark sequences inserted across the sixteen synthetic chromosomes to distinguish them from their natural counterparts.

The first synthetic yeast chromosome, chromosome III, was completed in 2014. Debugging the sequence – the painstaking process of error correction – required almost as much effort as the synthesis itself. In 2023, nearly two decades after the project's inception, the team announced a major milestone: they had completed the synthesis of all sixteen synthetic chromosomes, as well as the seventeenth artificial neochromosome. They had also assembled 6.5 of the synthetic chromosomes into a new synthetic yeast strain to build a 'half-synthetic' yeast, whose genome sequence was close to 50 per cent synthetic.

The artificial yeast cells functioned remarkably well, signalling what Boeke described as the 'end of the beginning'. The final step would be to bring all sixteen synthetic chromosomes and the neochromosome together within a single cell, to create the world's first fully synthetic eukaryotic organism.

The laborious, handcrafted manner in which the synthetic yeast genome was built – using the standard construction method of assembling small fragments of DNA into progressively larger pieces – is reminiscent of the early days of automobile manufacturing, when cars were built by hand, rather than on an industrial scale. Our ability to write DNA remains far more rudimentary than our ability to read it. But it is only a matter of time before it catches up. When it does, chemical genome synthesis may eventually replace editing as the preferred method for modifying genomes. The efficient and targeted delivery of large synthetic DNA sequences into genomes, however, remains a substantive ongoing issue.

Unlike genome editing, which modifies existing sequences, synthetic genomics enables the design and construction of novel genomes that have never previously existed in nature. The Sc2.0 project has shown that extensively engineered synthetic genomes can generate viable synthetic multicellular organisms. As a result, biology now stands at the threshold of transitioning from a largely descriptive science to a generative one. In the future, we won't just catalogue species, we will create them.

9
The End of Illness

> For if it be true that in every phenomenon of Nature there is something of the marvelous, surely that factor is nowhere more evident than in the workings of the metabolic processes of living things.
>
> – ARCHIBALD E. GARROD, 1909

In his treatise *De ratione motus musculorum* (The Reason of the Movement of the Muscles) published in 1664, the English physician William Croone – one of the founding fellows of the Royal Society – outlined a rational explanation for how muscles function. Croone framed his ideas in terms of the chemical and mechanical principles that were, at the time, becoming increasingly influential in the natural sciences. After his death in 1684, an annual lecture series, the Croonian Lectures, was established in his honour. The individual tasked with presenting the 1908 series was another English physician, Archibald E. Garrod. The landmark lectures that he delivered at the Royal Society that June would establish him as both the father of chemical genetics and the first individual to demonstrate that Mendel's laws of inheritance generalised beyond peas to humans.

Garrod proposed that the biochemical differences between individuals – their 'chemical merits' – influenced the characteristics of human nature that distinguish one person from another. Conversely, 'inborn errors of metabolism', or chemical abnormalities, could predispose to disease. The invisible structures underlying these chemical abnormalities, or 'freaks of metabolism', were, he argued, 'analogous to structural malformations' seen, for example, in anomalous body forms.

While such chemical 'shortcomings' predisposed individuals to disease, biochemical individuality and variability could also confer 'immunities from the various mishaps which are spoken of as diseases'.

Despite lacking any insight into the chemical nature of the genetic material, Garrod had established the foundations of medical genetics. He inferred the existence of biochemical pathways whose components were prone to abnormalities that resulted in bodily malfunction. The chemical components of metabolic processes had a machine-like nature, being in some ways analogous to mechanical parts. The enzymes that controlled the chemical pathways of living systems formed the cogs and wheels of biochemistry. Diseases, Garrod suggested, were 'merely extreme examples of variations of chemical behaviour, which are probably everywhere present'. There was, furthermore, 'no rigid uniformity of chemical processes in the individual members of a species'. They differed 'slightly in their chemistry as they do in their structure'.

Just over four decades earlier, an obscure Austrian monk called Gregor Mendel had uncovered the nature of the rudimentary hereditary principles that would later underpin Garrod's concept of biochemical individuality. Though this work went largely unrecognised during his own lifetime, Mendel's insights – obtained through his breeding experiments in a monastery garden – would go on to establish the foundations for a genetic understanding of disease.

Having failed his high school teacher training examinations, and keen to avoid the inconveniences associated with earning a living, in October 1843 the cash-strapped Austrian farmer's son Johann Mendel – who subsequently adopted the monastic name Gregor Mendel – joined the Augustinian monastery of St Thomas in Brünn (now Brno in the Czech Republic). At a certain point he was pressured to become a parish priest, but the abbot, Cyril František Napp, eventually relieved him of his ecclesiastical duties owing to his unsuitability for that vocation. This was due to his 'unconquerable timidity' when visiting the sick, and 'anyone ill and in pain'. Meanwhile, he continued in his capacity as a friar, which gave him greater freedom than the monks to pursue his academic studies and other intellectual pursuits.

Unfortunately, following a papal investigation into the monastery led by the Bishop-Prince Philipp Gotthard von Schaffgotsch in Brünn, followed by a site visit in June 1854, Mendel was accused of studying 'profane sciences at a worldly establishment in Vienna at the expense of the monastery'. Schaffgotsch had discovered that Mendel had been attending lectures at the University of Vienna to prepare for a teaching career. As a result of this and other evidence of secularism and neglect of religious piety among his colleagues, Schaffgotsch filed a report to the Vatican recommending that the monastery should be dissolved. Fortunately, this did not happen.

Around that time, at the age of thirty-two and while working in the monastery's five-acre garden, Mendel set about performing the first of a protracted series of breeding experiments aimed at gaining insights into the heredity of the garden pea *Pisum sativum*. His cross-breeding work, using twenty-two true-breeding varieties of *Pisum sativum*, began in earnest in 1856 and continued for eight years, until 1863. He eventually demonstrated that heritable factors, now known as genes, controlled the characteristics of pea plants, such as flower colour, seed shape and height. He did this by identifying plants with mutant versions of particular traits and then crossing them. For example, upon discovering dwarf mutants of pea plants 'suffering from a heritable lack of tallness', he was able to infer that tallness was controlled by a hereditary factor.

Mendel's genius lay in his realisation that it was possible to record the behaviour of hereditary information across generations without knowing anything about its material basis. This insight led him to the revolutionary proposal that the information describing the different features of plants – be they flowers, seed shape, or colour – was specified by independent 'elements' operating according to a distinct hereditary mechanism.

Mendel detailed his findings in a paper *Versuche über Pflanzen-Hybriden* (Experiments on Plant Hybrids), which he presented to the Natural Science Society of Brünn on 8 February 1865 and again on 8 March. He subsequently published his paper in 1866. The results of his meticulously conducted breeding experiments led him to infer that the characteristics he had studied were controlled by factors (*Merkmal*) that occurred in pairs and segregated independently of one another. While some were dominant, others were recessive. Purple flower colour was, for example, observed if at least one of the two factors in a plant was dominant, while white colour was only observed if both were recessive.

Unfortunately, Mendel's presentations and paper on the subject failed to generate any interest. After confidently informing his friend Gustav Niessl von Mayendorf that his time would eventually come, he swiftly abandoned his experiments, and in 1868 was appointed to the position of abbot at the monastery. Around that time, in a letter written to his friend, the Swiss botanist Carl Wilhelm von Nägeli, Mendel lamented, 'I am really unhappy about having to neglect my plants and my bees so completely.' One of his contemporaries at the monastery noted how Mendel's theory had been suspected of being 'contrary to the revealed truths of the Christian religion'.

Although unappreciated during his lifetime, Mendel's laws of heredity were independently and simultaneously rediscovered in 1900 – more than thirty years after he had first formulated them – by three unconnected individuals: the Dutch botanist Hugo de Vries, the German botanist Carl Erich Correns and the Austrian agronomist Erich von Tschermak-Seysenegg. Hugo de Vries referred to the presumed hereditary particles as 'pangenes'. This was later shortened to 'gen' or 'gene' by the Danish botanist Wilhelm Ludvig Johannsen in his 1909 book *Elements of Exact Heredity*.

The origins of Garrod's insights may be traced to 1897 – eleven years prior to presenting his 1908 Croonian Lectures – when a mother arrived with her infant at Great Ormond Street Hospital in the Bloomsbury area of London, where he was working as an assistant physician. She was disconsolate because her child's nappies were inexplicably stained black. When the same mother returned with another infant three years later, Garrod's hunch that the sibling would also produce black urine proved correct.

Recognising the potential importance of this phenomenon, and eschewing his colleagues' disdain for his preoccupation with the biochemistry of urine, Garrod began searching for further case reports documenting this condition. He soon identified a sizeable collection in the literature describing instances where a 'peculiar coloration of the urine' had developed on standing. The case of a monk with black urine had, for example, been described by the German physician Johannes Schenck von Grafenberg in his 1584 treatise *Observationes medicae de capite humano* (Medical Observations on the Human Head).

When in 1902, the biologist William Bateson, working in the Department of Zoology at the University of Cambridge, published *Mendel's Principles of Heredity* – which explained how mating between first cousins allowed autosomal recessive Mendelian diseases to manifest – Garrod connected these principles with the cases of black urine he had observed in the offspring of consanguineous partnerships. He realised that the transmission of the disease must be due to a 'peculiarity of the parents', which could 'remain latent for many generations'. The disease only occurred following a union between two individuals carrying a recessive gene. Mating within consangineous marriages between first cousins provided the condition necessary for the rare recessive characteristic to reveal itself, as it allowed two rare forms of a gene to be brought together in a single individual. The inheritance of human metabolic diseases recapitulated the recessive inheritance pattern Mendel had observed in peas. It seemed, at least in this sense, that humans were 'peas writ large'.

The rare autosomal recessive disease causing black urine, known today as alkaptonuria, became the prototype for other inborn errors of metabolism. We now know that alkaptonuria is a multisystem disease resulting from the deficiency of an enzyme called homogentisate 1,2-dioxygenase (HGD). HGD is responsible for the breakdown of homogentisic acid, a metabolic intermediate produced by the breakdown of the amino acid tyrosine. In healthy individuals, HGD cleaves the aromatic benzene ring of tyrosine to produce a metabolite called 4-maleylacetoacetate. The accumulation of homogentisic acid with unbroken benzene rings in urine gives it its characteristic black colour, as it is oxidised by air on standing, polymerising to form a black pigment.

Affected individuals are generally well for the first three decades of life, but pigment deposits gradually accumulate in the joints of their bones and in particular in the hip, knee, and lumbar and thoracic vertebrae of the spine, where they cause severe arthritis, which in some cases requires joint replacement. The pigment also accumulates in the earlobes, skin, muscles, tendons, and in the whites of the eye – known as the sclerae – giving them a distinctive blue hue. Deposits of homogentisic acid may also lead to the production of kidney stones and prostate stones, damage the leaflets of the aortic valve in the heart, and occlude the coronary arteries supplying blood to the heart. In 1979, deposits of homogentisic acid were identified in the cartilage of a 1,500-year-old Egyptian mummy called Harwa, indicating that alkaptonuria was present in antiquity.

The machine-like conception of human biochemistry, in which individual, genetically encoded components of biochemical pathways may break and be fixed through their repair, removal or replacement, formed the philosophical foundation of molecular medicine. It also played a critical role in establishing the modern medical paradigm focused on identifying molecular targets for intervention. Yet, as science has advanced, this Mendelian framework has revealed its limitations.

Not all apparently Mendelian diseases are, for example, monogenic – influenced by a single gene. Mendelian inheritance in some seemingly monogenic diseases may be more complex than the traits of peas, with the effects of genes being modified by several factors. Furthermore, not all the components of human biochemical networks are proteins. Many are made from RNA. If, as has historically been the case, we focus our attention on the 1.5 per cent of the human genome that codes for proteins, we risk overlooking the full complexity of human biochemistry and the substantial collection of non-protein components. It is estimated that at least 80 per cent of non-coding DNA is transcribed into RNA, with the mode of inheritance for non-protein-encoding genes appearing to recapitulate that of protein-encoding genes. Moreover, many characteristics are influenced by multiple genes, and individual genes may perform more than one function.

Traditionally, the search for the gene responsible for a simple Mendelian disease involved tracing affected individuals through well-documented family lineages. The profound impact that alterations to individual letters in DNA sequences may have on human biology is extraordinary. As of June 2024, the Online Mendelian Inheritance in Man (OMIM) database had catalogued 8,511 distinct human Mendelian diseases caused by mutations in single genes. Of these, 4,405 follow an autosomal recessive inheritance pattern, 3,733 are autosomal dominant, and 373 are X-linked.

Occasionally, nature accelerates the laborious process of identifying the causative genes of Mendelian diseases, offering researchers a rare shortcut. One such instance may be found in Ricaurte, a small town nestled in the foothills of the Andes in Colombia. Long known as the home of *los bobos*, 'the foolish ones', Ricaurte has the largest known cluster of individuals with fragile X syndrome in the world. Fragile X is the most common global cause of intellectual disability, affecting around 1 in 2,000 males, and 1 in 4,000 females. It is characterised by cognitive issues such as memory impairment, as well as behavioural and emotional disorders such as Attention-Deficit/Hyperactivity Disorder (ADHD), depression, and autism. Other features include infertility, convulsions, tremors, speech disorders, and distinctive physical features such as large ears and a prominent jaw and forehead. The gene responsible for fragile X is located on the X chromosome and inherited in an X-linked dominant manner. This means that a single copy of the mutant gene is sufficient to cause all the observed features of the disease.

Remarkably, around five per cent of Ricaurte's residents have either the full-blown disease or a milder version. Protestant missionaries once attributed the disorder to divine punishment, blaming the townspeople's veneration of *El Divino*, an image of Jesus in Ricaurte's white A-frame church. Others suspected sorcery, while some, more rationally, implicated contamination of the groundwater from nearby magnesium mines. However, in 1991, Stephen Warren and Ben Oostra at Erasmus University Rotterdam identified the true cause. The disease resulted from a faulty gene known as Fragile X Mental Retardation 1 (*FMR1*).

The protein encoded by this gene, Fragile X Mental Retardation Protein (FMRP), is an important 'hub' that interacts with around 100 other proteins. This helps explain why the disorder impacts so many different body systems. While most healthy individuals have between six and fifty-four CGG trinucleotide repeats – a short pattern of three nucleotide letters repeated again and again – those with Fragile X syndrome have up to 200 copies. These additional repeats result in the silencing of the gene's activity by a process called methylation. The excess DNA disrupts the structure of the X chromosome, which can be seen under a microscope as a small constriction near the end of its long arm, creating a tiny island that appears to break away from the rest of the chromosome, making it appear 'fragile'.

Individuals with mild forms of the disease, called premutation carriers, have lower numbers of repeats. Their X chromosomes lack the characteristic fragile appearance.

The medical geneticist Wilmar Saldarriaga-Gil, at the University of Valle in Cali was able to trace the founder mutation back to an individual called Manuel Triviño, one of Ricaurte's original settlers in the 1880s. He also found that individuals with the same *FMR1* mutations had very different levels of disease severity, suggesting the presence of as yet poorly defined genetic or environmental modifying factors that influence the extent of the mutant gene's pathological effects. Paradoxically, townspeople with fewer numbers of repeats showed more severe signs and symptoms of the disease than might otherwise be expected.

Fragile X's pattern of inheritance differs from that of other Mendelian diseases, as the defective gene resides on the X chromosome – one of the two sex chromosomes – rather than on one of the autosomes. Affected males can only transmit the disease on to their daughters, while females may pass the disease to either sex, since women have two X chromosomes. One of the two X chromosomes randomly shuts down in each cell in the body in a process known as X chromosome inactivation (XCI). If a woman carries just one abnormal copy of the *FMR1* gene, there is a 50 per cent chance that the mutant rather than the healthy copy of the gene in any given cell will be shut down. The cells in the body of affected individuals are thus 'mosaic', with some inactivating the mutant X chromosome and others the healthy one. This mosaicism, combined with other genetic, environmental, and epigenetic factors, adds significant complexity to the disease's expression.

In a different part of the world, the history of phenylketonuria (PKU) offers another illuminating case in Mendelian genetics. In 1934, the Norwegian physician, biochemist, and farmer's son Ivar Asbjørn Følling, working at the Oslo University Hospital, met four-year-old Liv and six-year-old Dag. They were the children of Harry and Borgny Egeland, who had been referred to him for a second opinion. The children had been healthy at birth but gradually developed severe cognitive impairment, as well as other issues, including difficulty sitting upright and feeding themselves, and pronounced involuntary eye movements. The parents had also noticed an unusual musty odour in the children's urine. Determined to identify the cause, Følling conducted a series of tests. As part of his investigations, he added iron chloride to a urine sample. Instead of turning the expected red-brown colour, it turned dark green. He had never seen anything like it before. He decided to investigate further, eventually showing that the unexpected colour change was due to the presence of a metabolite called phenylpyruvic acid.

Liv and Dag became the first recorded cases of phenylketonuria (PKU), a rare, autosomal recessive inherited Mendelian disease affecting roughly one in 10,000 children. This inborn error of metabolism is caused by mutations in the *PAH* gene, which encodes the enzyme phenylalanine hydroxylase. A deficiency in this enzyme prevents affected individuals from breaking down the amino acid phenylalanine, found in protein-rich foods. This leads to the accumulation of phenylpyruvic acid, which Følling's colleagues insensitively referred to as 'idiot acid'. Elevated levels of this are toxic to the brain and responsible for the urine's distinctive odour. By 1953, researchers had shown that low phenylalanine diets could mitigate the disease's effects. In 1984, the *PAH* gene was identified as the root cause.

More recently, whole-genome sequencing has revealed that humans differ by approximately 0.5 per cent at the genetic level. Around 95 per cent of these differences lie outside the regions containing the 19,969 identified protein-encoding genes in the human genome, and instead map to the non-coding regions. While PKU has long been regarded as a quintessential example of a single-gene autosomal recessive Mendelian disorder, in 2021 Liuqing Yang and Yajuan Li at the University of Texas MD Anderson Cancer Center in Houston conducted a detailed analysis of *PAH* mutations and disease severity. They found that the *PAH* mutation alone could not account for the variability in disease severity. Individuals with identical mutations differed significantly in their cognitive outcomes. This suggested that the *PAH* mutation alone could not predict the severity of the disease. Other genetic or regulatory factors were clearly at play.

The discovery of a long non-coding RNA (lncRNA) molecule encoded by a gene called *Pair*, which, when deleted, mimicked the disease presentation caused by a *PAH* mutation, suggested that Pair lncRNAs are integral components of the metabolic networks contributing to the disease, and play a key role in phenylalanine metabolism. Malfunctioning or absent lncRNAs may impair these metabolic functions in a manner similar to protein abnormalities. Synthetic lncRNA mimics, therefore, hold promise as therapeutic agents to restore these metabolic functions in situations where they are faulty or missing.

On 15 September 1657, the English diarist John Evelyn made his way to London to view a performance featuring a Turkish rope dancer renowned for performing stunts blindfolded. While attending this event, he encountered the international celebrity Barbara van Beck, known as the 'bearded woman'. Evelyn described how 'the hairy maid' had 'a most prolix beard, & mustachios, with long locks of hair growing on the very middle of her nose, exactly like an island dog'. He managed to secure an audience with her after the show, observing that she 'plaied well on the harpsichord'. Sometimes called 'werewolf syndrome', hypertrichosis can result from at least ten near-identical syndromes. Each has its own distinct pattern of inheritance. Congenital hypertrichosis lanuginosa, for example, follows an autosomal dominant pattern of inheritance, while cataract-hypertrichosis-intellectual disability syndrome is autosomal recessive. Despite their different genetic origins, these conditions converge on a similar appearance.

The 8,511 known Mendelian diseases provide striking examples of human genetics operating in a machine-like manner. The repair of Mendelian diseases and the adjustment of monogenic traits could, in principle, be achieved with relative precision. In this analogy, the chemical components of human biochemistry behave like the parts of a motor car, with genes representing the different mechanical elements. A malfunction, such as a broken clutch or worn brake pad, can be addressed by taking the vehicle into a garage and replacing or repairing the faulty component without disrupting the rest of the system. Thanks to the curiosity and astute observations of a young Spanish researcher studying an obscure bacterium, we now possess a technology – gene editing – that can repair the broken components of genetic machines.

From a distance, the arid landscape of Las Salinas, located in the pink salt flats of Alicante on Spain's south-eastern Mediterranean coast, resembles a Martian landscape. This environment is so extreme that almost nothing can survive. The lagoons cutting into the marshland are artificial, created to produce salt that crystallises along their edges. This salt-encrusted periphery is peppered with pinkish-red seams of sediment teeming with microbes. The colouration arises from chemicals called carotenoids that protect microbes from the sun and change colour according to the degree of salinity. It was hard to imagine that this obscure and inhospitable landscape held the key to the development of one of the most consequential tools in genetic engineering.

In 1989, Francisco Mojica, a 28-year-old graduate student who had recently completed his military service, began studying *Haloferax mediterranei* – a rod-shaped, salt-loving microbe isolated from the salt flats near Alicante a few years earlier. This obscure organism became the focus of his research, which would ultimately have a profound impact on medicine and biology. Yet at the time, its extraordinary significance was far from clear.

While examining the microbe's genome, Mojica noticed something unusual: clusters of nearly identical DNA sequences, each about thirty letters long, repeated hundreds of times. These sequences had a striking feature – they read the same way backwards and forwards. Sandwiched between these palindromic repeats lay highly variable sequences, each around thirty-five letters long, which he called spacers. Mojica named the entire structure CRISPR, short for 'clustered regularly interspaced short palindromic repeats'. The purpose of these enigmatic sequences, however, remained a mystery.

Over the following years, Mojica and other researchers discovered CRISPR's function – it was part of a primitive immune system that bacteria use to protect themselves from their bacteriophage adversaries. CRISPR functioned as an immunological memory bank, storing fragments of viral DNA within its spacers – like a molecular 'most-wanted list'. These fragments, or 'fingerprints', help bacteria to recognise and destroy the same virus if attacked again.

CRISPR works together with a protein called Cas9, which functions like 'molecular scissors' to cut DNA at precise locations specified by the CRISPR sequences. The Cas9 protein consults its spacer sequences like an oracle. When a virus invades, CRISPR uses RNA copies of its spacers, known as CRISPR RNAs (crRNA), to guide Cas9 to specific viral DNA sequences, which it then cuts and destroys. In so doing, it neutralises the viral threat. Remarkably, CRISPR arrays are able to continuously update their spacers, adding new viral fragments following each bacteriophage encounter, thereby learning from each experience.

The French biochemist Emmanuelle Charpentier later discovered another crucial component of the CRISPR defence system, called trans-activating CRISPR RNA (tracrRNA). This weaponises Cas9 by acting as a handle that connects it to the crRNA. Together with her collaborators Martin Jínek and Jennifer Doudna at the University of California, Berkeley, Charpentier simplified the system by fusing the crRNA and tracrRNA together to form a single 'guide' RNA (sgRNA). The team then attached this to Cas9 and showed that this streamlined system could target specific genomic sequences with exquisite precision.

Together, they had shown that an esoteric microbial immune system could be repurposed into a programmable tool for editing genomes. Shortly afterwards, Feng Zhang and George Church, working independently in Boston at MIT and Harvard, modified the technology for use in human cells. This required the addition of a DNA sequence that helped transport CRISPR-Cas9 into the nucleus. It soon became apparent that CRISPR-Cas9 could be used to fix the broken components of the human genomes responsible for the 8,511 known monogenic diseases.

The transformative potential of CRISPR-Cas9 was illustrated on 2 July 2019, when Victoria Gray, a mother of four from Mississippi who had been diagnosed with an autosomal recessive Mendelian disorder called sickle cell disease (SCD) at three months old, became the first person to be treated with a CRISPR-based gene-editing therapy.

Sickle cell disease, first described in 1910 by the Chicago physician James B. Herrick, affects around 20 million individuals worldwide. It causes severe pain and dramatically shortens life expectancy. Herrick documented the case of a 20-year-old African American whose oxygen-carrying red blood cells had a 'peculiar' appearance. The cells were 'very irregular', with 'large numbers of thin, elongated, sickle-shaped, and crescent-shaped forms' visible. The change in red blood cell morphology results from a change to a single nucleotide – an A-to-T substitution in the gene encoding one of the two protein chains of haemoglobin – which causes the molecule to become misshapen. In 1949, the biochemist Linus Pauling showed that mutant sickle cell haemoglobin, HbS, behaved differently to healthy haemoglobin HbA. Sickle cell became the first 'molecular disease'. By that time it had been shown that individuals with sickle cell anaemia inherited an abnormal gene from both parents. Those with just one abnormal gene, known as the 'sickle cell trait', were carriers of the disease, developing only mild symptoms.

During a sickle cell crisis, the abnormally shaped red blood cells cause excruciating pain as they obstruct tiny blood vessels. These blockages may also result in strokes and life-threatening organ failure. The misshapen red blood cells are also prone to rupture, causing a haemolytic anaemia. Victoria Gray described a sickle cell crisis as 'like getting struck by lightning': 'Your bones feel as if they have been crushed', as if you have been 'hit by a car'. As a result of those crises, 50 to 90 per cent of affected infants in sub-Saharan Africa and India with sickle cell disease die before their fifth birthday.

In December 2023, the drug exa-cel (Casgevy) received approval from the FDA. This confirmed the feasibility of using CRISPR gene editing to treat a wide range of monogenic human diseases and raised the possibility of obtaining cures for the remaining component-like Mendelian diseases. The drug is a 'one and done' treatment administered via intravenous infusion. Stem cells are removed from the patient's bone marrow, edited using a virus containing a CRISPR construct, and then transplanted back into the patient.

Sickle cell disease is especially prevalent in sub-Saharan Africa and in Mediterranean countries such as Turkey, Italy and Greece. The reason for this was revealed by the Oxford physician Anthony (Tony) C. Allison. He observed that there was considerable geographical variation in the prevalence of the sickle cell trait in Kenya. In tribes living near the Mombasa coast or Lake Victoria, the prevalence of the trait was as high as 20 per cent In the arid, mosquito-free regions of central Kenya and the highlands, however, just one per cent of the population had sickle cell trait. Allison had an epiphany. Malaria was endemic in the wetlands of the Mombasa coast and Lake Victoria, where the *Anopheles* species of mosquito was rife. Might the sickle cell trait in some way confer some resistance to infections with the *Plasmodium falciparum* malarial parasite carried by infected mosquitoes? The parasite was known to reproduce in red blood cells during its lifecycle.

In 1953, Allison set out on a 'sickle cell safari' in Uganda to test this hypothesis. Setting up base in Kampala, he measured the levels of malarial parasites in the red blood cells of children at the weekday markets in the Buganda region near the city. Those with sickle cell trait had levels of malaria parasites at least half of those children with two copies of the sickle cell gene. He then extended his analysis across East Africa, travelling to the forests of western Uganda and coastal regions of Tanzania and Kenya. He collected around 5,000 samples in total, representing three countries and thirty tribes. The results confirmed his hunch. In non-malarial regions virtually no sickle cell genes were present, while in malarial regions the frequency was as high as 40 per cent.

Allison showed that while two copies of the gene cause severe sickle cell disease, single copies cause only mild disease but provide protection against the mosquito-borne parasite *Plasmodium falciparum*, which causes malaria. Despite the fact that individuals with two copies of the mutant gene develop life-threatening sickle cell disease, the protection conveyed by a single copy of the mutant gene led to its maintenance in populations living in endemic regions. This was the first example of how evolution maintains disease-causing genetic variants in populations if they confer a survival advantage. There is a trade-off between the liability of the sickle cell mutant gene when two copies are present and the adaptive advantage conferred by a single copy in regions where malaria is endemic. Sickle cell trait carriers are relatively asymptomatic.

Allison published his results in 1954, concluding that 'the proportion of individuals with sickle cells in any population' is 'the result of a balance between two factors: the severity of malaria, which will tend to increase the frequency of the gene, and the rate of the elimination of the sickle-cell genes in individuals dying of sickle-cell anaemia'. One of the mechanisms by which sickle cell trait protects carriers from malaria was elucidated by Milton J. Friedman, who described in 1979 how the sickling process leads to oxidative damage that impedes the growth of the *Plasmodium falciparum* parasite.

While developing in the uterus, the foetus produces a unique form of haemoglobin called HbF, which binds oxygen more tightly than the adult variant. This allows the foetus to extract oxygen efficiently from its mother's blood. After birth, however, the gene responsible for HbF production is turned off by the production of a protein regulatory element called the BCL11A repressor. The observation that the few individuals who keep foetal haemoglobin switched on in adulthood have less severe symptoms suggested an innovative therapeutic approach. By using CRISPR gene editing to disable the *BCL11A* gene encoding the BCL11A repressor, HbF production could be switched on again in adults. Remarkably, HbF works almost as well as adult haemoglobin. In this instance, the HbF gene, which is usually silenced following birth, can be made to function as a 'spare tyre' by using CRISPR editing to reactivate it in adults.

Although all patients with sickle cell disease share an identical mutation, many other monogenic diseases exhibit considerable genetic diversity. The dystrophin gene *DMD* causing Duchenne muscular dystrophy (DMD), for example, has more than 3,000 variants, each of which occurs in different individuals. At 2.3 million base pairs long, it is also the largest gene in the human genome. This X-linked recessive condition leads to debilitating muscle degeneration in affected boys, typically claiming their lives by the time they are thirty years old. It would be impractical to develop a CRISPR-based therapy for each mutation, but gene editing is not the only technology that can be used to address Mendelian disorders. Gene replacement therapy, which uses viruses to ferry healthy replacement versions of whole genes into cells, has also proved highly effective in treating a variety of monogenic disorders.

On 22 June 2023, Elevidys, the first gene therapy for DMD, was approved by the FDA for boys between four and five years of age. Due to the huge size of the dystrophin gene, the drug was designed using a truncated synthetic gene encoding a smaller version of the dystrophin protein, known as microdystrophin. This innovation was inspired by a patient who, despite missing half of the dystrophin protein, displayed only mild clinical signs of muscular dystrophy.

Gene replacement therapies are now approved for the treatment of several other rare Mendelian diseases, including an inherited form of progressive blindness called retinal dystrophy (caused by mutations in the *RPE65* gene), haemophilia B (a blood-clotting disorder caused by a defective factor IX gene called *F9*), spinal muscular atrophy 1 (a progressive, muscle-wasting disease) and epidermolysis bullosa (a debilitating and excruciatingly painful blistering skin disorder). While highly effective, the cost of these drugs – typically several million dollars for a single treatment – puts them far out of reach for most healthcare systems.

It is important to note that all these gene therapies are 'somatic', meaning that the changes they introduce into genomes are not inherited. They do not affect the so-called 'germline' egg and sperm cells, whose genomes transmit hereditary information.

On 4 October 2023, Aissam Dam, an eleven-year-old boy from Morocco, was referred to the US for treatment, where he became the first patient to receive gene therapy for congenital deafness. He had a type of deafness caused by a mutation in a gene called *otoferlin*, which plays an essential role in the function of hair cells in the cochlea of the inner ear. The gene therapy he received at the Children's Hospital of Philadelphia placed a healthy copy of the gene into the genomes of the cells in his ear. It was a huge success. Following treatment, Aissam joyfully exclaimed, 'There's no sound that I don't like.'

While gene therapy and gene editing demonstrate how the individual Mendelian protein components of biochemical systems can be fixed in a curative manner by repairing or replacing faulty genes, these therapies are still being optimised. One challenge with gene replacement therapies is maintaining the long-term activity of the introduced gene. Over time, the patient's cells can shut down these artificial genes through 'epigenetic' regulation, causing incremental 'drop-off' of gene expression year by year. CRISPR gene editing is not, furthermore, always as precise as one would like it to be. 'Off-target' cuts created by Cas9 invariably occur at one or more undesired locations and have the potential to cause significant safety issues. It is notable that Sarepta Therapeutics, the company that developed the approved DMD drug Elevidys, announced on 22 July 2025 that it had paused all US shipments of the drug. This was due to three patient deaths caused by acute liver failure. It remains to be seen whether the observed toxicity was directly related to the drug.

Yet another obstacle results from the fact that patients may mount immune responses to proteins on the coat of the adeno-associated viruses (AAVs) that are commonly used to introduce replacement genes or CRISPR editing components into the cells of a patient. Another issue is that AAVs can accommodate only relatively small genes. Their restricted 'payload capacity' limits the size and complexity of the genetic material that can be introduced into patients' cells. The extent of the changes that gene editing can introduce into genomes is, at present, consequently limited.

CRISPR-Cas9 is ideally suited for the treatment of Mendelian diseases caused by mutant versions of small genes, but it cannot be used to rewrite genomes at scale, other than by introducing multiple parallel edits in a process known as multiplexed editing. However, this is very different from being able to completely redesign genomes, which is possible using the genome design and writing strategies of synthetic genomics.

Alternative delivery methods, such as the herpes simplex virus – which has up to thirty times more payload capacity than the AAVs that are typically used to ferry DNA cargoes into cells for gene therapy – offer the promise of delivering more DNA into genomes. Similarly, tiny microspheres capable of ferrying genes into cells, called lipid nanoparticles, are being explored as a 'non-viral' delivery method. Emerging technologies based on mobile genetic elements, such as transposons and retroelements, are also likely to play a role in reprogramming genomes, as they can potentially integrate much larger pieces of DNA, allowing genome engineering at scale.

The CRISPR–Cas9 gene-editing system is very versatile, and may be adapted to produce innovative types of editing tools. For example, the cutting activity of Cas9 may be replaced with a mutant version of Cas9 called dCas9 (dead Cas9), which lacks the DNA-cleaving ability but retains the ability to be guided by a single RNA to bind a specific DNA sequence with high precision. Cas9 may then be used as a programmable platform to which an enzyme called DNMT3A, a DNA methyltransferase can be attached. This enzyme adds methyl groups to CpG sites – the key targets for methylation in the genome – thereby silencing gene expression at the position it is guided to by a mechanism called epigenetic repression. The chemical modification process that allows the activity of specific genes to be silenced without altering their nucleotide sequences, is known as 'epigenetic editing'.

In 2024, a team led by Angelo Lombardo in Milan successfully used epigenetic editing to reduce the activity of the *PCSK9* gene, which controls the production of low-density lipoprotein (LDL) – 'bad' cholesterol – in the human liver. Abnormally high levels of LDL predispose individuals to cardiovascular disease, including heart attacks and strokes.

One potential advantage of epigenetic editing is that it may be safer than conventional editing, as it does not involve cutting the genome. Whenever conventional CRISPR-Cas9 editing is used, unintended cuts may be introduced into DNA sequences resembling the target sequence. This may result in unintended insertions and deletions into select regions of the genome, including genes and the sequences that regulate them. It may also lead to translocations, in which segments of the genome rearrange themselves incorrectly, either within or between different chromosomes. As well as disrupting normal gene function, off-target editing may also activate genes that cause cancer.

Another type of genome modification, known as base editing, rewrites the code of genes at the level of single nucleotides. Like epigenetic editing, base editing also reduces the risk of off-target effects. It has been used successfully to deactivate the *PCSK9* gene as well. Base editing also employs sCas9, but instead of attaching DNMT3A, as is the case in epigenetic editing, it couples Cas9 with a deaminase enzyme. As with epigenetic editing, base editing reduces *PCSK9* gene activity sufficiently to remove the need for a long-term cholesterol-lowering therapy.

Gene therapy may also be performed *in utero* on a developing foetus, in a way that corrects the somatic cells, but does not introduce heritable alterations into the germline. The key advantage of this approach is that it can prevent the accumulation of irreversible, disease-related damage early on in life. In Gaucher's disease, for example, mutations in the *GBA* gene, which encodes an enzyme involved in breaking down a lipid called glucocerebroside, leads to the accumulation of fatty molecules in brain cells and other cells in the body. This causes devastating progressive brain damage, as well as damage to other organs, which is irreparable once it has occurred. Many affected children do not live beyond two years.

It therefore makes sense to correct the faulty gene as early as possible to prevent the accumulation of damage. In 2018, Simon Waddington at University College London showed that the injection of a correct version of the gene into the foetal brain of a mouse with a mutant version of *GBA* prevented the onset of the neurodegeneration that invariably accompanies the disease. Tippi MacKenzie, a foetal medicine specialist at the University of California, San Francisco, has called foetal gene therapy 'the next frontier'.

While the long-term safety of gene replacement therapy, conventional gene editing and epigenetic editing remains to be seen, one of the most significant issues at present is its cost. Most monogenic diseases are rare, making them commercially unattractive for biotechnology and pharmaceutical companies. Because of this, it is hard – at least for the time being – to see how these therapies could ever become widely available on a global scale. This is especially true because many of the estimated 300 million individuals worldwide affected by Mendelian diseases, such as sickle cell disease, live in economically challenged regions of the world.

In a significant breakthrough and landmark moment for patients with ultrarare genetic diseases, in May 2025 researchers at the Perelman School of Medicine at the University of Pennsylvania and the Children's Hospital of Philadelphia (CHOP), announced that they had successfully treated a child called JK with a personalised form of CRISPR gene editing therapy. It enabled a single nucleotide mutation to be changed to the healthy version. Patients with ultrarare mutations are typically disenfranchised as biopharmaceutical companies derive no economic benefit from treating them and are consequently not incentivised to invest the necessary resources to develop appropriate therapies.

Born with a disorder called carbomyl phosphate synthetase 1 (CPS1) deficiency – which occurs in just 1 in 300,000 individuals, may cause brain damage due to increased levels of ammonia in the blood, results in the mortality of up to 50 per cent of affected children, and invariably leads to a liver transplant – JK was successfully treated with the drug in February 2025. Remarkably, the team managed to manufacture the customised gene therapy and dose the patient in just seven months. With the cooperation of regulatory agencies, it should be possible to generalise this bespoke gene editing approach to allow for the treatment of many other ultrarare genetic diseases caused by single mutations.

It is possible to imagine a future in which – resources permitting – it should be possible to permanently cure most or all of the 8,511 known human Mendelian diseases. These diseases could, in principle, be removed as a recurring feature of human nature. The cost to global healthcare systems would, however, be prohibitive, as if corrected at the somatic level rather than the inherited germline, every affected child would need to be corrected, either *in utero* or in early childhood. The genes causing the Mendelian diseases would continue to persist in the germline, and contribute to a process of heredity, meaning that every new generation would require treatment.

It may, therefore, for purely economic reasons, at some point be necessary to consider the plausibility, safety and ethics of correcting the genomes of affected individuals at the inherited germline level. In the absence of new disease-causing mutations, this would eradicate such conditions from human populations on a permanent basis, much in the same way that the infectious disease smallpox was eliminated on a global scale, as a result of a World Health Organization eradication programme. Introducing permanent changes into the human genome would, however, be far more consequential than eradicating a virus like smallpox, and the ethical and societal implications of this would need to be carefully examined.

While gene editing and gene therapy may eventually enable us to prevent, cure or ameliorate most or all monogenic human diseases in which faulty genes behave like the broken components of human-made machines, most genetic human diseases are not caused by single genes. They are 'polygenic' rather than 'monogenic' – caused by small contributions from multiple genes – and present a major conceptual and technological challenge.

It is when addressing these types of complex diseases caused by large numbers of genes scattered across the genome, which are also impacted by environmental and lifestyle factors, that the human body-as-machine metaphor begins to break down.

Examples of highly prevalent complex diseases of global significance include: hypertension, coronary artery disease, diabetes, asthma, epilepsy, obesity, Parkinson's, Crohn's, multiple sclerosis, rheumatoid arthritis, cancer and psychiatric diseases like depression. Unlike Mendelian diseases, these conditions arise from the contributions of numerous small effects across multiple biochemical pathways. This makes it difficult to identify the individual contributions of each gene, let alone repair them. The genes implicated in complex diseases are, furthermore, typically entangled in a host of unrelated cellular processes, making it hard to identify specific therapeutic targets. These diseases cannot, therefore, be repaired in the same way as Mendelian diseases. They highlight the limitations of the existing component-focused molecular medicine paradigm and the apparent duality of biological systems, whose components can be modelled in a machine-like manner in some instances, but not in others.

This complexity extends beyond diseases to complex traits like human height. Collaborative genome-wide association studies (GWAS) conducted at Harvard Medical School and MIT and published in October 2022, analysed data from nearly 5.4 million individuals to identify DNA variants associated with height. GWAS track millions of one-letter variations in DNA called single-nucleotide polymorphisms (SNPs), to identify markers linked to specific traits. The SNPs serve as markers both for themselves and for the DNA sequences in close proximity to them.

An astonishing 12,111 different genetic variants were implicated in the control of human height, each of which could potentially provide insights into the underlying biological mechanisms. These variants clustered within genomic segments covering about 21 per cent of the genome. Individually, the impact of each variant was very small, but together they accounted for most of height's heritability. In prior attempts to use GWAS to identify genes contributing to height, there had always been a substantial degree of 'missing heritability' that could not be accounted for, but which was known to be associated with specific gene variants. This is because the minuscule effects of some genes could not be statistically resolved in smaller datasets but became visible when larger ones were examined. It turns out that a significant proportion of the human genome appears to be involved in determining height – the exact opposite of a Mendelian trait.

By compiling this substantive dataset, the team achieved 'saturation' for the trait, and pinned down almost every source of variability. It is hard to do this for most complex diseases, however, owing to their lower prevalence and the variability of their diagnostic criteria, compared with easy-to-measure traits like height. GWAS demonstrate that rather than acting independently like the orthogonal components of Mendelian diseases, most genes appear to work together as parts of an interconnected network. If the genetic factors determining a trait like height are so numerous, one can only imagine how complex the causal structures underlying traits like creativity, consciousness, musical ability, morality and the human imagination must be.

For complex diseases like type 2 diabetes, hundreds of GWAS hits have been identified. Schizophrenia's genetic contributions are so widely dispersed that any randomly selected million-nucleotide-long region of the human genome is likely to contain at least one contributing risk locus. Intriguingly, up to 95 per cent of GWAS hits map to non-coding regions. This suggests that most traits and diseases are driven by changes in gene regulation, rather than by alterations to protein-encoding gene sequences.

In 1826, Maria Halm, the wife of Anton Halm, a Viennese pianoforte player and composer, asked her husband – who at the time was completing a four-handed piano transcription of Ludwig van Beethoven's *Große Fuge* (Great Fugue), Op. 133 – whether he could procure a lock of the composer's hair as a memento. Halm approached Beethoven's friend, the violinist Karl Holz, with the request, but unbeknown to Beethoven, Holz mischievously sent Halm's wife some goat's hair.

When Halm wrote to Beethoven on 24 April 1826 to inform him that he had completed the transcription, he told Beethoven that his wife was aware of the practical joke Holz had played on them. Beethoven immediately sent her an authentic lock as recompense, along with a note requesting her forgiveness and affirming, 'That is my hair!' Around thirty years after Beethoven's death, Halm gave this lock to Beethoven's biographer, Alexander Wheelock Thayer, who wrote an accompanying note on 25 April 1826 that read: 'Hair from Beethoven's head received from himself by Anton Halm.'

On 24 April 2023, almost two centuries later, Johannes Krause and his colleagues at the Max Planck Institute published a paper titled 'Genomic Analyses of Hair from Ludwig van Beethoven'. The analysis was based, in part, on a genome sequence derived from DNA prepared from the Halm/Thayer lock of Beethoven's hair. Beethoven had, in a document addressed to his brothers Carl and Johann dated 6 October 1802 – subsequently known as the *Heiligenstadt Testament*, after the small village where he wrote it, located just north of Vienna – requested that his physician Dr Johann Schmidt make the details of his illness known after his death. The testament, which was found in a concealed compartment in Beethoven's desk the day after his death on 26 March 1827 by his lifelong friend Stephan von Breuning and former associate Anton Schindler, details Beethoven's increasing despair at the prospect of his impending deafness, 'a permanent malady' whose cure, he believed, would 'take years or even prove impossible'.

While failing to find the genetic cause of Beethoven's progressive and eventually complete deafness, or his chronic gastrointestinal problems, they were able to show that Beethoven was infected with hepatitis B and had a strong genetic predisposition to liver disease, which might explain his documented liver issues. The difficulty in predicting the genetic basis of complex traits, both disease-related and otherwise, is, however, demonstrated by the fact that when Beethoven's genome was assessed for beat synchronisation – considered to be a proxy for musical talent, and which may be assessed by the question 'Can you clap in time with a musical beat?' – he scored very poorly.

This illustrates the challenges of determining the genetic contributions to traits such as musical ability and creativity, which likely emerge from the interplay between genetics, environment and culture. It may turn out that the direct genetic contribution to these types of human capabilities is negligible. There is likely no simple, reductionist genetic calculus to explain and predict complex traits like creativity. But genes clearly play an important role in generating the foundational structures that facilitate the broad potential for the development of such phenomena.

To understand the genetic basis of complex diseases, human traits and human biology in general, it will be necessary to assemble a comprehensive compendium of sequencing data. This would exhaustively catalogue genomic variation across a broad swathe of ethnically diverse individuals in both health and disease. Advances in whole-genome sequencing (WGS) technology are making this increasingly feasible. Comprehensive mapping projects – aimed at defining the genetic architecture of complex diseases across diverse populations – are already establishing the paradigm for this monumental task.

The US-based *All of Us* research programme, launched in 2018 by the National Institutes of Health (NIH), exemplifies this ambition. The programme seeks to gather genetic information on an unprecedented scale, aiming initially to compile the WGS data of one million participants that reflect the genetic diversity across the US. This genetic data is linked to electronic patient records. Upon completion, it will form the most diverse and extensive genetic resource of its kind, creating an encyclopaedic catalogue of human genetic disease variants. Notably, half the participants in the programme self-identified as being of non-European ancestry. On 14 March 2024, the project published its preliminary results. After sequencing the complete genomes of the first 245,388 volunteers, a total of more than one billion genetic variants had been identified. These types of large, curated databases, along with the ability to interrogate them using deep learning AI to identify underlying patterns, will help determine the genetic basis of complex diseases.

GWAS have contributed to our understanding of how DNA sequence variants inherited from interbreeding events occurring tens of thousands of years ago between our ancestors and now-extinct species of archaic humans such as Neanderthals and Denisovans, have had an outsized impact on our biology. The genomes of most Europeans and Asians contain around 2–4 per cent Neanderthal DNA, while Melanesians and Aboriginal Australians have up to 5 per cent Denisovan DNA.

Some Neanderthal variants of human genes contribute to health vulnerabilities. GWAS data have linked Neanderthal versions of genes in our genomes to depression, obesity, osteoporosis, pre-cancerous skin lesions, coagulation disorders leading to blood clots and strokes, urinary tract disorders and various skin conditions. Their contributions, however, are generally very small. Other effects, however, have proven to be beneficial. Modern Tibetans, for example, have a Denisovan-derived version of the *EPAS1* gene, which helps them to live at altitudes of up to 4,000 metres by preventing excessive blood thickening. Other Neanderthal and Denisovan genes have been shown to enhance the innate immune system, providing increased protection against infections with organisms such as the bacterium *Heliobacter pylori*, which can cause stomach cancer.

But it is not just ancient genetic legacies that have contributed to disease risk in modern humans. Scandinavians are more than seventeen times more likely to develop the autoimmune disease multiple sclerosis (MS) than individuals in sub-Saharan Africa. Genetic detective work has shown that this disparity may be traced to the Bronze Age around 5,000 years ago, when nomadic cattle herders known as the Yamnaya, who lived on the steppes near the Black Sea, migrated across central and northern Europe. As they swept through the continent, they brought genes with them that predisposed to MS. In some northern European regions, these large-scale migrations completely replaced local hunter-gatherer populations, leading to an exceptionally high prevalence of the MS-related disease gene variants in those geographical areas.

The international Human Variome Project (HVP), based in Victoria, Australia, aims to catalogue human diversity by mapping the uncharted variation in the estimated 0.5 per cent of the human genome that differs between individuals. These small differences contribute to inter-individual variability and disease susceptibilities. There is no single standard 'healthy' version of the human genome. As Somerset Maugham observed in his 1915 novel *Of Human Bondage*, normalcy is 'the rarest thing in the world'. The UK Biobank project, which plans to generate genome sequence data from 500,000 individuals of primarily European descent, is focused on the exome – the part of the genome coding for proteins. While the scale of the project is impressive, its focus on whole-exome sequencing, rather than whole-genome sequencing, excludes critically important information contained in the non-coding regions of the genome.

Should the data from these extensive WGS projects be combined into a single database representing all global ancestries, it will enable the creation of a fully inclusive human 'pangenome'. This cataloguing of the global diversity occurring at every position within the human genome would provide a definitive genetic description of the human species, and detail the genetic causes underlying the variability of all human traits. The landscape of human variability will have been comprehensively charted, creating a definitive resource for understanding the genetic basis of being human.

The foundations for this kind of genetic exploration were constructed more than a century ago by the statistician and geneticist Ronald Aylmer Fisher, when he was twenty-eight. His 1918 paper 'The Correlation Between Relatives on the Supposition of Mendelian Inheritance' established the statistical basis for quantitative genetics – the study of the variation responsible for the emergence of complex traits under the influence of multiple genes. Fisher proposed that such traits result from the action of 'cumulative Mendelian factors'. His key idea was that continuous traits like height could be explained by a model in which a large number of variable genetic factors, each making only a small contribution to the trait, were independently inherited according to Mendel's laws. This provided the theoretical bridge between monogenic and polygenic traits and established the foundations for modelling the polygenic inheritance of most human diseases.

However, focusing exclusively on genetic information written into the human genome risks what has been called 'human genome exceptionalism' – the tendency to attribute a unique and special status to genetic information that elevates its importance above all the other sources of information that contribute to human nature. To counterbalance this, it is helpful to introduce the concept of what we may call the human 'informiome'. This term captures all the different types of information contributing to human nature, regardless of the substrate or physical medium in which they are represented. Each organism is defined by its own distinct 'information cloud' that defines its informiome. This includes, for example, in social species, those aspects of behaviour that are learned – and therefore represented in the brains of individual organisms – and not geneticaly encoded. Such behaviours are transmitted both within and across generations through extra-genetic forms of cultural inheritance.

While the human genome forms the foundation of the human informiome, other layers of extra-genetic information are equally important. These include the epigenetic information – the chemical modifications to DNA that alter gene expression – which is superimposed on our genomes (the epigenome) through methylation and other forms of chemical modification that silence or activate genes. The concept of the informiome also encompasses the genomes of the various bacteria, fungi, viruses and parasites that live in, on, and all around us, as well as the influence of development, learning, culture, social factors such as lifestyle, nutrition, and the physical and biological environments in which we are embedded. It serves to democratise the diverse sources of information that make us human.

The concept of an informiome extends to the information defining non-human species as well. In their 2021 study, Erol Akçay and Amiyaal Ilany showed how the inherited social networks of wild spotted hyenas shape their social behaviour. The offspring of higher-ranking mothers inherit their mothers' social networks, and this is predictive of their increased survival.

Microorganisms comprise a key component of the informiome and impact metabolic pathways. The complete inventory of the microorganisms associated with an organism (including bacteria, viruses and fungi) is known as its microbiome. The microbiome has substantive effects on the biology of a species.

The nature and extent of the human microbiome – the ecosystems of microorganisms living in and on us – remain unknown. If, as a back-of-the-envelope calculation, we estimate that the human gut contains around 1,000 unique bacterial species with approximately 2,000 genes each, the number of different bacterial genes in a human host is around two million, or roughly one hundred times the estimated 20,000 protein-encoding genes in the human genome.

The complete set of genes from the various species comprising an individual's microbiome may be thought of as its 'extended genome', which is integral to and functionally connected with the human genome. Microorganisms in the human intestine, for example, regulate the absorption, metabolism and storage of lipids. In 2023, Lora Hooper and Yuhao Wang at the University of Texas, Dallas, showed that microorganisms could repress the expression of a long non-coding RNA encoded by a gene called *small nucleolar host gene 9* (*Snhg9*) in the cells lining the intestine of mice. This, in turn, repressed the activity of a gene called *PPAR-gamma*, resulting in an increase in lipid absorption and metabolism. It showed how bacteria can change the behaviour of genomes and alter their activity.

Microorganisms can also have a profound impact on the behaviour of organisms. Wolves that have been infected by the parasite *Toxoplasma gondii*, for example, are more likely than uninfected animals to become leaders of their packs. There is also an increased chance that infected animals will leave the pack to strike out on their own. The infection alters their behaviour, making them bolder through an as-yet-unknown mechanism. The microbiome may also impact a range of other behaviours. In 2022, Christoph Thaiss and Lenka Dohnalová at the University of Pennsylvania, Philadelphia, showed that the microbiome of mice can determine whether they are active or lazy. It can influence other features too. The bacterium *Lactiplantibacillus plantarum* has, for example, been shown to promote growth in undernourished mice.

Alterations to the microbiome have been linked to several human diseases, ranging from neurodegenerative diseases such as amyotrophic lateral sclerosis and Alzheimer's disease to inflammatory diseases such as Crohn's. Some bacteria even appear to play a role in malignancies, like prostate cancer. Indeed, tumours often contain large numbers of bacteria. While their role remains poorly understood, it appears that they can impact the function of immune cells and convey resistance to cancer therapy. Other bacterial species may also impact the response of cancer patients to immunotherapies. There is tremendous diversity in the microbiomes of different human populations, and this remains a subject of ongoing research, as is the heritability of microbiomes among human family members. Close relatives typically have microbiomes that are more similar to one another than to non-relatives. The basis of this similarity, however, is unclear. It may relate to their shared genetics, environment, or both.

In *The Nature of Knowledge* (1994), the biologist Henry Plotkin, working at University College London, explored an area of the philosophy of biology known as evolutionary epistemology, which views biological information as a form of knowledge. Plotkin envisaged the various informational mechanisms that organisms employ at different levels as tracking environmental change of different frequencies. While genomes track informational change of a very low frequency – being modified only very slowly across generations – development and learning track change of intermediate frequency. Culture, on the other hand, tracks the highest frequency of change, and furnishes immediate, contextually relevant adaptations. Social contributions to the informiome include phenomena such as loneliness, social status and inequality. None of these are directly represented in the code script of genomes, but each may make significant contributions to health.

Ultimately, human health must be perceived as being embedded within a complex web of distinct informational systems. While grounded in genetics, these are distributed across multiple different layers, which include social and environmental factors. Fruit flies and mice provide compelling examples of how behaviour can impact organismal function.

The prolonged absence of social contact in fruit flies leads to dysregulated sleep and feeding patterns, while a loss of social status in mice results in changes in the medial prefrontal cortex of their brain. Disrupted sleep is also a risk factor for Alzheimer's disease. Lifestyle factors such as exercise have been shown to stimulate the production of immune cells in bones, which help protect against the amyloid build-up in the brains of patients with Alzheimer's.

Public health, social stability and access to essential resources also play a key role in disease susceptibility. Without affordable, widely available and comprehensive healthcare – alongside access to basic resources such as food security, a balanced diet, clean drinking water, mosquito nets, vaccines, housing, sanitation, education, employment and adequate social networks – health inevitably suffers. Worldwide, more than two billion people lack access to safe and reliable drinking water. Political and economic stability, social equality and the absence of warfare are also of critical importance for health maintenance. The World Health Organization has estimated that around 3.8 million people die each year from illnesses related to household air pollution, caused by open fires and stoves burning kerosene, coal, wood or animal manure. The gene variants present in staple crops such as wheat and rice, and in livestock such as pigs and cows, likely exert a greater impact on global human health than rare genome variants that affect only small numbers of individuals.

The emerging field of 'systems genetics' seeks to determine how genetic information is generated within and propagated across molecular and cellular networks, and how these interactions result in higher-level biological functions. These arise through self-organising and emergent behaviours that influence critical processes. There are doubtless different and less intuitive types of potential therapeutic targets located within gene networks that influence the dynamics of network behaviour. An awareness of this could lead to a new type of 'network medicine'.

Dennis Bray, a mathematical biologist at the University of Oxford, has proposed that enzymes and receptors function as 'readers' of molecular signals, functioning like computational elements. While enzymes read the concentration of their substrate, receptors read the levels of their ligands. Similarly, individual cells function as nodes within cellular networks. These networked computational units generate emergent, self-organising network behaviours that are not directly genetically programmed.

We may need to recast our understanding of biology within a new conceptual framework, defined by interconnected gene networks.

In 2007, Kwang-Il Goh and Albert-László Barabási proposed that so-called 'disease genes' tend to reside not at the central hubs of these networks, but at their functional periphery. Given that many of the genes associated with disease also perform essential roles, it may be time – at least in some cases – to relinquish the notion of individual 'disease genes' altogether, and instead reframe our thinking in terms of local network dysfunction.

Network medicine promises to move beyond the reductionism of molecular medicine, replacing its component-based logic with a richer understanding of the emergent behaviours across the human interactome. This expansive landscape includes not just the products of protein- and RNA-encoding genes, but also the dynamic populations of cells. Such systems of highly interconnected proteins and regulatory RNA molecules behave less like machines and more like neural networks, which have been trained by evolution to produce particular outputs. Unlike the rigid circuitry of AI, the wiring of biological networks is shaped by chance encounters between molecules, constrained by the laws of diffusion.

The transition from a component-based paradigm to a network-based one demands that we think about biological systems in a fundamentally different way. In his September 2022 paper entitled 'Can a Biologist Fix a Radio? – Or What I Learned While Studying Apoptosis', the biologist Yuri Lazebnik of Cold Spring Harbor Laboratory demonstrated how biology's reverse-engineering approach is ill-suited to the task of repairing even a simple transistor radio. With this analogy, he showed that we currently lack an appropriate engineering language for biology that is capable of capturing the complex behaviours of biological networks and their non-orthogonal parts.

To develop effective treatments for complex, polygenic diseases, we will need to adopt this network-based paradigm and move beyond the human-as-machine conception. In a landmark computer experiment in 2008, Barabási took the first step towards imagining what a network medicine intervention might look like. He outlined the basis for a strategy to treat a monogenic disease – not by fixing the defective part directly – but by rewiring the system in which it is embedded. Using a computational approach, he devised a study to test whether the loss of enzyme function caused by a mutant gene could be offset by manipulating the structure of the surrounding metabolic network – a process he called 'synthetic rescue'. The idea was to 'restore biological function by forcing the cell either to bypass the functions affected by a defective gene, or to compensate for the loss of function'. Astonishingly – and counterintuitively – he found that some of these functions could be rescued not by repair, but by introducing further damage through additional gene deletions. The removal of a second enzyme-encoding gene was, paradoxically, able to compensate for the malfunction of the first. Future therapeutic interventions are likely to increasingly reference network medicine approaches. The capacity for such interventions will be facilitated by ABI, which will help identify the required network modifications.

The emerging science of synthetic genomics will, in time, allow more extensive forms of network manipulation, enabling the reconfiguration of molecular systems from the ground up. This forward-engineering approach will be a powerful addition to the traditional reverse-engineering strategies used to investigate biological function. Substantive rewiring of molecular systems is beyond the reach of gene editing, which can amend pre-existing genomes but not write them from scratch. The 'compensatory rescue' that Barabási achieved in his pioneering network experiments represents the simplest form of this approach, and will likely provide the conceptual foundation for a new kind of network-based therapeutic science. Ultimately, however, the future of network medicine will rest on our ability to draft, revise and author the metabolic circuitry of life, not gene by gene, but system by system.

Large-scale genomics projects, like the study of consanguineous Pakistani populations led by Sekar Kathiresan at MIT, have revealed the surprising robustness of metabolic networks and may help establish some of the key tenets of network-based approaches. Kathiresan and his colleagues trawled through WGS databases to identify individuals who had inactive versions of genes but lacked corresponding deficits. Their landmark study of 10,500 individuals of Pakistani origin, a population in whom first cousin marriage is relatively common, showed that more than 1,300 genes could be 'knocked out' of the human genome without causing obvious consequences. This suggests that many genetic functions are either redundant or readily compensated for. Were it possible to establish the molecular basis by which such compensations occur, we might devise new therapeutic strategies that leverage the intrinsic capacity of networks to preserve their core functions. A more extensive 'human knockout project' would extend these findings, identifying other non-essential human genes and exploring, at the network level, how each gene deletion may be counterbalanced.

As we begin to contemplate the possibility of engineering biology at scale, we are inevitably drawn back to the question of what constitutes a disease – and, more provocatively, what we mean by the term 'natural'. Take ageing, for example. Is it a disease that needs curing, or a defining feature of being human? In 1900, the average life expectancy worldwide was thirty-two, but by 2021, it had more than doubled to seventy-one. To someone living a century ago, dying in one's thirties might have seemed normal, if not expected and inevitable. A hundred years from now, it may seem equally absurd that people once died at the 'young' age of ninety. Our sense of what is 'natural' is, to some degree, socially and culturally constructed. The nature of organisms is itself the product of a long and haphazard evolutionary process, and changes over time.

Disease and illness have played an important role in human cultural history. The evocative self-portraits of the Mexican painter Frida Kahlo, such as her iconic *The Wounded Deer* (1946), capture the artist's physical and emotional suffering after contracting polio and being severely injured in a bus accident. The same could be said of Vincent Van Gogh, who wanted his portraits to show 'the soul of the model' and to 'appear as revelations to people in a hundred years' time'. In 1819, the poet John Keats wrote 'Bright star, would I were stedfast as thou art' while dying of tuberculosis, contrasting the desire for permanence with the reality of mortal existence. Yet romanticising illness is a luxury typically afforded in hindsight. It is one thing to appreciate its contribution to art, music and literature, but quite another to endure it. Few would willingly trade their health for artistic inspiration.

Ideas about disease are profoundly shaped by the prevailing 'tastes and preoccupations of society'. One can distinguish between 'illness', which references the subjective experience of suffering, and 'disease', which reflects the ideas we structure around that illness. Lovesickness, once regarded as a legitimate medical disorder, marked by symptoms ranging from anorexia to insomnia and melancholy, today lives on purely as a metaphor. The 1774 oil painting *Erasistratus Discovering the cause of Antiochus' Disease*, attributed to the French neoclassical artist Jacques-Louis David, immortalised the diagnosis of lovesickness and the futility of seeking a cure. Battista Fregoso, the fifteenth-century Genoese nobleman and son of the Doge, described lovesickness as love that is 'not merely behaviour resembling sickness' but 'a true disease, virulent, and dangerous'. One imagines that in potential future authoritarian societies, free will and the capacity for dissent might be construed as an inconvenient illness. In such dystopian social orders, dissent itself may be classified as a disorder, with even minor transgressions pathologised and transformed into new diagnostic categories.

The inventor of the telephone, Alexander Graham Bell, devoted his life to developing a machine that could faithfully transmit sound across great distances. This ambition culminated on 7 March 1876, in the granting of a US patent for 'the method of, and apparatus for, transmitting vocal or other sounds telegraphically'. Bell's fixation with sound and hearing likely had personal origins: his mother – an accomplished pianist – lost her hearing in adulthood and taught him British Sign Language. His wife, Mabel Hubbard, also lost her hearing after contracting scarlet fever at the age of five. Yet, rather than embracing sign language, Bell became an opponent of its use. He championed an oral language for the deaf, called Visible Speech.

However, Bell's concerns went further. He actively canvassed against the marriage of two deaf individuals, fearing the emergence of a so-called 'deaf variety' of humankind. His misguided advocacy lent support to a series of unethical legislative efforts in the US, some of which criminalised marriage between deaf people.

Today, deafness is understood very differently, and such views are rightly condemned. Many within the Deaf community – who capitalise the 'D' in Deaf to signify that deafness is a distinct cultural and linguistic entity – see deafness not as a disability, but as a natural variation of the human condition. For them, sign language is not just a means of communication, but an integral part of their cultural identity: something to be celebrated and protected, rather than pathologised and corrected.

There are over 300 sign languages across the world, each a richly expressive linguistic system with its own grammar, syntax and vocabulary. In his 1996 poetry collection *Dandelions*, the American Sign Language (ASL) poet Clayton Valli articulated the quiet defiance of Deaf culture. His work celebrated the resilience of sign language in the context of the prevailing global cult of 'oralism', which sought not only to undermine and marginalise sign language, but to erase it. Valli likened Deaf people to dandelions – forced to take root in inhospitable ground and to 'grow' on the terms of those inhabiting the world of hearing – and yet, despite this, still persisting as the validity of their own world was marginalised. Meaning in his poems is conveyed through a rich array of handshapes, body movements, facial expressions, rhythm and spatial cues. There is a particular type of social intimacy and of shared space and language within Deaf culture that is not typically found among individuals with intact hearing. Some Deaf parents, embracing this identity, actively hope to have Deaf children and view prenatal testing with unease. Bell himself, in a final moment of reconciliation, appears to have capitulated, apparently signing into his wife's hand on his deathbed.

Cancer, in contrast, is a condition that evokes no such cultural ambiguity. It is, universally, regarded as a disease. It is also the field in which molecular medicine has made some of its most remarkable advances. It has allowed us to glimpse the possibilities for improving human health and for eventually curing most, if not all, human diseases. Notable successes – such as the attainment of a functional cure for many patients with chronic myeloid leukaemia (CML) using an oral tyrosine kinase inhibitor (TKI) drug called imatinib – indicate how devastating diseases, once thought to be untreatable, can now be functionally cured. This would have been unthinkable only a few decades ago. Previously intractable cancers like CML have become the poster child for mankind's desired attainment of a disease-free future. They are emblematic of the types of remarkable therapeutic successes that we might reasonably expect to achieve in the future.

The story of Doug Olson, a patient who developed chronic lymphocytic leukaemia (CLL), provides another example of this emerging medical frontier. Diagnosed in 1996 after a protracted duration of experiencing increasing fatigue, by the summer of 2010 he had run out of options and became resigned to his destiny. His cancer had mutated so much that it had become insensitive to all known approved medicines. When he heard about a new experimental treatment option called chimeric antigen receptor T-cell therapy, or CAR-T, he felt that he had nothing to lose. The therapy, a 'living drug', was being tested at the Abramson Cancer Center and Perelman School of Medicine at the University of Pennsylvania, in a study led by Carl June and David Porter. It involved removing some of the patient's T-cells, reprogramming them using gene editing, and then infusing them back into the patient's body.

The results were, quite simply, miraculous. Just weeks after receiving the treatment, Olson's doctors could not find 'a single cancer cell' in his blood. This outcome was beyond even their wildest expectations. More than a decade later, Olson's blood remains cancer-free. In 2022, Carl June stated that 'we can conclude that CAR-T cells can actually cure patients with leukemia'.

Cell therapy is, however, still in its infancy, and is not yet fully optimised. Some patients experience life-threatening side effects or relapse, and the therapy has not yet managed to make substantive inroads into solid tumours, such as breast cancer, pancreatic cancer or lung cancer. But it is likely just a matter of time before that happens. Cell therapy has already been expanded beyond T-cells to include different cell types, such as natural killer cells. Some approaches – so called 'in vivo' cell therapy – have dispensed with the infusion of engineered cells altogether. Genetic material and gene editing components are instead introduced directly into the cells of the human body. This allows the patient's cells to be directly reprogrammed without having to remove them. But the next major technological leap is likely to come from the rewriting of cellular genomes, rather than just editing them. This offers the promise of constructing synthetic human cells that have been specifically designed and optimised for cell therapy.

Such genomically rewritten cells promise to have extraordinary safety and efficacy, not just in the context of cancer, but in the treatment of autoimmune disease and beyond. The ability to design and synthesise human genomes using ABI, will transform cells into miniature, programmable biocomputers. Programmable engineered cells – their synthetic genomes filled with applications, much like the software applications of smartphones – are poised to play a central role in the future of human healthcare. The emerging ability to rewrite and construct the genetic code script of immune cells – and that of other cell types – at scale, for therapeutic purposes represents one of the most potentially significant emerging frontiers in clinical medicine.

Complex diseases do not readily conform to the conventional reductionist molecular model that has underpinned the successes of molecular medicine over the past half-century. If we wish to eliminate such diseases, humanity faces some daunting options. To do so we must either learn to interpret and manipulate the behaviour of biological networks, or contemplate the audacious and ethically challenging notion of rewriting the human genome – regionally or in its entirety – to reconfigure its architecture in ways more conducive to health.

The first option, although difficult, is intellectually and ethically straightforward. The second, however, would propel humankind into completely uncharted territory. The prospect of rewriting any portion of the human genome raises profound ethical and societal questions. Could this ever be justified? How far are we, as a species, prepared to go to eliminate disease and foster healthy longevity? And to what extent is the potential for death and disease itself an essential part of our humanity?

Were we to permit regions of the human genome to be refactored, like the revisions that Drew Endy introduced into the code of bacteriophage, to remove billions of years of accumulated 'technical debt', we might inadvertently alter facets of our humanity that we do not yet fully comprehend. Should the optimisation of health prove a sufficient rationale for such an enterprise, it could set a precedent that places other aspects of human nature at risk. Heritable genome engineering is not something that should be undertaken lightly. Nor are we at present even close to being in a position to contemplate it.

It is, in truth, hard to imagine a future in which large-scale reconfigurations of the human genome could ever be ethically justifiable. Among many other things, the genome and the informiome to which it belongs, may only be partially computable. Our attempts to engineer biological systems in a fully predictive manner will likely always be frustrated by unknowable and undecidable factors. Some of these relate to the environmental factors that impact human biology. Other elements may be intrinsically undecidable. There will also inevitably be limits to the extent to which biological functions can be disentangled, necessitating compromises to accommodate conflicting constraints.

If an exhaustive understanding of biological systems remains beyond our reach, then the only way to assess the outcome of different versions of rewritten genomes would be to 'run' the programmes, by generating genomically rewritten individuals. This would clearly be ethically indefensible. Moreover, one is obliged to consider whether there is something intrinsically sacred – if not religiously, then biologically, or historically – about the naturally evolved human genome. Might it simply be wrong to attempt to permanently redesign, replace or repair anything other than the smallest regions of it?

If the ultimate goal of clinical medicine is to liberate humankind from disease, then large-scale genomic reprogramming may ultimately be the only way to achieve this. There may, consequently, be a naturally and ethically defined limit to how far we can and are prepared to go to end human disease, and alleviate suffering at the level of genomes. It may be necessary, in the end, to accept that some suffering cannot be eliminated without compromising our deeper sense of what it means to be human. We may, additionally, prefer not to set a precedent for the refactoring of human genomes at scale, regardless of any potential health benefits. The balance of risk and benefit in this regard will doubtless form a central focus of future philosophical debate. We will likely need to look elsewhere for many of our cures – evolution's profligate and tangled spaghetti code continuing to confound and thwart us at every step.

10

The Future of Species

> In the years ahead, genes are going to be synthesized. The next steps would be to learn to insert them into and delete them from genetic systems. When, in the distant future, all this comes to pass, the temptation to change our biology will be very strong.
>
> – HAR GOBIND KHORANA, 1968

One of the history of science's most enduring mysteries is the disappearance, without trace, of a unique manuscript. The legendary star map created by the ancient Greek astronomer Hipparchus was the earliest known attempt to chart the cosmos with numerical precision. The missing star catalogue eluded discovery for nearly two millennia. Its existence was known only through scattered references embedded in later works, such as Cicero's *Aratea* – a Latin rendering of Aratus's 275 BCE astronomical poem *Phaenomena* (Appearances).

The lost manuscript holds great significance, as it represents humankind's first known attempt at the 'mathematisation' of nature – the systematic application of mathematical principles to the prediction of natural phenomena. It marked a turning point in the history of science, transforming astronomy from a descriptive, observational enterprise into a predictive one.

Hipparchus composed his star map at least three centuries before Claudius Ptolemy published his astronomical treatise *Almagest* (The Great Treatise) in the second century CE. Hipparchus's compendium detailed the positions of nearly every star and heavenly body in the cosmos visible to the naked eye. It blended Babylonian celestial modelling with Greek geometry and was a work of extraordinary consequence. The search for this missing star map has been the holy grail of historians of science for centuries. The consensus, however, was that the map had been lost for ever.

But this assumption was re-evaluated in 2012, when Peter Williams, a biblical historian at the University of Cambridge, was looking for a summer project for his student, Jamie Klair. He suggested that Klair examine a manuscript known as the *Codex Climaci Rescriptus*. This Christian manuscript, written in the ancient Semitic language of Syriac by scholars at the Greek Orthodox St Catherine's Monastery in the Sinai Peninsula of Egypt between the sixth and eleventh centuries CE, comprises a total of 146 folios. Assembled over time from various sources, in 2014 the folios were transferred to the Museum of the Bible, in Washington D.C.

The *Codex* is a palimpsest – a manuscript in which earlier 'lower text' writing is scraped away and overwritten with 'upper text'. In this case, Syriac text was written over recycled parchments that had originally contained Christian Palestinian Aramaic and Greek script dating back to the sixth century CE. This lower text had been mostly removed when the parchment was scraped clean to make way for the new Syriac upper text. Yet during his examination, Klair noticed faint traces of Greek astronomical material concealed beneath the Syriac layers – remnants of a text that had not been fully removed at the time the parchment was being prepared to receive the new Syriac writing.

Five years later, with the assistance of Victor Gysembergh and Emanuel Zingg at the Centre National de la Recherche Scientifique in France, the folios underwent multispectral imaging. This process uses different wavelengths of light in conjunction with computational methods, to recover hidden text. It was then that Williams noticed something remarkable. Concealed beneath the surface text were star coordinates, including the following: *Ο στέφανος ἐν τῷ βορείῳ ἡμισφαιρίῳ κείμενος κατὰ μῆκος μὲν ἐπέχει μ° θ̄ καὶ δ´ ἀπὸ τῆς ᾱ μ° τοῦ σκορπίου ἕως ῑ <καὶ> δ´ μ° τοῦ αὐτοῦ ζῳδίου*, which translates as: 'Corona Borealis, lying in the northern hemisphere, in length spans 9° 1/4 from the first degree of Scorpius to 10° 1/4[8] in the same zodiacal sign (Scorpius).'

The precision of the recovered coordinates, and the way they reflected the Earth's axial wobble, which causes stars to slowly shift their positions in the sky over time, allowed the text to be dated to 129 BCE, the exact same time that Hipparchus was known to have compiled his star map. It confirmed Williams's original suspicion that they had unwittingly stumbled upon the holy grail – an original fragment of Hipparchus' elusive star catalogue.

While Hipparchus' mathematical imagination was fixed on the stars, the cosmos of modern biology is populated not with heavenly bodies but with genomic sequences – the biological code scripts that compute the nature of species. This universe cannot be charted with the naked eye or the lens of a telescope. Instead, it is rendered visible through two emerging instruments: AI-augmented genome design and genome synthesis. Together, these technologies promise to achieve something previously impossible – to establish a mathematical grammar of species creation, with a predictive acuity reminiscent of Newton's celestial mechanics. If successful, biology would be transformed into a predictive science. Life would become, at least in part, computable, and perhaps even programmable.

One might imagine a biological analogue of Hipparchus's *Star Catalogue*: a comprehensive record of the genomes of all existing species on Earth, made visible through DNA sequencing technologies. This 'Species Catalogue' of biological possibility would also include all extinct species that might once have been visible but have disappeared from view. Were we to extend this concept to include not only the past and present, but the future as well – to encompass all possible species based on the rules of life as we currently know them – the Species Catalogue begins to resemble what we have previously called Fred's Library. It accommodates an endless sea of unrealised species that fly, glide, whirr, whizz, run, walk, waddle, hop, skip, jump, climb, float, and swim across its iridescent mathematical landscapes.

The current problem, however, is that biology lacks the kinds of universal mathematical laws that structure the physical sciences and allow them to make predictions. Biology does not, for example, have laws corresponding to Gauss's law, Faraday's law of induction, or Newton's laws of motion. Darwin's theory of evolution by natural selection, while providing mechanistic explanations for how natural species originate and change over time, is a descriptive theory rather than a predictive one. It tells us how species have evolved but cannot anticipate the types of species that might emerge in the future. While rich with data, biology lacks a coherent mathematical framework for organising, generating, and forecasting biological knowledge.

But let us now take the idea of the Species Catalogue one step further, and imagine a version that transcends Fred's Library. This is because it includes not only all known and future forms of life constructable using existing biology, but also life built from entirely new chemistries. The code scripts of these hypothetical forms of life would encode their information not in DNA and RNA, but in unfamiliar genetic alphabets. The artificial biological scripts generated from these alphabets would be profoundly alien to life's natural biology. A completely new astronomical chart – not of stars but of species – would need to be configured to accommodate them.

These may, in time, supplement life's familiar natural architecture, using, for example, non-natural amino acids to expand the natural set and to assemble organisms from repertoires of amino acids that extend beyond those currently used by living organisms. Some may not use amino acids at all, building life's machinery from completely new classes of polymers. It would be a miserable universe indeed that was unable to accommodate such diversity. Fred's Library – the theoretical space of organisms constructed from conventional building blocks – occupies just a tiny corner of this mathematical hyperspace, which accommodates organisms built using alternative chemistries and molecular scripts.

When Hipparchus looked up at the night sky, he was able to discern only the faintest trace of the complexity of the known universe. Likewise, as we peer into the vast space of biological possibility, we are able to observe just an infinitesimal fraction of its expanses. Within this framework, our own species is an unremarkable and irrelevant detail. As the philosopher Bertrand Russell stated in *The Impact of Science on Society* (1952): 'Life is a brief, small, and transitory phenomenon in an obscure corner, not at all the sort of thing one would make a fuss about if one were not personally concerned.'

The parallels between astronomy and biology are more than just metaphorical. In 1609 Galileo Galilei, modifying a rudimentary design that had been invented a year earlier by the Dutch eyeglass maker Hans Lippershey, constructed the first astronomical telescope. It enabled him to visualise worlds extending far beyond the immediately visible skyline documented by Hipparchus. With the launch of the Hubble Space Telescope centuries later in 1990, the extent of the visible universe dramatically expanded. Previously invisible regions of sky were now seen teeming with sumptuous, spiralling, sparkling galaxies. The deployment, in 2021, of the James Webb Space Telescope – the most powerful telescope ever built – unveiled further breathtakingly expansive seas of multicoloured galaxies lurking in the invisible corners of space–time.

To map the expanses of the Species Catalogue, comparable technological leaps will be required. Today's tools allow us to construct the rudimentary beginnings of a Hipparchian-style chart of biological diversity: a genomic map encompassing the estimated 1.9 million known existing species catalogued in the Encyclopedia of Life (EOL) database, as well as the 10,000 or so new species discovered each year. But like Hipparchus's star chart, this contemporary compendium of the biological 'night sky' captures only what is visible.

Known species are numerically dwarfed by those that have been lost. Between four and thirty billion species are estimated to have gone extinct over the course of life's history. Only a fraction of these have left physical remnants that can be sequenced. Were we to add in the estimated four million known bacterial species, the trillions of viral species thought to exist, and the likely astronomical number of extinct viral and bacterial species, the Hipparchian map of potentially observable biological possibility would buckle at its seams.

If we could define the generative rules of biology as it is – its natural grammar – and eventually extend that understanding to encompass the rules governing hypothetical species built using alternative chemistries prescribing life as it might be, we would possess not merely a theory but a new kind of predictive instrument. A visualisation device powerful enough to gaze further into the depths of the Species Catalogue than ever before. It would enable us to furtively glimpse the teeming worlds of possible species, the blueprints of living forms that have never been, and likely never will be, realised. Each of these latent species has a concrete mathematical reality, and could be transformed into the flesh and blood of a living species. The eventual mastery of life's generative grammar, enabled by the telescopic capacities of artificial intelligence applied to vast biological datasets and genome synthesis, may come to represent humanity's crowning scientific achievement.

Should we wish to journey even deeper into the Species Catalogue to explore potential species made from substances unlike anything used historically by life on Earth, we will need an even more powerful kind of instrument. One capable of visualising not only genetic variation within known types of biology, but of venturing into new worlds of biochemical innovation. Such a device would allow us to scan the alternative futures of biology itself. To envision life not as it is, but as it could be. Life forged from foreign molecular logics that might have arisen elsewhere in the universe, or may perhaps one day exist here on Earth. Such a device would be capable of effortlessly transporting us across the unimaginably expansive and timeless landscapes of the universe of all possible species.

We are currently in the Hipparchian phase of biology, staring up at the edges of the known visible biological universe, aware only of what we can see. But we stand on the threshold of a magnificent journey of boundless extent, one that could bring the invisible universe of biological possibility into full resolution.

What we call 'natural' today refers only to those biochemistries that have originated as a result of evolution by natural selection. But this distinction is, in many ways, a product of history. Had life evolved under slightly different conditions, the chemistries that today we think of as 'artificial' might have formed the basis of life on Earth. The distinction between 'natural' and 'artificial' is therefore somewhat arbitrary, an artefact of history rather than an inevitability.

As we embark on this cartographic endeavour of almost unfathomable scale, we find ourselves – like Antonio Pigafetta, the chronicler of Magellan's voyage in the sixteenth century to circumnavigate the Earth – commencing our journey without the benefit of maps. Our voyage through the Species Catalogue to explore its previously uncharted galaxies of unknown species will be a spectacular endeavour. The denizens of this mathematical space of possibility will reveal life's deepest mysteries. This fantastic journey will transform humankind's understanding of its place in the universe, extending the reach of human possibility in ways hardly imaginable. How we choose to apply the unique map that we create in the new age of synthetic biology will shape the future trajectory of our species. It will test the limits of our judgement and imagination, and strain our moral compass.

We should recall, however, that the Species Catalogue does not stand alone. It is embedded within and inseparably connected to a series of higher-dimensional informational spaces. These include not just genetic blueprints but the superimposed layers of epigenetic, developmental, learning and cultural information, as well as the code written into the genomes of the microbial and viral entities that form an informational shadow world, which impacts us in complex ways. The full mathematical space capturing every informational feature of every possible species is the catalogue of all possible informiomes, which we might call the 'Informiome Catalogue'. This metacatalogue of the mathematical possibility for all possible species predates biology itself, originating at the time of the creation of the universe. The mathematical structures in the Informiome Catalogue describing potential organisms are both timeless and immutable.

It is to the dazzling complexity of the finite world of living species, as we know it, that we must first turn before embarking on any exploration of the infinite. For it is the natural life around us that sustains and gives meaning to human existence. In order to go beyond history, we must not erase it.

In his apocalyptic 1922 poem *The Waste Land*, T. S. Eliot depicts the ruins of a confused and fallen world, undermined by the devastation of the First World War. It is a place of fractured memories and obliterated aspirations, where a tormented nature baulks at the incivilities of human behaviour. A portrait of a broken world, violated by the excesses of modernity. 'The dead tree gives no shelter,' Eliot wrote, 'the dry stone no sound of water.'

And yet, failing to learn from the past, we continue along the same path, unable or unwilling to change. The consequences are no longer distant or abstract. We are rapidly destroying our world, undermining the conditions that make life possible, and hurtling towards a global catastrophe of our own making. The Earth's climate – driven by the relentless build-up of carbon dioxide and methane emissions – is now warmer than at any point in the last 125,000 years. This has led to an increase in the frequency and severity of floods, droughts, and other extreme weather conditions. Sea levels are rising, glaciers are disappearing, and wildfires are devastating landscapes across the world – from the Australian bush and the Amazon canopy to the hillsides of California.

Meanwhile, atmospheric carbon dioxide has reached levels not seen in more than two million years. The situation beneath the Earth's surface is no less troubling. Wells around the world are in danger of running dry. If groundwater levels decline by just a few more metres, it will threaten the drinking water supply and irrigation necessary for agriculture for billions of people worldwide.

There can be no infinite without the finite. The worlds we imagine – or may someday engineer – depend on the continuity of the one we currently inhabit. And yet the natural world is faltering. Wild species are vanishing, supplanted by their domesticated counterparts. Forests are receding. Coral reefs are bleaching. Wildernesses are fracturing. Permafrost is thawing. Non-native species are spreading. Habitats are destabilising. Snowmelt is accelerating. Migration corridors are narrowing and closing. Across the world, entire ecosystems are unravelling.

The effects of this are visible everywhere. As their habitats shrink, wild animals are increasingly being forced into densely populated areas at the margins of human settlements. Oceans are warming to levels that, even recently, were unthinkable. The polar ice caps are melting – none more ominously than the Thwaites Glacier in Antarctica, known as the 'Doomsday Glacier'. Across the world, animals are acting strangely. Swarms of jellyfish are infiltrating the Sea of Japan. Lyme disease-bearing ticks are migrating northward to Canada. Bats are dying en masse in Australia.

But climate change is not just an isolated environmental phenomenon; it is destabilising societies and undermining the fragile livelihoods of already vulnerable communities. Acting as a catalyst for poverty, it also fuels warfare through political unrest. All this is unfolding while the global human population continues to expand relentlessly. It is projected to reach 9.7 billion by 2050, further exacerbating the problem and placing an even greater strain on the already fragile ecosystems that sustain us.

The uncontrolled combustion of fossilised organisms on a planetary scale; methane emissions from cattle and flooded rice paddies; the proliferation of plastic pollution, microplastics, sunscreens, insecticides, and agrochemicals; the wanton destruction of habitats; and the accelerating loss of biodiversity, together are propelling Earth's species into a death spiral from which they may never return. The root cause lies not in biology or geological events, but in evolution's most dangerous invention: the informational structures generated within, transmitted between, and parasitic upon human minds. We call these 'ideas' and 'ideologies'.

These mental constructs – shaped by ambition, propelled by corporate greed, and accelerated by political expediency – have driven an unprecedented assault on the natural world. Our species' intellectual competence, it seems, has not been accompanied by a corresponding measure of wisdom, restraint, or common sense.

Each of Earth's species, however inconspicuous or seemingly irrelevant – from the ungainly flight of a heron to the most translucent, gelatinous algae – deserves our attention and protection. They are woven together into ecosystems of staggering complexity that we barely understand. Together, they form the essential fabric of life on Earth.

In his poem 'An August Midnight' written in 1899, Thomas Hardy recounts an encounter with four insects that flit across his desk one summer night, including 'A sleepy fly, that rubs its hands'. Initially dismissive of these diminutive 'winged, horned, and spined' creatures, he gradually comes to see them differently. His uninvited companions are keepers of unique and valuable knowledge inaccessible to him. This change in perspective culminates in a moment of revelation: '"God's humblest, they!" I muse. Yet why? They know Earth-secrets that know not I.'

Like Hardy, we too must come to see that the genome of every species contains unique Earth-secrets – code scripts stretching far back in time that are worthy of preservation in their own right. Some of these secrets may, by chance, prove invaluable to humanity. The discovery of CRISPR, a revolutionary gene-editing system derived from an obscure bacterium found in the salt flats of Alicante, shows how unlikely and unexpected Earth-secrets may have transformative potential for humankind. Who could have imagined that a technology capable of modifying the human genomes would emerge from such an apparently inconsequential microorganism? One can only wonder what other unknown treasures reside within the code scripts of Earth's species, waiting to be discovered. The loss of even a single species is not just an ecological and cultural tragedy but a reckless forfeiture of knowledge and of untold possibilities.

The path forward lies not in the subjugation of nature, but in its stewardship. Protecting Earth's climate, habitats, microenvironments, and the fragile structures of its diverse, interconnected species requires urgent action. This includes conservation, habitat preservation, rewilding, and the development and adoption of sustainable technologies. Synthetic genomics has the potential to play a significant role in this effort, offering new tools for repair and resilience and providing sustainable alternatives to existing energy sources and materials.

Saleemul Huq, the Pakistani-Bangladeshi climate visionary, devoted his life to the pursuit of climate justice and global climate action. He was an advocate for the world's most vulnerable, and consistently highlighted the disproportionate impact that industrialisation and climate change have on low-income countries. He was also a champion of inclusive advocacy, believing that the communities most affected by climate change should be empowered to help define the solutions.

It has been estimated that reducing global beef consumption by just 20 per cent could cut deforestation related to cattle grazing by half. But no single nation state can solve the climate-change problem alone. Climate action requires alignment, consensus, and a sense of shared responsibility. Governments must make a sustained and coordinated effort to end poverty, to prevent and reverse global warming, and to protect the ecosystems upon which life depends. Climate change is the most urgent issue of our age. It is of existential importance, both for our species and for all others. Without a thoughtful and definitive resolution of this issue, the future of species will be bleak indeed. Nature is not a commodity to be endlessly exploited. Neither is it inexhaustable. It too has a breaking point.

Throughout Earth's history, five mass extinction events have led to a succession of different worlds, each with radically different species structures. The most catastrophic of these – the Permo–Triassic boundary extinction event occurring some 250 million years ago – was the closest that life has ever come to total annihilation. For some time, scientists believed that the event, much like the conspiracy in Agatha Christie's *Murder on the Orient Express* (1934), in which all twelve suspects turned out to have perpetrated the crime, had multiple causes: widespread ocean anoxia, acid rain, runaway climate change, a sudden and massive release of methane from the ocean floor, and even a meteorite strike. But recent evidence suggests that the most likely cause was a terrestrial one – a massive volcanic eruption in the Siberian Traps of eastern Russia.

We are now in the early stages of what many scientists believe is the sixth mass extinction. Unlike those that preceded it, this one is exclusively human-generated, and is occurring at a rate that may be as much as eighty times higher than the geologically defined extinction rate. Historically, fewer than two mammalian species have gone extinct per million years. But in the last five centuries, at least eighty mammalian species have disappeared. While extinction is a natural phenomenon, the speed and scale at which it is now occurring exceeds anything in the palaeontological record.

In 2011, the palaeontologist Anthony David Barnosky warned that up to three-quarters of animal species alive today could vanish within the next three centuries. Evolution, it seems, is struggling to keep pace with the rate of the environmental destruction we have created. When we ruthlessly and indifferently crush nature, we simultaneously undermine our own humanity.

The genome of the Scottish wildcat, *Felis silvestris*, the last remaining wildcat species in Great Britain, has become so corrupted by interbreeding with domestic cats that it is now considered functionally extinct. The magnificent ivory-billed woodpecker, *Campephilus principalis* – once relatively common in an area stretching from the Carolinas through the south-east of the US to Texas, until habitat loss and hunting led their numbers to sharply decline – was, by 2021, teetering on the edge of extinction. There have, however, since then been a few isolated sightings in Louisiana, which have been documented with grainy photographs. In China, the dhole *Cuon alpinus*, also known as the red wild dog, lingers in dwindling pockets of habitat with only a few hundred individuals remaining. As populations shrink their genetic diversity is eroded, which accelerates the cycle of inbreeding and extinction.

Nowhere is the loss more pronounced – or more neglected – than in the world of insects. We are living through the most significant insect die-off since the Cretaceous–Paleogene extinction event that wiped out the dinosaurs. Insects account for more than half the known species on Earth. Their disappearance, driven by habitat loss, pesticides, and pollution, should concern us all. The American biologist Edward O. Wilson once stated that it's 'the little things that run the Earth'. Unlike the calamities of the ancient past – including massive volcanic activity – the catastrophic loss of species is not inevitable and may be prevented if we take immediate action. We will need to reimagine our relationship with the living world. The future of life on Earth will depend on this.

In his 2016 book *Half-Earth*, Wilson proposed an intriguing solution: that at least half the Earth's land and seas should be returned to nature. This rewilding effort, he argued, could preserve over 80 per cent of Earth's species. But to achieve this, we will need to substantially reduce Earth's population, shift to more sustainable modes of energy and food production, eat less meat, and reduce our reliance on materials like plastics, which are contaminating ecosystems across the world. Importantly, governments will need to cooperate to prioritise nature's well-being by promoting biodiversity and habitat preservation.

Should we wish to get some idea of the kinds of remarkable and unworldly potential species that populate the Species Catalogue, we need only examine the strange and exotic creatures already known to science – both living and extinct. The giant penguin *Kumimanu biceae*, for example, which stood 1.77 metres tall and weighed over 100 kilograms, roamed the waters around New Zealand some 60 million years ago. Then there's *Heracles inexpectatus*. This flightless parrot lived around 15 million years ago and, at almost a metre high, was taller than a St Bernard dog and weighed twice as much as the largest existing parrot species.

At the opposite end of the spectrum is *Brookesia micra*. This Lilliputian lizard, measuring just twenty-four millimetres in length, inhabits the forests of northern Madagascar and is about the size of a matchstick head. The 'uncrushable' diabolical ironclad beetle, *Phloeodes diabolicus* is another improbable curiosity. Native to North America, it is scarcely larger than a grain of rice. Thanks, however, to the remarkable design of its wing case – whose components lock together 'like a 3D jigsaw puzzle' – it can withstand crush forces up to 39,000 times its body weight.

Other oddities include the pea-sized frog *Microhyla nepenthicola*, which is just 11 millimetres long, and found in Borneo and parts of Asia, Africa and Europe. Then there is *Clitarchus hookeri*, a New Zealand stick insect capable of reproducing in the presence or absence of a mate, and *Dryocopus pileatus*, the pileated woodpecker, which hammers its beak against tree trunks at rates of up to twenty times a second – equivalent to slamming your face against a wall at twenty-five kilometres per hour without sustaining concussion.

The oceans also harbour eccentric and unexpected organisms. The other-worldly weedy seadragon *Phyllopteryx taeniolatus*, for example, drifts across Australia's shallow coral and rocky coastlines. Its long, leaf-shaped appendages, tapered tail, and elongated snout give it an appearance reminiscent of both a miniature mythical beast and an animated Victorian ornament.

The invisible world discernible through a microscope is no less fantastical. The microbe *Idionectes vortex*, for example, resembles a microscopic flying saucer. And sometimes the invisible can become visible. The giant, sulphur-oxidising bacterium *Thiomargarita magnifica* (meaning 'sulphur pearl' in Latin), with an average length of a centimetre, is the largest bacterium ever described. Visible to the unaided human eye, it was discovered in the early 2010s growing underwater on rotting red mangrove leaves on the French West Indies island of Guadeloupe. Being more than 4,000 times larger than a standard microbe, it challenges our preconceptions about the theoretical limits of bacterial cell size.

This menagerie of bizarre creatures – extending across every scale – with their improbable anatomies, eccentric behaviours, and enigmatic lifestyles, offers a fleeting glimpse into the range of organismal possibility existing at the edges of our imagination. But they are only tiny specks in the secret ocean containing all possibility.

Humans, too, are represented in this compendium, which accommodates a vast mathematical space containing the exhaustive sequences defining all human possibility – only the tiniest fraction of which has ever been realised. Each potential human species has its own distinct anatomy, behaviour, and potential culture. While the fossil record provides rudimentary insights into some of the different anatomical ways of being human, it tells us little about the emotional, psychological, or cognitive make-up of alternative human species, or other aspects of their nature. There are over 350,000 known species of beetles, while currently there is only one living species of human. But potential human diversity is just as dazzling. Imagine the exhilaration of perusing a library of 350,000 different species of human.

In 1994, a team led by the anthropologist Tim White at the University of California, Berkeley, uncovered the fragmented 4.4-million-year-old skeleton of *Ardipithecus ramidus*, known as 'Ardi'. Ardi – likely the precursor of all known human species – represents one of the earliest blueprints for a human-like creature. Painstakingly extracted from the reddish-brown sediment of Ethiopia's Afar Rift Valley, a once-wooded landscape teeming with monkeys, centipedes and antelope, Ardi, with its paradoxical jumble of anatomical characteristics, looked more like a character from *Star Wars* than an ancestor of *Homo sapiens*.

We now know that as recently as 50,000 years ago, we shared the planet with at least three other human species. They were not subtle variations on a theme, but wholly distinct – as different from one another as a pigeon is from an ostrich, or a flamingo from a pelican. Among them was *Homo floresiensis*, also known as 'hobbit man'. Living until around 18,000 years ago, and overlapping with *Homo sapiens*, this unique species of human was just over three feet tall, chinless, and large-toothed. It lived on the Indonesian island of Flores, alongside Komodo dragons, giant rats, and pygmy elephants.

Unlike the homebody hobbits who seem never to have strayed far from their island idyll, the Neanderthals – *Homo neanderthalensis* – with their pronounced brow ridges, protruding jaws, broad ribcages, and low, sloping foreheads, were the backpackers of the archaic human world. They ranged widely until some 40,000 years ago, when they abruptly disappeared from the fossil record. The Denisovans, the most enigmatic of the human species who overlapped with us and disappeared around 50,000 to 20,000 years ago, are known from just a few fragments of finger bones and teeth.

Had things been different, it might have been a descendant of a hobbit man, Denisovan, Neanderthal, or some other extinct member of the hominid family writing this book, speculating about the sudden disappearance of a curious species known as *Homo sapiens*. It is hard to know what each of these alternative species of human might have gone on to achieve. Would 'hobbit man', for example, eventually have been capable of composing an opera like Mozart's *The Magic Flute* (1791) – conjuring up a fictional character like the Queen of the Night to life? Might they eventually have formed rock bands and written a surrealistic, drug-inspired song like the Beatles' 'Lucy in the Sky with Diamonds' (1967) or Lady Gaga's catchy and theatrical pop anthem 'Poker Face' (2008)? Could Neanderthals, given time, have invented digital computers, aeroplanes, and smartphones? Might they even have surpassed us, both intellectually and morally, and generated more nuanced and compassionate cultures?

These questions are unknowable. But thanks to unprecedented developments in ancient DNA sequencing, we can now peer through the keyhole of deep time to gain some insights into these vanished species and explore their mysteries. Complete, and in some cases fragmentary, genome sequences have been retrieved from several extinct human species. Most notably, in 2014, Svante Pääbo published a complete, high-resolution genome sequence of a Neanderthal, opening a window into a vanished world. Beyond our human relatives, ancient DNA also allows researchers to virtually reconstruct entire ecosystems.

While extinction has, for most of natural history, been an irreversible event, the recovery of genetic sequence information from long-extinct species has led to the suggestion that there may be value in resurrecting – de-extincting – and rewilding species such as the woolly mammoth, Pyrenean ibex, dodo, great auk, and Tasmanian tiger (thylacine). But this prospect remains fraught with ecological and ethical uncertainties, and requires debate and circumspection.

The biotechnology company Colossal Biosciences, based in Dallas, Texas, and founded in 2020 by the Harvard geneticist George Church and tech entrepreneur Ben Lamm, is in various stages of resurrecting extinct species. Among the creatures they hope to bring back are the woolly mammoth and the dodo, as well as the critically endangered tooth-billed pigeon, native to the Samoan rainforest. Backed by several Hollywood film executives and a roster of high-profile investors, Colossal claims that it will 'make extinction a thing of the past'.

They plan, for example, to take a mammoth genome and implant it into a living Asian elephant via a synthetic embryo – created using pluripotent stem cells capable of developing into any cell in the elephant's body. This is expected to lead to the birth of creatures resembling mammoths, which have not existed on Earth for more than 4,000 years. According to Colossal, these will 'walk like a woolly mammoth, look like one, sound like one, but most importantly [they] will be able to inhabit the same ecosystem previously abandoned by the mammoth's extinction'.

On 7 April 2025, accompanied by an international media fanfare, Colossal announced that they had 'successfully restored a once-eradicated species through the science of de-extinction'. The species in question was the dire wolf, whose genome had been partially reconstructed using DNA extracted from two Ice Age fossils. A distant relative of modern wolves, the dire wolf vanished from the grasslands and woodlands of North America around 10,000 years ago.

Sceptics were quick to denounce the work, claiming that Colossal's dire wolves were imposters – little more than grey wolves in dire wolves' clothing. One critic argued that Colossal had shown a 'casual disregard not only for truth, but for life itself'. Carl Zimmer, writing in *The New York Times*, expressed a very matter-of-fact view: 'It's a grey wolf clone with 20 dire-wolf gene edits and some dire wolf traits.'

Beth Shapiro, Colossal's chief executive officer, defended the endeavour by invoking the 'morphological species concept' – the idea that a species can be defined by its physical characteristics – arguing that the de-extincted dire wolves were 'functional copies' of the historical version. 'It's not possible', she explained, 'to bring something back that is identical to a species that used to be alive. Our animals are gray wolves with 20 edits.' But there are in fact thousands of genetic differences separating grey wolves from their long-extinct dire wolf relatives. While gene editing can produce convincing facsimiles, it cannot resurrect the actual historical species.

There was, additionally, disagreement about whether the resurrected gene-edited wolves even looked like historical dire wolves. The coats of the originals are thought to have been reddish, in contrast to the sparkling white fur of the dire wolf pups – Romulus, Remus and Khaleesi – created by Colossal. Currently housed on a 2,000-acre reserve at an undisclosed location, their purpose and future remains uncertain. Phie Jacobs, writing for *Science*, noted that while dire wolves were once vital components of their ecosystems, 'the habitats they lived in – and many of the animals they predated – no longer exist'.

Aside from their molecular and anatomical characteristics, it is hard to imagine how aspects of historical dire wolves that are likely not genetically encoded, such as their hunting methods, pack dynamics, and vocalisations, could ever be recreated. The same holds true for other de-extinction candidates. The historical 'song' of the dodo, for example, has been lost for ever.

In light of such limitations, critics argued that Colossal risked misleading the public by suggesting that true de-extinction is possible. There was, additionally, concern that their work might inadvertently undermine the sense of urgency surrounding conservation efforts focused on the protection of endangered species.

Current technology does not yet allow the genomes of extinct organisms to be written out from scratch. Instead, the genomes of extinct organisms are recreated using a method called comparative genomics. In the case of the woolly mammoth, for example, DNA is extracted from preserved remains and then compared with sequences obtained from its closest living relative, the Asian elephant. These are then used to fill in any missing gaps resulting from the degradation of the mammoth's sequences. Once the genetic differences have been mapped, the Asian elephant genome is then modified at each of the relevant sites using gene editing to generate a 'proxy mammoth' genome.

Evolution has imposed considerable constraints on human possibility. The developmental pathways selected millions of years ago have homed in on a relatively narrow repertoire of form and function. And yet, the depth of the possibility inherent within the Species

Catalogue – visible in the eccentric forms of the fragmented remains of extinct human species – suggests that there are countless different ways of being human: anatomically, biochemically, cognitively, behaviourally and culturally. Some familiar; others fantastical.

Our current body size, for example, precludes the possibility of flight. And yet it is easy to imagine gnat-sized humans hovering in the air, or miniature insect-like humans skipping across the surface of water like water boatmen. The impulse to blend the anatomical traits of other species with those of humans appears to be hard-wired into our imagination. Such themes occur repeatedly in literature, mythology and in the religious iconography of Judaism, Christianity, and Islam.

In the Old Testament of the Bible, the prophet Ezekiel described a vision in which he saw four living creatures. Each of these cherubs had the likeness of man, but with four faces – that of a man, a lion, an ox and an eagle – four wings, calf-like feet, and a body that sparkled like burnished bronze. The Prophet Mohammed is similarly said to have been carried on the Night Journey (*al-'Isrā' wal-Mi'rāj*) from Mecca to Jerusalem and then to the heavens by the supernatural hybrid creature Burāq. Burāq is described in Hadith literature and later Mughal and Ottoman depictions as variously half-mule, half-donkey, with a human face and wings at its sides.

The hilltop Cathedral of San Giusto in Trieste, Italy, contains two intricate Byzantine mosaics decorating the eastern apse of the cathedral. These depict the diaphanous bodies of the archangels Gabriel and Michael. Sumptuous, elegantly curved and delicately feathered wings of generous proportions have been appended to their lithe figures. These fabulous images of angels, created in a pre-scientific age – when genetics was still unknown – anticipate the possibility for the infinite malleability of organismal form.

The Species Catalogue also accommodates different potential versions of humans who, while looking and behaving like us, are wired differently at the molecular level. These are of particular interest, as they include a subset of humans in whom disease is absent or much attenuated. They also contain versions with greatly extended healthy lifespans. The genomes computing these individuals hold the deepest secrets of human health and longevity – the keys to the future of clinical medicine.

To explore these hidden possibilities, we will need to move beyond reading genomes to modelling them computationally using artificial intelligence. To test these design principles, perhaps by making modest modifications to human genomes to increase healthy longevity and reduce the potential for disease, will require the ability to construct new versions of genomes. But while our current genome-writing capabilities allow us to write out the genome of a yeast, a human genome – which is 266 times larger – remains a formidable challenge. And yet it will eventually be possible to construct not only a single human genome from scratch, but also vast libraries of human genomes containing hundreds of thousands, if not hundreds of millions, of alternative human genome sequences. The challenge will be how to test the impact that each of these variants might have on human health in an ethical manner.

While synthetic human genomes could be studied in isolated cells, the information obtained from such systems would be limited. Testing rewritten, or reimagined genomes in actual humans, on the other hand, would be ethically indefensible – as well as being impractical, given our protracted lifespan. One of the most profound questions facing the future of human biology and medicine, therefore, relates to how we might ethically investigate the effects of alternative versions of genomes that have the potential to impact health and disease.

Digital simulations of biological systems – known as 'digital twins' – will likely provide the most practical way forward. Yet, for the time being, virtual biology remains an imperfect surrogate for the real thing. It is not even clear that the human genome is fully computable. For now, 'wet biology' – the study of actual living organisms – remains essential. If we wish to understand how modified human genomes operate, we will need to design, construct, and test them in living systems.

The best existing surrogates are so-called 'model organisms', which replicate aspects of human biology with varying degrees of success. Their ability to function as proxies is based on the fact that many pathways and biochemical mechanisms have been conserved between species across evolutionary time. Of the various models tested to date – from bacteria and yeast to flies and worms – it is the house mouse, *Mus musculus*, that has emerged as the most robust model for studying human health and disease. When its full genome sequence was published on 5 December 2022, the astonishing similarity to our own was immediately apparent – despite the fact that the two species last shared a common ancestor around 75 million years ago.

Approximately 95.5 per cent of the protein-coding sequences in mice and humans are identical. Yet, at the level of individual genes, the extent of the similarity between corresponding genes is more variable, ranging from 60 per cent to 99 per cent. In contrast, only about 28 per cent of the non-coding sequences are shared between the two species, indicating that the regulatory regions of the human genome have changed extensively. It is these differences in the regulatory, non-coding regions that, for the most part, explain why a mouse is a mouse and a human is a human.

While in liver cells the pattern of gene expression is nearly identical in mice and humans, in other cell types the patterns diverge. For example, around half of the mouse genome's transcription factor binding sites – critical for regulating gene activity – are absent in humans. Another quarter have migrated to different regions of the human genome. This phenomenon is most pronounced in enhancers, DNA sequences that initiate gene activity and are often located far from the genes they control.

One way of addressing these differences is to transform the mouse into a more compelling surrogate for human biology. Given the broad architectural similarities between their genomes, this might be achieved by selectively 'humanising' defined coding and non-coding regions of the mouse genome with their human equivalents. In its most straightforward implementation, regions of the mouse genome known to be linked to human disease would be overwritten with human versions. The replacement of mouse sequences with human versions would result in partially 'humanised' mouse models.

In 2023, Jef Boeke and Weimin Zhang at the Institute for Systems Genetics in New York demonstrated the feasibility of overwriting mouse genome sequences with human versions. The technique they described, called mSwAP-In (mammalian switching antibiotic resistance marker progressively for integration), enables extensive rewriting of mouse embryonic stem-cell genomes. When these rewritten cells are injected into early-stage embryos, they can develop into living mice with genomes that contain human genetic code script.

To demonstrate the utility of this method, the team engineered a mouse in which a 180,000-nucleotide region including the ACE2 gene and its associated regulatory sequences, was overwritten with the human equivalent. This gene encodes the ACE2 protein, which the SARS-CoV-2 virus uses to enter human cells. When exposed to the virus, these partially humanised mice responded more like infected humans than unmodified mice. By removing some of the evolutionary barriers that prevent mouse genomes from mimicking human biology, such models are likely to make substantial contributions to our knowledge of biology and drug development.

Entire systems, such as the mouse immune system or even whole chromosomes, could, in principle, be replaced with human versions. A mouse with a human immune system, for example, would be far better suited to testing therapies that manipulate human immunity. One could imagine a mouse genome in which every gene is replaced with its corresponding human version. Sydney Brenner once showed that if you put pufferfish genes into a mouse, the mouse can 'read' them correctly. This suggests that a mouse genome might similarly be able to read human versions of its genes, provided that the non-coding sequences retain some degree of similarity.

A mouse operating with human versions of each of its genes, while retaining its natural non-coding regions, would provide an opportunity to study the function of human non-coding DNA, which could then be systematically inserted into the hybrid genome. The extent to which a mouse genome should be humanised, however, is not just a technical issue but also a philosophical and ethical one. Given that we do not yet understand the basis of conscious awareness, the extensive humanisation of both coding and non-coding sequences risks creating a mouse with some degree of consciousness awareness – the presence of which may not be readily detectable.

Meanwhile, in what synthetic biologist Tom Ellis at Imperial College London described as 'a wake-up call for those who think synthetic genomes are only for microbes', a team in China has turned its attention to the plant kingdom. Led by Junbiao Dai at the Agricultural Genomes Institute at Shenzhen and Yuling Jiao at Peking University, in 2023 they began the first ever attempt to synthesise the genome of a multicellular organism. Their focus: *Physcomitrium patens*, a moss with a genome comprising twenty-six chromosomes. Starting with its approximately 500-million-base-pair genome, which is around forty times larger than that of yeast and 6.4 times smaller than the 3.2 billion base pairs of the human genome, the aim of the synthetic moss genome project, also known as SynMoss, is to engineer a synthetic, streamlined version of the natural moss genome.

By rewriting the genome of a multicellular organism, the team aims to identify the DNA sequences essential for growth and survival, while introducing the possibility of new capabilities relevant to industrial, agricultural and medical applications. Moss is an attractive multicellular system for genome engineering, as a single cell can grow into a complete organism. As a result, editing the genome of just a single cell can reprogramme the entire organism.

The team began by synthesising and replacing the 155,000-nucleotide-long short arm of moss chromosome 18, reducing its size by 56 per cent. Surprisingly, removing large sections of its non-coding 'dark matter' DNA had little impact. Although the refactoring disrupted some of the genome's three-dimensional chromatin structure and altered the activity of certain genes, the overall functionality of the organism remained intact. By 2024, synthetic moss cells carrying the engineered genome had been shown to grow normally and produce spores.

Success in synthesising an artificial genome that encodes a multicellular organism will have far-reaching implications both the future of both plants and animals. It will open up the possibility of creating plants with designer genomes able to withstand pests and diseases, that have increased yields and productivity, reduced fertiliser dependence, are capable of nitrogen fixation, have reduced water requirements, and that can survive the increasingly harsh global climate conditions. It will also facilitate the cultivation of plants indoors and in extreme environments like space. Using conventional breeding methods, it would typically take around eight years to develop a new crop variety. The use of synthetic biology greatly accelerates this process, while also providing possibilities that go beyond what is currently possible in nature.

The Synthetic Plants Programme, initiated by the UK's Advanced Research + Invention Agency (ARIA), aims to redefine the future for agriculture by building programmable plants. As one of the investigators, Jake Harris at the Department of Plant Sciences, University of Cambridge, stated, 'instead of modifying an existing chromosome' it will be possible to 'design it from the ground up'. This capability will move agricultural science 'beyond the limitations of natural genomes'. It is 'about designing entirely new capabilities in plants – from the molecular level up'. The plan is to build synthetic chromosomes in moss and transfer them into potato plants. The techniques and learnings from the streamlining and engineering of plant genomes and those of other simpler species – including viruses, bacteria and yeast – have direct relevance for the rewriting of the human genome.

A minimal, engineered human genome could serve as a universal platform for the development of next-generation cell therapies, with potential utility in cancer, autoimmune diseases and beyond. Like a smartphone operating system, it would contain core biological functions onto which specific therapeutic 'applications' could be uploaded. Removing non-essential sequences, such as those involved in organ formation, which are not required by individual cells, would enhance the efficiency, engineerability and cost-effectiveness of artificial human genome synthesis.

Once the generative rules of biology have been elucidated and ABI comes of age, it might be possible to synthesise the genomes of crops that have never previously existed. These could surpass the importance of traditional staples like maize, wheat, rice, potatoes and soybeans. Guided by AI generative foundational models, the Species Catalogue could be searched to identify optimal solutions. These would be based exclusively on required utility, such as the ability to withstand extreme climate conditions, without needing to reference historical designs.

A compelling example of this potential is the wild South American rice *Oryza alta*, which is highly resistant to drought, salinity and diseases. But it is not currently suitable for cultivation, as its nutritious grains cannot be harvested. This is due to a phenomenon known as seed shattering, in which seeds fall to the ground when they ripen. Were its biology better understood, it should be possible to rewrite its genome to domesticate it – something that would take thousands of years to achieve by selective breeding. In 2021 Jiayang Li, at the Institute of Genetics and Developmental Biology in

Beijing, sequenced the *Oryza alta* genome and compared it to that of domesticated rice. This revealed a potential route for its artificial domestication. It suggested the possibility of transforming *Oryza alta* into a new harvestable cereal that could enhance global food security.

As the world's population continues to grow and the climate becomes increasingly unstable due to global warming, food security remains one of humankind's most compelling challenges. Estimates indicate that by 2050, global food production will need to increase by 70 per cent to meet demand. While agricultural productivity in regions such as sub-Saharan Africa continues to accelerate, around 75 per cent of this growth has come from the expansion of the area of land under cultivation rather than through gains in crop yields. The escalating global demand for food remains a key driver of the destruction of wild habitats and loss of biodiversity across sub-Saharan Africa and beyond. Agriculture is one of the areas where the ability to design and synthesise genomes is likely to have the greatest impact. But the availability of fresh water is a critical limiting factor for agricultural productivity, as agriculture consumes around 90 per cent of the world's freshwater resources. Reducing water requirements could increase the photosynthetic efficiency of plants, and help limit the appropriation of wilderness.

In 2018, a research team led by Stephen Long and Katarzyna Glowacka at the University of Illinois at Urbana-Champaign, showed that the efficiency of water use in tobacco plants could be substantially increased by introducing extra copies of a gene called *Photosystem II subunit S* (*PsbS*) into their genomes. This modification reduced the opening time of their stomata – the tiny pores in leaves that, like the adjustable aperture of a camera lens, allow carbon dioxide to enter the plant while releasing water vapour.

Krishna Niyogi at the University of California, Berkeley, explored whether similar results could be achieved by modifying the regulation of the endogenous *PsbS* gene. In 2024, he showed that this could be achieved by modifying a non-coding sequence in the plant's own *PsbS* gene. While the alterations to the tobacco plant genome involved just small edits, a more extensive redesign using genome synthesis could modify multiple plant features simultaneously. Substantial alterations to the nature of plant and animal species are likely to be achievable through reprogramming non-coding regions, but we don't yet understand their function well enough to systematically manipulate them beyond a few well-defined examples. But the significant impact of these types of minor edits suggests the possibility of transforming plant biology on a more substantive scale.

There is, in theory, almost no area of human endeavour that could not in some way be impacted by the ability to design and construct the artificial genomes of species from first principles. Microbial genomes, for example, could be engineered to convert industrial carbon dioxide emissions into feedstock for jet fuel production. Food could be packaged using biodegradable seaweed-derived protein films that replace plastic. Organisms that lack natural photosynthetic capabilities could be reprogrammed to capture sunlight and produce sustainable energy. Plant immune systems could be augmented with antibody-like 'nanobodies' – using genetic sequences borrowed from camels and llamas – to enhance their resistance to pathogens.

Synthetic biology also promises to transform the architecture of plants. It could be used to redesign plant root systems to improve water and nutrient uptake during droughts. Cereal crops like wheat and barley may be reimagined to produce heavier seeds or to increase the number of seeds in each spike. Nature's tricks – such as the blue mussel's ability to produce powerful vanadium-based adhesive proteins, or the ability of spiders to manufacture silk with a tensile strength comparable to that of steel – could inspire new types of biomaterials. The natural properties of these materials could be modified and augmented using AI-guided protein design. Their systematic combination could produce novel biological composite materials, leading to a new biological materials science.

The implications extend to space exploration as well. Plants could be engineered to operate using alternative metabolic pathways that support growth in the reduced light conditions of extraterrestrial locations. This would make it possible to cultivate food on spacecraft during the anticipated three-year journey to Mars, during which a crew of six astronauts would consume an estimated 10,000 kilograms of rations. It would also enable food to be cultivated year-round in regions such as the South Pole, where light is scarce. In 2022, Feng

Jiao at the University of Delaware demonstrated that mushrooms and green algae could be engineered to grow in the dark, sustaining themselves on acetate, the principal component of vinegar.

Liberating agriculture from its dependence on the sun would have a profound impact on global food production. Artificial photosynthesis is potentially eighteen times more efficient than natural photosynthesis and could provide an orthogonal way to generate foodstuffs. The metabolisms of plants could also be rewritten to produce pharmaceuticals. Microorganisms could be engineered to degrade microplastics in Earth's ecosystems, to clean up oil spills, or to convert greenhouse gases into nutrients. Biological innovation could also extend to computing. Living brain organoids, comprising collections of genomically rewritten living cells, could one day be used to build biocomputers. The genomes of animal cells, such as those of pigs, could be partially overwritten with human sequences to facilitate human organ transplantation – work that is already underway.

In January 2024, a team led by Abraham Shaked at the University of Pennsylvania Transplant Unit in the US, announced that they had successfully connected a gene-edited pig liver – made by the company eGenesis and perfused using a device manufactured by the company OrganOx – to a brain-dead patient. The organ functioned for the full 72-hour duration of the experiment and showed no signs of rejection.

Rewritten organisms could be used to make biosensors, and to 'grow' self-assembling living structures such as houses. Materials based on natural polymers, such as biodegradable cellulose-derived plastics, could replace traditional plastics. Smartphones might one day become living, self-replicating devices. The harnessing of DNA's information-storing capacity could revolutionise global information archiving. A gram of DNA can theoretically store up to 215 petabytes (a petabyte is a million gigabytes) of information. It has been estimated that the entirety of the world's data will have reached 175 zettabytes (a zettabyte is a million petabytes) by 2025. All of it could, in principle, be stored in around eighty-one kilograms of DNA. In 2019, George Church and Henry Lee outlined the design of a prototype for a universal DNA information-storage system. Biology may eventually become the default engineering material for most applications.

The ability to design and synthesise genomes, especially in the context of 'strong ABI', which allows for the full mastery of biology's generative grammar, will have a transformative impact on society and turn biology into a predictive engineering discipline. The use of sustainable, biologically-inspired technologies and artificial species to address everyday needs promises to reduce pollution and combat global warming. Biological devices will become a pervasive feature of everyday life. One potential frontier application is the development of living quantum computers that use biological components and are capable of functioning at room temperature. Such devices would have the potential to revolutionise AI and increase the likelihood of generating conscious awareness in machine architectures.

One can imagine a conversational interface integrated with ABI, in which a user types in a request specifying a desired species or phenotype, and the system responds with a functional genome sequence 'printout'. A query might be: 'Design a cash crop that can survive extreme temperatures', 'Design a minimal human genome', 'Design the genome of a completely new immune cell type', 'Create an organism capable of synthesising waterproofing materials for the clothing industry' or 'Design a microorganism that can synthesise rocket fuel'. It might alternatively be: 'Construct an organ that performs the functions of the human liver, but is half its size', or 'Design a microbe that is capable of purifying water'.

Once we are in a position to colonise habitable planets beyond Earth, we might ask it to suggest the design of suitable ecosystems for planetary colonisation. As human biology becomes increasingly computable, commands might also become more abstract, such as 'Design a synthetic emotion that evokes qualitatively new mental states'. Whether we would wish to do so, however, is another question.

The instruction might, however, be more open-ended, encouraging the AI device to 'hallucinate' novel types of species, with a request such as 'Provide the designs for some artificial species capable of solving challenges that are of critical relevance to humankind', or 'Design new forms of intelligence using biological architectures distinct from those generated by conventional nervous systems'. We cannot, however, overlook the possibility that an operator might type unethical commands such as 'Configure a human genome that removes the capacity for free will', or 'Design a mind that lacks empathy and moral reasoning'. While offering significant potential benefits to society, the ability to speak, write and author the language of biology will be accompanied by significant risks. It is especially vulnerable to misuse by totalitarian regimes. Human nature, as we currently know it, is at risk of being irreversibly undermined.

But let us now return to the speculative domain of the Species Catalogue, and in particular to the region existing beyond biology as we currently know it: a mathematical frontier that we can discern, but that is largely uncharted. It contains species built from chemistries fundamentally different from those with which we are familiar. The borders of this land, where existing biology is marginalised or irrelevant, were first explored by the American biochemist Seymour Cohen in his 1963 paper 'On Biochemical Variability and Innovation'. Pioneering this uncharted world of alternative biology, Cohen allowed his imagination to extend outside the known building blocks of living things and the biochemical unity of all existing life on Earth.

While some viruses encode their genetic information in RNA, most organisms use sequences of DNA. The proteins encoded by protein-encoding genes are, in most cases, built from combinations of a standard set of twenty 'canonical' amino acids. The first of these to be discovered, asparagine, was isolated from asparagus in 1806 by the French chemist Louis Nicolas Vauquelin and his assistant Pierre Jean Robiquet. After the full set of twenty commonly occurring amino acids had been identified, two additional amino acids were discovered in proteins: selenocysteine and pyrrolysine. In humans, selenocysteine is used to make just twenty-five proteins, but it is more widely used in the proteins of bacteria and archaea. Pyrrolysine is not used to make any human proteins, but is used in the proteins of bacteria and archaea. Neither selenocysteine nor pyrrolysine, however, is directly specified by the genetic code.

Cohen speculated that it might be possible to encode genetic information in molecules other than DNA and RNA, and to devise genetic codes that accommodate the use of non-natural amino acids capable of generating new functionalities. He noted that the chemical possibility for naturally occurring analogues of nucleotides, such as nebularine, tubercidin and cordycepin, suggested the potential for 'biosynthetic mechanisms that are almost totally unexplored'. He wondered whether the existing biological world, in which genetic information is encoded in DNA and RNA, could represent 'the best of all possible biochemical worlds', or whether we might encounter or design life that utilises different chemistries to encode its genetic information. In short, could a 'coding, replication and information-transfer system be configured that lacks nucleic acids'?

All life on Earth utilises a narrow repertoire of twenty amino acids. Yet, as Cohen observed, biochemical pathways exist for the synthesis of at least eighty additional types of amino acids. It has been estimated, however, that there are more than 500 different naturally occurring amino acids, of which 140 have been artificially incorporated into proteins through various means. So why does nature use a set of just twenty amino acids to build natural proteins? Cohen noted that existing organisms 'are the products of selection', which reflect historical biochemical choices. Nature has utilised just the tiniest fraction of the biochemical possibilities for building living things. Cohen speculated that the use of alternative building blocks to encode the information of living things and construct organisms may result in behaviours and functionalities that are more than 'mere modifications' of existing biology. The deep seams of chemical possibility inherent in nature offer the potential for introducing 'great complexity' into species.

The term Artificial Biological Intelligence (ABI) has been used earlier to define a hypothetical form of artificial biological intelligence and generative competence that is equivalent to – or closely approximates to – that of natural biology. It is possible, similarly, to envisage what we may call Artificial Biological Superintelligence (ABS), a form of biological intelligence that surpasses the capabilities and competence of natural biological possibility through its ability to construct organisms by means of alternative chemistries. By deploying such unnatural chemistries for both information storage and expression – of which unnatural amino acids may be viewed as being the most rudimentary rendition – ABS may be expected to describe regions of the space of biological possibility that cannot be articulated by natural biological languages, and which ABI must therefore pass over in silence.

To appreciate the constraints imposed on biological languages by their current limited vocabulary, it is helpful to look at how letter usage influences the structure of written language. A notable example is the lipogram (from the Greek *leipen* 'to remove' and *gramma*, 'letter'), a form of writing in which a linguistic constraint is imposed that forces a particular letter to be omitted. For instance, in 1968, Georges Perec wrote an entire novel, *La Disparition* (A Void), without using the letter 'e'. Perec was a member of the experimental literary Oulipo group of mathematicians and authors who gathered at the Café de Flore in Paris and attempted to produce art that was derived from the application of arbitrary rules and algorithms. The exclusion of 'e' – the most commonly used letter in the French alphabet – imposed considerable limitations on what could be articulated.

Similarly, constraining the language of life to protein sequences that are restricted to using just twenty different amino acids and the five informational nucleotides deployed in DNA and RNA, limits life's expressive possibilities. Living organisms could potentially be released from such constraints, and built from different chemistries. Cohen's visionary ideas have, in recent times, been realised experimentally. It has been shown, for example, that it is possible to synthesise proteins that incorporate 'non-canonical' amino acids, which provide them with distinct chemical properties that are not found in nature.

The biochemist Steven A. Benner, based at the Foundation for Applied Molecular Evolution in Florida, set about trying to redesign genetic molecules. He was famously dismissive of DNA's structure, calling it a 'stupid design'. Why, he asked, would you entrust the 'valuable genetic inheritance that you're sending on to your children to hydrogen bonds in water'? In an attempt to expand DNA's alphabet beyond the four natural nucleotide base letters and to transform it from its 'lipogrammatic' state into something more versatile, Benner set about designing and constructing new kinds of nucleotides. He synthesised exotic new biological alphabets in the hope that the foreign letters might be incorporated into existing genetic material.

But it turned out to be far more challenging than he had expected. Ultimately, however, he succeeded, doubling the size of the natural genetic alphabet by producing a synthetic type of genetic material called Hachimoji DNA (*hachi* meaning eight and *moji* meaning letters in Japanese), which used eight nucleotide 'letters', rather than four. This significantly increased the density of the information that can be stored in synthetic hereditary material.

Philipp Holliger, at the LMB in Cambridge, also succeeded in creating synthetic genetic polymers that are not found in nature. He called them xeno nucleic acids (XNAs). He showed that heredity and evolution – the fundamental hallmarks of life – are not exclusive to DNA and RNA molecules. There could, in principle, be huge numbers of alternative genetic systems capable of encoding the information of living organisms. These unconventional biological architectures remain largely unexplored. They form part of the currently barely visible array of galaxies of future biological possibility that are not yet resolvable. The study of such systems is known as xenobiology.

To explore whether the basic logic of the genetic code – the rules that cells use to turn DNA sequences into proteins – might be reimagined, in 2019 Jason Chin, Kaihang Wang and Julius Fredens at the LMB redesigned and synthesised a completely artificial version of the 4-million-nucleotide-long genome of the bacterium *E. coli*. They called this synthetic organism Syn61. The logical structure of its genetic code was radically different from anything previously produced by natural evolution.

Nature uses a near-universal genetic code comprising sixty-four three-letter combinations, or codons, to specify each of the twenty naturally occurring amino acids that most cells use to make proteins. A few species have minor variations of the code, indicating an intrinsic potential for malleability in the code's structure. Three of the sixty-four codons specify punctuation marks that demarcate the ends of genes – so-called stop codons. The remaining sixty-one codons, known as 'sense' codons, specify amino acids. Because there are more codons than amino acids, most amino acids are represented by more than one codon. These are known as synonymous codons. The code is thus degenerate, with eighteen of the twenty natural amino acids being coded by more than one synonymous codon. The choice of the particular synonym coding for an amino acid can impact biological outcomes. Substituting one for another can result in deleterious consequences, manifesting as alterations to gene regulation or growth deficiencies.

The team wondered whether it was necessary to have so many synonymous codons. Might it be possible to compress the naturally occurring genetic code to reduce the number of synonymous codons specifying each amino acid? If the number of codons specifying genetically encoded amino acids could be reduced, it would provide a way to protect the genomes of organisms from one another with a biological firewall – for example, allowing cells to be virus-proofed. It might also allow unnatural amino acids to be incorporated into proteins, or enable the genetically encoded synthesis of polymers made from building blocks other than amino acids.

They showed it was possible to remove three of the sixty-four codons (two coding for amino acids, and one for a stop codon) and rewrite every occurrence of those codons across the entire genome. The result was a unique sixty-one-codon genome.

Chin, Wang and Fredens had created a customised organism with a genetic logic that was distinct from that of all known biological entities. They had shown that rather than being fixed, life's constructional logic, is, instead, flexible, and can operate with a reduced set of synonymous codons. History has locked in just one of the many possible alternatives, each with its own possibilities and constraints. There are a vast number of theoretically possible recoding schemes, and it is unlikely that the one used by the natural genetic code is optimal.

This work was extended by Michael Grome and Farren J. Isaacs in 2025, who recoded the genome of *E. coli* in a way that compressed its three synonymous stop codons into a single stop codon. This required the recoding of 1,195 synonymous stop codons. The two stop codons that had been 'freed up' were now available to be reassigned to code for unnatural amino acids. It built upon a landmark paper that Isaacs published in 2011 with George Church, in which they showed that the genome of a living cell could be treated as 'an editable and evolvable template.' Using a genome engineering method called Conjugative Assembly Genome Engineering (CAGE), they were able to modify the genetic code in a living *E.coli* cell on a genome-wide basis, changing 314 instances of TAG codons to a corresponding – so-called synonymous – codon, TAA.

That same year, Wesley Robertson and Jason Chin announced that they had authored a synthetic, recoded *E. coli* organism with a heavily compressed fifty-seven-codon genetic code, called Syn57. They eliminated seven codons in total – six sense codons and one stop codon. The bacterium still used the full set of twenty natural amino acids, but encoded them using just fifty-five codons rather than the usual sixty-one. This suggested that the genetic code could tolerate an even more extreme deviation from its natural configuration. The design required more than 100,000 alterations to the code of the natural bacterial genome.

By 'freeing up' unused codons in this way, it is possible to reprogramme the genetic code to allow proteins to incorporate non-canonical amino acids. For example, Syn61 has the potential to use up to three unnatural amino acids, while Syn57 could incorporate seven. The utility of introducing artificial amino acids into proteins was demonstrated in 2013, by Ryan Mehl and Jennifer Jackson at the Franklin and Marshall College in Pennsylvania. They showed that the addition of an unnatural amino acid into an enzyme called nitroreductase, derived from *E. coli*, was able to improve its performance by a factor of thirty fold – something that was not achievable using natural amino acids.

In the future, unnatural amino acids will routinely be specified at the genetic level through genome recoding. The expansion of the repertoire of twenty commonly used natural amino acids to include unnatural ones offers the possibility of expanding nature's capabilities, by altering individual proteins and the nature of species in ways that we have only just begun to explore.

While DeepMind's AI protein structure prediction model AlphaFold, revolutionised structural biology by enabling atomic-level prediction of the structures of proteins constructed from the twenty naturally occurring amino acids, it was not configured or trained to predict the structures of artificial proteins that incorporate unnatural 'non-canonical' amino acids. But in 2025, Qiuzhen Li,

Diandra Daumiller and Patrick Bryant, at the Wenner-Gren Institute in Sweden published RareFold, a deep learning AI model that is similar to AlphaFold, but expands the chemical alphabet available for protein design to include twenty-nine non-canonical amino acids.

This opens up the possibility of designing artificial proteins with completely new properties and functions unavailable in nature. Some artificial proteins containing unnatural amino acids are likely to have novel properties – including the ability to target previously inaccessible surfaces, increased thermo-stability and reduced immunogenicity. Synthetic, recoded organisms could be engineered to incorporate the metabolic machinery necessary to synthesise unnatural amino acids within cells, thereby creating organisms with the ability to construct artificial proteins. The genetic code could also be compressed even further to allow for a more diverse repertoire of unnatural amino acids to be utilised within synthetic life.

Might it be possible, to go beyond natural and artificial amino acids, to create life that is not based on amino acids at all? In 2024, Jason Chin and Daniel L. Dunkelmann showed that they could expand the chemical scope of the genetic code in *E. coli* to allow for the incorporation of non-canonical monomers – chemical building blocks that are not amino acids. These included ß-hydroxy acids, macrocyclic peptides and depsipeptides (unique polymers containing both amino acids and hydroxy acids). They had, for the first time ever, succeeded in breaking the 'evolutionary deadlock' that had, until now, restricted the machinery of life to using just a single class of monomer – amino acids.

This raises the possibility of building living things with unnatural chemical 'hardware' that goes beyond the historical protein technology used by all existing life on Earth. Life's alphabet could not only be expanded, but completely reimagined. Life would be guided into uncharted landscapes, endowed with entirely new properties that would reinvent the way organisms function. It would enable the creation of entirely new kinds of life.

11

Authoring Human Genomes

> It is likely that we will be able to completely redesign [the human genome] in the next thousand [years].
>
> – STEPHEN HAWKING, 1998

On Tuesday, 10 May 2016 – sixteen years after the announcement of the completion of the sequencing of the human genome in the East Wing of the White House – a disparate collection of around 135 individuals gathered in a conference room at Harvard University. The invitation-only meeting was shrouded in secrecy. A few attendees admitted to being unsure why they had been invited. Journalists were barred from attending, and participants were explicitly instructed to avoid engaging with the press or posting about the event on social media.

The meeting marked the beginning of a bold new endeavour. Initially conceived by the California-based geneticist and futurist Andrew Hessel as a 'grand challenge' for humankind, Hessel and his co-founders of the Human Genome Project-Write (HGP-write) initiative – the geneticists Jef Boeke and George Church and the attorney Nancy J. Kelley – had decided to chemically synthesise a human genome. The HGP-write initiative was presented as the natural successor to the Human Genome Project, which had successfully mapped humanity's genetic blueprint.

The Human Genome Project was itself first conceived at a U.S. Department of Energy meeting in Utah, in December 1984. After fifteen years of effort, the first draft of the human genome sequence was announced on 26 June 2000. This was followed by the preliminary completed human genome sequence on 1 April 2003. Francis Collins, who headed the government-led portion of the HGP, sometimes referred to in this context as HGP-read, described the draft human genome sequence as providing 'the first glimpse of our own instruction book'. Bioethicist Sheila Jasanoff called it 'a powerful new way to represent human identity'. It provided the first insights into the nature of human individuality, and a basis for mapping inter-individual differences to specific genetic sequences.

While the plan to sequence the human genome had, at the time, been derided by some – including Sydney Brenner – as being premature, the ability to sequence healthy and diseased human genomes went on to revolutionise science and medicine. 'The most wondrous map ever produced by humankind' offered the promise of providing unprecedented insights into human biology and disease pathology. It led to the development of targeted therapeutics informed by sequence variations shown to be associated with particular human diseases. It also helped define individual risk and facilitated the development of personalised medical treatment plans. The information specifying the basis of human life had been democratised, and the entire world was now free to explore its previously uncharted landscapes.

The Human Genome Project catalysed investment in DNA sequencing technology, which eventually led to a dramatic decrease in the cost of DNA sequencing and the time taken to complete a genome sequence. As a result, the whole-genome sequences of hundreds of thousands of individuals, encompassing a broad range of ethnicities, have now been completed and continue to accumulate at an ever-increasing rate. They represent the multifaceted nature of all humanity and will, over time, through the analysis of individual variability, help teach us who and what we are as a species.

While covering most of the regions coding for proteins, the two 'completed' human genome sequences presented in 2000 lacked information from around 200 million nucleotides – representing roughly 8 per cent of the human genome – and were replete with errors. This was because some sequences were especially hard to decipher. The highly repetitive sequences that cap the ends of chromosomes, known as telomeres, for example, were particularly difficult to sequence. It was not until 2022 that most of the gaps were filled and a close to fully complete 'end-to-end' human genome sequence was presented by the Telomere-to-Telomere (T2T) Consortium. Once the human genome had been sequenced, it was perhaps unsurprising that the idea of writing a human genome was raised.

In his 2012 publication 'Time for another human genome project?', in which he first suggested synthesising a human genome, Andrew Hessel drew parallels between the worlds of biology and computing. He argued that the Human Genome Project had 'transformed biology into a digital information science'. Having shown it was possible to read DNA code script, the next step – beyond understanding its meaning – was to learn how to write it.

If genomes could be drafted like works of literature, humanity would be able to leverage the learnings from sequencing human genomes more effectively. It might, for example, be possible to identify, repair or rewrite the corrupt code that causes human diseases. Biology would, at that point, have been transformed into a specialised form of software engineering. Life would literally become programmable. Synthetic human genomes could be tested in cells by replacing natural genomes with artificially designed sequences. The ability to synthesise genomes from the 'bottom up' would far exceed the capabilities of gene-editing technologies, offering the opportunity to redraft and reimagine genetic text, as opposed to simply tinkering with it. The operating systems of species would be fundamentally redesigned, rather than simply adding or removing specific applications.

The HGP-write team set themselves a ten-year goal of chemically synthesising a human genome – the instruction manual for building and operating a human being. But there were some technical barriers that would first need to be overcome. Synthesising DNA was expensive, laborious and error-prone. It was also immensely difficult to write out the highly repetitive sequences that make up almost half the human genome. It was decided that the primary goal of HGP-write would be to 'reduce the costs of engineering and testing large genomes' by a thousandfold, paving the way for the development of ultra-low-cost, rapid technologies for constructing genomes at scale. The goal of synthesising a human genome, the team argued, would serve as a central challenge to drive progress in genome design and synthesis technologies.

George Church optimistically stated that 'we have had a revolution in our ability to read genomes' and 'the same thing is happening now with writing genomes'. The reality, however, was very different. The anticipated revolution in writing genomes did not materialise. The prospect of writing a human genome over a reasonable time frame and at an acceptable cost was barely feasible. Attempting to do so with existing technology would be immensely time-consuming and costly. The efforts of the Sc2.0 consortium to synthesise the genome of yeast gave some indication of the magnitude of the challenge. By the time of its completion, it would have taken the international consortium nearly two decades to build the synthetic, redesigned genome. At three billion nucleotides long, the human genome is around 260 times larger than that of yeast. It also contains more chromosomes and is filled with highly repetitive sequences that are especially challenging to synthesise.

While the $3 billion price tag for sequencing the first human genome has, in just two decades, been reduced by a factor of six millionfold – to around $500 for a high-quality genome completed in less than four hours rather than in two decades – comparable improvements in genome-writing technology have lagged far behind. It has proved challenging to move beyond the foundational construction methods based on the chemical synthesis of small fragments of overlapping DNA (typically up to 300 nucleotides long), which are then assembled into larger fragments.

In addition to its scientific importance, the HGP-write team believed that the chemical synthesis of an artificial human genome would capture the public's imagination. They saw it as a way to 'galvanize the scientific community' and rally support for synthetic biology by attracting the necessary resources from government agencies, philanthropists and investors. This would raise global awareness of the field while stimulating public debate on the related ethical, legal and societal implications.

Genome synthesis was viewed as 'a logical extension of the genetic engineering tools' that formed the basis of the 1970s biotechnology revolution led by Herbert Boyer, Stanley Cohen and Paul Berg. Despite some initial concerns, recombinant biology – the science of cutting and pasting genes – had proven to be safe and consequential. It had, for example, resulted in the synthesis of the first artificial genes, which were integrated into bacterial genomes and used to make therapeutic human proteins, such as insulin for the treatment of diabetes.

In their landmark 2016 *Science* magazine paper, 'The Genome Project-Write', the HGP-write team outlined the framework for a new kind of constructional biology. They envisaged a 'learning-by-building' approach for the natural sciences, where synthetic versions of natural genomes would be designed, built and tested in model organisms. The nature of organisms would be inferred by recreating them from scratch. Biological designs would be converted into genetic code scripts and then assembled and tested. This new constructional approach to unpicking life's secrets would offer profound insights into the 'operating systems' of living things.

In so doing, it would pioneer a new era of programmable biology, capable of generating organismal novelty in a predictable manner. Life would be engineered like a machine. It would also become possible to consider overhauling the 'human operating system' to make it less complex, and more predictable and engineerable. The ability to synthesise genomes would allow engineering principles to be systematically applied to entire genetic systems, rather than just to their components.

The project inevitably raised considerable ethical concerns, not least because it was not clear why the team wanted to synthesise a human genome in the first place. There appeared to be no direct justification for doing so. To address these concerns, which the team had anticipated, and to ensure HGP-write proceeded responsibly, the organisers established plans to incorporate appropriate legal and ethical oversight, public engagement and societal decision-making. They also made it clear that they had no intention of making synthetic humans lacking biological parents. Their activities would be focused on generating cell lines containing synthetic human genomes and self-organising 'organoids' derived from these cells.

As a result of their shared evolutionary history, the genetic language of all organisms is broadly conserved across the diverse branches of the tree of life. All life uses the same programming language. If you can programme a bacterium, you can programme a human. But while the chemical language used by life on Earth is universal, the semantic component of this language – the rules of its grammar – varies across species. One of HGP-write's key goals was to decode these variations, especially within the substance of the human genome's 'dark matter' – the non-coding regions responsible for key regulatory functions that are frequently implicated in disease.

To address this, a series of small pilot projects would be initiated alongside the synthesis of the human genome, to explore the re-engineering of the human genome on a regional basis. The development of methods to introduce large pieces of DNA into cells, and even entire genetic loci containing key genes and accompanying stretches of non-coding DNA, would allow the regulatory genome to be interrogated.

The generation of artificial chromosomes called 'neochromosomes', would enable the relocation of metabolic pathways and gene circuits to new genomic regions. Sophisticated types of gene therapy involving regulatory switches and control circuits would also be designed. One of the more ambitious parts of the initiative was the planned creation of a reference human 'pangenome' embodying the global average 'baseline' of human genetic variation, with every variable position across the human genome being represented by the most common variant.

Perhaps unsurprisingly, news of the inaugural Boston HGP-write meeting leaked out, generating considerable global attention. *Time* magazine headlined its story as 'The Brave New World of the Synthetic Human Genome', while *The Washington Post* headline was 'After Secret Harvard Meeting, Scientists Announce Plan for Synthetic Human Genomes'. *The New York Times* led with 'Scientists Talk Privately About Creating a Synthetic Human Genome'. Concern was expressed that the meeting was held behind closed doors, and that this may have been done deliberately, to avoid public scrutiny.

The New York Times noted that the prospect of creating 'human beings without biological parents' was 'spurring both intrigue and concern'. George Church was swift to correct the misunderstandings, making it clear that the project had no intention of making synthetic people, stating that the media were 'painting a picture' of the project that did not reflect its aims. There was nothing in the project that advanced the agenda of making a synthetic, parentless human being. The ability to write a human genome would, however, take humankind a step closer to that possibility, and there was significant potential for the technology to be misused.

Some of the criticism originated within the synthetic biology community itself. In a critique published on 10 May 2016, titled 'Should We Synthesize a Human Genome?', the synthetic biologist Drew Endy and bioethicist Laurie Zoloth argued that synthesising a human genome would constitute an 'enormous moral gesture' that lacked a clearly defined purpose. Endy also stated that he felt the human genome was 'special' and should not be industrialised or commoditised. It was, furthermore, thought to be unnecessarily provocative to advance a project that might be perceived as undermining the creation narratives of several global religions. Endy and Zoloth suggested that the genome of a less controversial species, with more immediate utility, would have been a better alternative. As Jef Boeke succinctly summarised, 'The notion that we could write a human genome is simultaneously thrilling to some, and not so thrilling to others'.

Endy's principal concern was that the project might lend spurious legitimacy to potential future efforts by individuals or organisations that might not share the same ethical perspective as the HGP-write initiative. The project raised the prospect of individuals or entities introducing synthetic human genomes into germ cells, potentially attempting to generate genetically enhanced, or synthetic, parentless humans. Despite the best intentions and strict regulatory measures, a rogue nation state with an unaligned ideology could easily deviate from internationally agreed norms and guidelines. Endy and Zoloth also criticised the HGP-write team for neglecting 'essential questions' regarding the legitimacy of creating synthetic humans. 'Would it', for example, 'be OK to sequence and then synthesize Einstein's genome?' for example.

It transpired that the reason for the Boston meeting's secrecy was to prevent the disclosure of the initiative before the publication of its mission statement in *Science* magazine on 8 July 2016. The founders of HGP-write had decided to remove the word 'human' from the project's name, prior to the furore caused by the 10 May 2016 Boston meeting, renaming it Genome Project-Write (GP-write).

The follow-on meeting held a year later in New York, was fully transparent and open to the press. While the renaming of the initiative suggested a shift in its objectives, the plan had always been to write the genomes of multiple species. To demonstrate the project's immediate utility and to address a scientific problem with nearer-term feasibility, the team scaled back their ambitions, making the virus-proofing of a human cell line their highest priority. This would be achieved by recoding the cell's genome. An 'ultrasafe' recoded cell line could enable the production of antibodies, vaccines and other biologics free from the risk of viral contamination.

But funding for GP-write was not forthcoming. No major government organisations, investors or philanthropists stepped forward to finance the project. Over time the initiative came to an abrupt end and the group of investigators disbanded. Their attempt to synthesise a human genome has for the time being been thwarted by controversy, and by technical, financial and public understanding issues. Visionary and scientifically astute, the GP-write team were almost a decade ahead of their time. While the project went into 'hold mode' for several years, it is currently in the process of regrouping.

The technological challenges of genome writing and assembly remained as formidable as ever, and the lack of any substantive knowledge of the underlying design rules governing the functional architecture of genomes continued to be problematic. Debugging even the simplest genomes to get them to function properly had proved to be as time-consuming and labour-intensive as constructing them. To fully master the art of genome synthesis, proficiency would be required across several relevant domains, namely reading, writing and a comprehension of the generative rules of biology. This would require, among other things, the type of machine-learning-enabled generative AI that had not yet become available.

Building on the foundations and learnings of GP-write, on 26 June 2025, a group in the UK – led by Jason Chin at the LMB in Cambridge and the Generative Biology Institute (GBI) at the Ellison Institute of Technology in Oxford, and involving groups in Kent, Manchester, Oxford and Imperial College London – announced a new synthetic human genome project, called the Synthetic Human Genome (SynHG) project. Funded by the largest medical charity in the world, the Wellcome Trust, SynHG's bold announcement managed to avoid the hostile reception that GP-write had received just nine years earlier. While aware of the fate of GP-write and the moral crisis and controversy it had invoked, Tom Collins at the Wellcome Trust, who approved the funding, hedged his bets, arguing that it was necessary to consider 'the cost of inaction'. The technology was 'going to be developed one day'. By 'doing it now', they were 'at least trying to do it in as responsible a manner as possible and to confront the ethical and moral questions in as upfront a way as possible'.

In the earliest stages of the project, the team planned to make sections of a human chromosome and then eventually a whole chromosome representing 2 per cent of the human genome. These, and the full synthetic chromosome would then be tested in human skin cells. The purpose of synthesising human genetic material from scratch was, they argued, to learn how DNA works, and in particular to unpick the function of the enigmatic non-coding 'dark matter' within the human genome, which programmes protein-encoding gene activity. 'The information gained from synthesising human genomes', Chin stated, 'may be directly useful in generating treatments for almost any disease.'

Rather than contemplating a 'brave new world' or crossing any moral boundaries, the team were focused on determining how the complete synthesis of a human genome could improve our knowledge of human health and longevity. The ability to synthesise human genetic material would go far beyond anything possible with gene editing, in terms of scale, precision and accuracy, and most importantly wrote on a 'blank slate' rather than being constrained by pre-existing genetic information. It was likely the only definitive way to understand how states of health and disease are computed by the human genome's sequences.

Synthetic genomes, programmed like a piece of computer software, could be inserted into cells and used to treat a host of medical conditions. Cells programmed to be disease-resistant could, for example, be used to repopulate damaged organs. The work would take engineering biology to a scale that far surpassed that achieved with the small genomes of viruses, bacteria and yeast. It would have consequences for the engineering of genomes across all species, including those of plants. Importantly, it would be accompanied by a parallel transdisciplinary initiative, named Care-full Synthesis, that would explore the socio-ethical, economic and governmental policy implications of the project across diverse global communities.

One potential application of synthetic human genomes is to produce a cell line that could serve as the Model T Ford equivalent for 'off-the-shelf' cell therapies. Manufactured by the Ford Motor Company from 1908 to 1927, the Model T was the first mass-produced car and became the 'industry standard'. A 'universal' cell, powered by a streamlined, refactored and minimised human genome, could be used to treat cancer as well as autoimmune diseases and other diseases.

Although engineered cell therapies have proven highly effective in treating blood cancers in patients who have failed multiple prior lines of therapy, they have not made a significant impact on the outcome of solid tumours. The microenvironment of solid tumours, where cancer cells reside, often excludes immune cells and generates factors that inhibit their function. One could, however, imagine a designed cell containing a human genome packed with technology that outsmarts tumour defence mechanisms. The chassis of this streamlined cell could include docking sites for uploading specific code relevant to each tumour type, as part of a system upgrade.

The efficient synthesis of large pieces of DNA is of critical importance to the construction of gigabase-sized (a billion DNA letters or more) genomes. These synthetic DNA fragments of up to 100,000-nucleotides-long function as the 'feedstock' for building large DNA molecules, all the way up to the size of genomes. Once these have been synthesised, the next step is to assemble them into larger fragments that can be used to hierarchically assemble entire synthetic chromosomes. The feasibility of synthesising human and other complex genomes in a scalable manner has increased significantly following a technological breakthrough in DNA assembly made in 2023 by a team led by Jason Chin, Julian Sale and Jérôme Zürcher at the LMB.

Their method, called bacterial artificial chromosome stepwise insertion synthesis (BASIS), is a bacterial-based platform that enables the rapid assembly of megabases (millions of nucleotides) of natural or synthetic DNA from smaller fragments. This could become one of the preferred platforms for the rapid and scalable assembly of large sections of synthetic human genomes. The team demonstrated that they could rapidly and with high fidelity, build a 1.1-megabase (one million nucleotides) region of human chromosome 21 by stitching together nine overlapping fragments.

The DNA fragments they planned to assemble were housed in donor bacterial artificial chromosomes (BACs), which were iteratively inserted into an assembly BAC. Nine iterations of the BASIS process were required to reconstitute the 1.1-megabase region of chromosome 21. The human DNA in both the assembly BAC and donor BACs existed as episomes – circular pieces of DNA that sit alongside, but are separate from, the bacterial genome. The 1.1-megabase fragment of human DNA they assembled contained many of the representative types of sequences found within the human genome, including protein-coding regions, introns and repetitive sequences, making it a good surrogate for the rest of the human genome.

While the first chemically synthesised *E. coli* genome assembled by Chin's team in 2019, took nearly two years to construct, the BASIS method is expected to allow synthetic *E. coli* genomes to be produced from functional designs in as little as two months. However, this assumes that the smaller chunks of DNA have already been synthesised. This DNA assembly method has the potential to revolutionise the biomanufacturing of 'big DNA'.

The Chin team also developed a method called continuous genome synthesis, which allows for the ongoing replacement of natural DNA sequences in human chromosomes and those of other species with synthetic sequences, in a scalable manner. This overwriting method enables the continuous replacement of 100-kilobase stretches of a genome with synthetic DNA sequences. They showed that, starting with BAC episomes containing around 100 kilobases (a hundred thousand nucleotides) of synthetic *E. coli* DNA, they could replace around one-eighth of the natural *E. coli* genome (around 0.5 megabases of synthetic DNA) at pre-specified positions in just ten days. By performing the various steps in parallel, the team were further able to accelerate the assembly of entire synthetic genomes.

Despite such tremendous advancements in the ability to manipulate very large pieces of DNA, the construction of whole genomes remains subject to a major rate-limiting step. Constructing the 50- to 100-kilobase chunks of DNA required as the 'feedstock' for higher-order genome assembly, is expensive, time-consuming and unreliable. The construction of these prefabricated DNA parts, which form the basic building blocks for assembling larger genomic structures, continues to present a significant challenge. Researchers such as Kaihang Wang at Caltech, with hands-on experience of synthesising a genome of the size and complexity of *E. coli*, readily describe the 'pain' they experienced in trying to source the multiple chunks of DNA required to build genomes.

The DNA synthesis companies tasked with producing the requisite 50- to 100-kilobase chunks of DNA would frequently fail to deliver the pieces that had been ordered, or take months to do so. The cost of ordering these parts was, moreover, very considerable. The total cost of the DNA parts necessary to build an *E. coli* genome could run into several million of dollars. It is therefore necessary to address this fundamental bottleneck and develop a method for rapidly, reliably, and accurately producing DNA building blocks of this size, regardless of the complexity of the sequence, while substantially reducing the cost. A method capable of achieving this, and enabling the routine construction of DNA of any sequence at scale, would democratise the process of genome construction, and pave the way for a global biorevolution.

We have, until recently, remained dependent on essentially the same DNA synthesis method pioneered by Har Gobind Khorana and later refined into the more efficient phosphoramidite chemical synthesis method by Marvin Caruthers. Synthetic fragments of DNA made in this way are assembled into larger pieces using methods such as Gibson Assembly, Golden Gate Assembly, and Polymerase Cycling Assembly (PCA). All such approaches, without exception, guide the assembly process using information encoded within the DNA sequence itself. The DNA sequence being assembled is therefore forced to perform two separate and often incompatible functions. It comprises part of the information that is incorporated into the final assembled sequence, while simultaneously also providing the information necessary for coordinating the assembly of the fragments into the correct order.

As a result, whether starting with chemically or enzymatically synthesised DNA fragments, all these methods struggle to assemble sequences that are highly repetitive or G- and C-nucleotide rich. Notably, around half of the human genome falls into this category, while in the genomes of plants this increases to approximately 85 per cent.

But in 2025, Kaihang Wang and Noah Robinson at Caltech in California, devised an entirely new and paradigm-changing method for constructing DNA called Sidewinder. It promises to fundamentally transform the way in which DNA is constructed in the future, by enabling DNA sequence space to be explored in an entirely unconstrained manner. Suddenly, and for the first time ever, the entire space was accessible. Sidewinder is philosophically distinct from all existing natural and artificial methods for joining pieces of DNA together. Wang and Robinson obtained their inspiration from the history of the book printing industry.

While in the later Medieval period a small number of scribes began to experiment with new ways of organizing their longer works – by, for example, adding catchwords at the bottom of each page, which were identical to the first word of the subsequent page – this practice was more of an oddity than mainstream. The great majority of manuscripts lacked any type of page numbers whatesoever. The first books printed on Johannes Gutenberg's press also had no page numbers. The famous Gutenberg Bible, for example, printed in 1455, lacked page numbers, as well as title pages. It was not until two decades years later, in 1470, that the German printer Arnold Ther Hoernen – operating out of Cologne – introduced a significant innovation into the design of printed books. He printed the first ever book containing page numbers titled *Sermo in Festo Praesentationis Beatissimae Mariae Virginis* (A Sermon for the Feast of the Presentation of the Blessed Virgin Mary).

The introduction of page numbers turned books into navigable physical spaces, enabling information to be effortlessly identified and retrieved. Wang and Robinson wondered whether it might similarly be possible to introduce page numbers into DNA. If successful, they would achieve a transformation in DNA construction every bit as consequential as that achieved by the introduction of page numbers into the printing industry. Sidewinder achieved exactly that, literally introducing molecular page numbers into the sequences of DNA. Unlike the page numbers of books, however, the page numbers in this case could be removed once they had arranged the DNA 'pages' into the correct order.

Sidewinder, for the first time in the history of natural and artificial DNA construction, enables the information of the sequence that is being assembled to be separated from the information of the assembly process itself. This is achieved by assigning synthetic 'page numbers' to each DNA fragment. These DNA page numbers are then used to programmatically assemble the DNA fragments into the correct order, without needing to reference the information in the DNA sequence itself.

Once the page numbers have performed their function, they are removed, vanishing without a trace, to create a 'scarless' contiguous piece of assembled DNA. As the assembly method does not reference the information in the DNA sequence to direct the assembly, the technique can construct DNA of any degree of complexity. Sidewinder's ability to dissociate these two distinct types of information – the sequence to be assembled and the page numbers required to direct their correct assembly – allows for the parallel assembly of DNA fragments in a single reaction.

As a result, Sidewinder can build DNA fragments of unprecedented size and complexity rapidly, efficiently and effortlessly, having a functionally limitless capacity. By allowing DNA sequences of any complexity to be rapidly and efficiently assembled, Sidewinder makes it uniquely possible to access the entirety of DNA sequence space. It is a universal time-travelling engine that can materialise anywhere within its vast expanses. It closes the gap between AI-informed computational creativity and synthetic feasibility, as unlike existing methods, it can assemble sequences of any complexity.

Once the Sidewinder technique has been fully automated, its ability to build DNA at scale will allow whole genomes to be synthesised in record time. Sidewinder's construction capability could, additionally, be combined with the ability to simultaneously sequence the newly constructed synthetic DNA, thereby allowing its accuracy to be monitored in real time. This type of 'read/write' device could be of transformative importance for genome writing, forming the basis of an integrated genome 'printing and copy editing' machine.

Sidewinder represents a step change in the ability to construct DNA at scale, and in addition to enabling the construction of large DNA building blocks of any degree of complexity, rapidly and accurately, it also promises to substantially reduce the cost of DNA construction. It will have a significant impact on our ability to synthesise the genomes of complex multicellular organisms – from fruit flies and nematode worms, to zebrafish, mice and humans – by allowing the smaller chunks of DNA from which genomes are built to be generated efficiently, at low cost, and on an industrial scale.

Sidewinder can also be used to construct complex DNA libraries, thereby allowing the protein design solutions identified by AI foundational models to be constructed swiftly and efficiently.

The conjunction of Sidewinder with genomic foundation models like Evo 2 that facilitate DNA sequence prediction and generation tasks at the molecular genome scale, will enable the full power of in silico AI-informed genome design and construction to be realised. Sidewinder is perfectly suited to generating the 50- to 100-kilobase building-block fragments that can – like the prefabricated concrete structures used to construct buildings – be used to write whole genomes. Once the 50- to 100-kilobase fragments have been constructed in a cell-free manner, they can then be assembled into much larger pieces of the DNA – up to genome size – within cells. The convergence of these two technologies – AI-informed genome design and DNA construction using Sidewinder – has the potential to turn DNA into a predictive engineering material of immense utility.

Attempts to develop enzymic methods for DNA synthesis at a scale reminiscent of nature's own strategy continue to be developed, and while these are now being used to make small pieces of synthetic DNA, they remain in their infancy. The enzymic DNA synthesis methods use an engineered version of an enzyme called terminal deoxynucleotidyl transferase (TdT), which is able to link nucleotides together without requiring a DNA template. While enzymic synthesis may be better suited to making the complex, repetitive sequences that challenge chemical synthesis methods, the fragments produced by TdT remain relatively short, and must be assembled using conventional technologies in the same way as chemically synthesised DNA.

Other innovative enzyme-related approaches are also being explored. Thomas Gorochowski at the University of Bristol and his collaborators have shown that DNA polymerase enzymes 'doodle' – joining nucleotides semi-randomly into long sequences of DNA in the absence of a template. Were it possible to control this process, nucleotides could be joined together in a programmable manner. This could, potentially, turn the DNA polymerase enzyme into a kind of nanoscale 'DNA typewriter'. Achieving this will likely require the use of AI-informed protein engineering to optimise the enzyme's behaviour, as well as the use of an electronic chip able to influence the doodling at an individual-nucleotide level, to achieve a prespecified outcome.

With the advent of Sidewinder and AI-informed foundational models like Evo 2, it is now possible to anticipate a remarkable leap forward in our ability to design and construct the artificial genomes of species. The rapid advances seen in DNA sequencing will likely, before too long, be recapitulated in the science of DNA writing. We can expect to reach a point where a genome sequence is typed into a computer and 'printed out' rapidly, accurately and inexpensively – much like a document printed out on a laser printer. Synthetic genomics will, at that point, have come of age.

The unprecedented power that these technologies confer – offering the prospect of redesigning and synthesising human genomes from first principles – raises some of the most challenging philosophical and ethical issues that humankind has ever had to consider. Issues that are far more fundamental and consequential than anything pondered by the great philosophers of history, such as Plato, Confucius, Na¯ra¯rjuna, Epicurus, Zarathustra, and Aristotle.

We will, collectively as a species, need to determine the extent to which we are prepared to modify our nature. Would redesigning the human genome ever, under any circumstances, be considered advisable, desirable, or permissible? This will involve achieving global alignment on what it means to be human, and weighing the relative importance of the benefits of a healthy and extended lifespan against the risks associated with altering our genetic material and introducing fundamental, and likely unpredictable, changes into core features of our nature. The natural limits of humanity – its beginning and end – will need to be formally delineated.

In the autumn of 1960, shortly after his arrival at Princeton, the Japanese biologist Osamu Shimomura met the biochemist Frank Johnson, who regaled him with descriptions of the brilliant and intense blue-green bioluminescence of the jellyfish *Aequorea victoria*, which pulsated rhythmically through the Pacific Ocean waters at Friday Harbor, Washington, like transparent dome-shaped umbrellas. Johnson invited Shimomura to join him in studying the chemical basis of bioluminescence. In the summer of 1961, they embarked on a 5,000-kilometre journey to Friday Harbor in Johnson's station wagon, loaded to the brim with laboratory equipment and supplies. They reached the West Coast seven days later, set up shop in a small laboratory with three other scientists, and then set about looking for jellyfish. Fortunately, these were abundant, and they scooped them into buckets using a shallow dip net before taking them back to the laboratory.

After extracting material from the light organs of over 10,000 jellyfish, Shimomura eventually discovered a protein with a brilliant green fluorescence, present in only tiny amounts. He called it 'green protein' – later renamed green fluorescent protein, or GFP. The protein comprised a 238-amino-acid chain that paired with an identical partner to form a dimer. Shimomura showed that the chromophore – the source of the brilliant green light – was formed by just three amino acids – serine, tyrosine, and glycine – located at its core. The intense green fluorescence of the protein was generated by chemical processes known as dehydration and dehydrogenation reactions, centered on these three amino acids.

GFP is unique in nature, as no other protein generates a fluorescent chromophore from its own internal structure without cofactors or substrates. This suggests it is a 'unicorn' protein – a very rare occurrence in protein sequence space, and difficult to find using natural evolution. GFP went on to have an important impact on biology, as demonstrated in work pioneered by Roger Y. Tsien at the University of California, San Diego. It enabled proteins to be fluorescently tagged, allowing them to be tracked in living organisms and providing major insights into cellular processes.

On 2 July 2024, a team led by Alexander Rives and Thomas Hayes at EvolutionaryScale, an AI protein-design research laboratory, published a preprint of a paper titled entitled 'Simulating 500 million years of evolution with a language model'. They set out to determine whether they could create an artificial version of GFP that had similar spectral features and fluorescence intensity, but very low sequence identity to the natural protein. Up to that point all variants of GFP had been made by mining proteins in the natural world. The question was whether artificial versions of GFP with entirely different folding geometries could furnish the same function.

Using GPUs and a multimodal generative AI transformer language model called ESM3, trained on sequence, structure, and function information from 2.78 billion different proteins, they succeeded in making an artificial version of GFP – called esmGFP. The total computing power used for the training was 1 trillion teraflops (an astonishing one trillion mathematical calculations per second); more than any other known model in biology. To find the protein, the model had to explore the natural protein universe – the space of all possible protein sequences. The remarkable thing about the unnatural GFP surrogate protein they discovered was that its amino acid sequence shared only 36 per cent identity with natural GFP. The new protein had been identified by searching a distant and previously unexplored region of protein sequence space that was densely populated with surrogates. They had identified a rich seam of fluorescent proteins hidden within theoretical protein sequence space that evolution had never previously discovered. They estimated that esmGFP was the equivalent of 500 million years of evolution distant from the parental version of GFP and the closest existing protein in nature. They had succeeded in teleporting across DNA sequence space.

The significance of this remarkable achievement is that it could lead to the design of artificial proteins with functionalities that go far beyond those found in nature, for use in the genomes of synthetic species. These artificial proteins will likely have little correspondence to the repertoire of proteins discovered by natural evolutionary processes, as natural proteins represent just the tiniest fraction of all theoretically possible proteins. The opportunities for designing new functionalities that this presents is staggering. While natural proteins utilise a relatively narrow set of topologies, which they reuse and elaborate, the potential set of all possible protein topologies is far greater.

The genomes of living species will no longer need to be constrained by the irrational searches of protein sequence space dictated by their Darwinian evolutionary history. Synthetic species will, furthermore, no longer be jeopardised by 'decisions' made by natural evolutionary processes billions of years ago, which have shut down huge regions of possibility within protein sequence space. They will be able to execute functions that transcend the natural repertoire discovered by evolution. This new method for making programmable proteins will have far-reaching implications – ranging from healthcare, for example, in the design of new cancer drugs, to biomanufacturing, with applications including carbon capture, and the prevention of coral bleaching.

If human genomes were refactored and minimised for therapeutic purposes, the proteins they encode could be redesigned or replaced with novel, AI-generated alternatives. These artificial proteins, originating independently of evolution by natural selection, may possess chemical features that greatly surpass those of natural proteins. Individual proteins within the human genome could, furthermore, be minimised by searching for alternative versions with similar performance but incrementally smaller size. The miniaturisation of proteins may enhance their ability to penetrate tumour microenvironments. Their genes would also be easier to deliver, and they would be less likely to invoke an immune response. By minimising all the proteins encoded by the human genome, a reduction in overall genome size could be achieved. The use of unnatural proteins in human cells could reconfigure their biology, generating new cell types of therapeutic significance and enhancing their ability to fight cancer.

David Baker, a biochemist working at the University of Washington, Seattle, who was awarded the 2024 Nobel Prize in Chemistry for his work on artificial protein design, has played a pioneering role in creating the science of unnatural protein generation – enabling us to go far beyond the protein repertoire provided by nature. In 2022 he showed how, starting with nothing other than the three-dimensional structure of a target, artificial proteins could be designed that are able to bind to specific sites on other proteins.

In 2024 he went a step further, showing that a deep learning generative AI model called RFdiffusion could make miniature artificial antibodies that can precisely recognise molecular components on disease-associated molecules. To date, the production of antibodies has relied on the immunisation of animals, or the screening of libraries of antibodies expressed in bacteriophages. The ability to design antibodies computationally – at the atomic level – that are able to recognise targets on diseased cells is revolutionary. It will usher in a new era of predictive, structure-based antibody design. Such designed antibodies – or other, similarly designed proteins – could be used as stand-alone reagents, or introduced into synthetic human genomes for use in cell therapy.

While all prior AI foundational models were constructed to explore DNA sequence space rooted in the sequence of the target gene, on 19 November 2025 Brian Hie and his colleagues at the Arc Institute at

Stanford published a landmark paper that achieved something that was both unprecedented and revolutionary. Inspired by the 'anthropological semantic' philosophy of the 'London School' of linguistics founder J.R. Firth who stated in 1957 that 'you shall know a word by the company that it keeps' and argued that words could only be understood by observing how they behave in real language, Hie wondered whether the same might be true for genes. In other words, might it be possible to understand a gene by the company that it keeps within a genome?

Using the biological language model Evo that they had previously developed, the team swiftly showed that they were able to learn the semantic relationships between multiple genes across hundreds of billions of bacterial DNA nucleotides. By exploiting this deeply embedded and otherwise invisible evolutionary information, they found that Evo could perform what they called a 'genomic autocomplete'. In other words, if Evo was prompted with DNA encoding a specific type of function, the model could travel across huge swathes of unrelated DNA sequence space with no evolutionary relationship to the target function and identify sequences that encoded similar functions.

This new type of function-guided design was able to venture far beyond natural sequences, and enabled the exploration of regions of DNA sequence space that existed beyond the observed evolutionary universe. By prompting the system with a desired function, it was now possible to 'semantically mine' for functions in alien regions of sequence space. Hie had managed to break free from historical evolutionary limitations on function. The team went on to use this approach to generate a unique database called SynGenome, containing 120 billion base pairs of AI-generated genomic sequences. These comprise raw material for semantic mining across multiple different functions.

It is evident that this approach could eventually be generalised to entire genomes. In this case, Evo would be prompted to generate the genome of an artificial species that fulfilled specific pre-defined criteria. The semantic mining approach pioneered by Hie represents a fundamentally different paradigm for exploring the expanses of biological possibility, and provides the rudimentary and tentative foundations of a remarkable and completely new way of generating the artificial genomes of entirely new species.

Were it desirable, safe, and ethical to do so, the simplest way to prevent or cure many human diseases or implement risk reduction might be to rewrite the source code of the human genome at the germline level – reconfiguring it in a more practical, optimal, rational, user-friendly and risk-adjusted manner.

The human genome could, in whole or in part, be refactored – much in the same way that Drew Endy refactored the T7 phage – while at the same time maintaining both its overall structure and individual variability. AI could be used to design the most optimal configurations of biochemical pathways to minimise the multiple conflicting compromises and constraints. There would not be a single optimal version, but multiple alternative versions, each satisfying a different set of conflicting constraints and resulting in different profiles of characteristics and vulnerabilities. While refactoring can help deconvolute a genome's complexity and promote modularity, it is unlikely that biological systems will ever achieve the kind of orthogonality present in human-designed machines.

But while the attainment of strong ABI – the ability to generate genome sequences that realise specific predefined outcomes, and to construct them – will, if achievable, furnish the required outcome, there will be a limit to our ability to disentangle the different functionalities and characteristics encoded by a genome's code script. This suggests that the future artificial, ABI-mediated design of synthetic biological systems will require a process for selecting the preferred and most optimal collection of suboptimal alternatives.

Many desirable phenotypes – including extended healthy longevity – that are potentially obtainable through the rewriting of genomes, will invariably be accompanied by unexpected, unwanted, and paradoxical effects. There is likely no 'best of all possible human genomes' from any particular design perspective – just an array of uniquely flawed alternatives. Each comprises just one of multiple possible imperfect computed solutions to a complex optimisation problem defined by user-specific parameters. Each has its own vulnerabilities and suboptimalities, falling short of its intended outcome in its own idiosyncratic way.

While rewriting the human genome is not, at present, a plausible or desirable strategy for addressing human disease in the genomes of living individuals, refactored human genomes may, in principle, provide the basis for new types of genomically rewritten cell therapies with significant therapeutic utility. Even if rewriting the genome of a living person were ethically permissible or scientifically desirable, many critically important aspects of the genome's behaviour remain unknown – including how it generates consciousness. As such, any extensive attempt to refactor a genome would risk inadvertently compromising such poorly understood phenomena.

To contemplate rewriting the human genome requires a comprehensive understanding of its functionality and of the contributions made by each of its components. This, of course, assumes that genomes are computable, even in principle – which may not be the case – and that suitable tools and model systems are available to discern the subtle effects that modifying genome sequences may have, for example, on the human cognitive and emotional repertoire. Without the ability to compute the results of modifications to the human genome through predictive computer simulations, the modified code would need to be run and tested in a living individual. But it is hard to see how this could ever be considered safe, ethical, or justifiable. Furthermore, we have already seen how the behaviour of genomes may be extensively modified by their informiomes – the informational context in which they operate – that add additional tiers of unpredictability and undecidability to genetically determined outcomes.

To accurately predict the consequences of extensive genome refactoring, the details of its interactions with the elements of the informiome residing outside the genome would need to be fully understood. If a microorganism can alter the expression of a regulatory long non-coding RNA in a human cell, then unexpected changes may be introduced into the behaviour of rewritten human genomes. The microbiome, for example, is not static: it undergoes continual turnover and there is considerable inter-individual variability. While the generic behaviour of genomes might, in principle, be computable, the dynamic variability inherent within individual informiomes makes it difficult to predict how individual genomes will respond in real-world contexts.

It is therefore instructive to consider some of the more tractable candidate anomalies within the human genome that might, in principle, be addressed by genome engineering. These exemplify the types of design features that may confer a disease propensity, and that could potentially be corrected using a version of 'genomic medicine' focused on limited, local redesign, rather than on more extensive, global rewriting.

On 24 April 1908, the Swiss psychiatrist Eugen Bleuler gave a lecture to the German Psychiatric Association – Deutscher Verein für Psychiatrie – in Berlin. He informed the audience that he was 'taking the liberty of employing the word schizophrenia'. In his 1911 monograph on the topic, *Dementia praecox oder Gruppe der Schizophrenien*, he wrote that in schizophrenia 'the personality loses its unity'. Schizophrenia is a devastating and highly prevalent psychiatric disorder, affecting nearly 1 per cent of the world's population. While environmental factors clearly play an important role in its aetiology, it is also known to have a strong genetic component.

In 2014, the Schizophrenia Working Group of the Psychiatric Genomics Consortium published a paper titled 'Biological insights from 108 schizophrenia-associated genetic loci'. They performed a GWAS (genome-wide association study) on a dataset comprising 36,989 cases of schizophrenia and 113,075 controls. The genetic variants identified as schizophrenia risk factors mapped to 108 different regions of the human genome that differed between those with and without the condition. The findings revealed the remarkable genetic complexity underlying schizophrenia and confirmed its genetic basis. The 108 variants identified contributed, collectively, to nearly every case. While no single obvious disease mechanism emerged from the data, the most significant association was with a region on chromosome 6 known as the major histocompatibility complex (MHC), which contains a cluster of genes involved in immunity. However, finding a locus is not the same as identifying a specific causal gene.

In 2016, Steven A. McCarroll and Aswin Sekar at Harvard Medical School took a closer look at the MHC locus to determine whether they could pin the GWAS causality down to a specific gene. They found that it mapped to two genes, C4A and C4B, which code for complement proteins that are part of the innate immune system. Suspecting that the gene may occur in different forms – one of which could predispose to schizophrenia – they analysed samples from schizophrenia patients and controls. They discovered that the number of copies of C4A and C4B varied between individuals, and that more copies correlated with stronger expression. Additionally, each gene came in long and short forms, with the long variant showing higher levels of expression. The long forms were longer because they incorporated the sequence of a human endogenous retrovirus

(HERV). HERVs are the remnants of ancient viral infections that occurred randomly in our ancestors millions of years ago. They make up around 8 per cent of the human genome. Remarkably, these chance, ancient viral infection events continue to influence genome behaviour and the disease propensities locked into human nature.

The team found that individuals with schizophrenia had, on average, a higher number of C4A copies and were more likely to carry the long form of the gene, leading to elevated C4A expression. But why would higher levels of C4A predispose an individual to schizophrenia? The answer lay in the versatility of C4A proteins. They are examples of the many proteins in the human genome that 'moonlight'. As we have seen, moonlighting proteins perform more than one function at different stages of development, or in different physical locations. In adults, the main role of C4 protein is in innate immunity, but during brain development it plays a central role in synaptic pruning – the process by which synaptic connections between neurons are eliminated. This process continues until early adulthood and is essential for producing a biochemically mature brain.

They hypothesised that increased C4A expression during brain development led to excessive synaptic pruning, a theory supported by the observation of synaptic thinning in the brains of patients with schizophrenia. In a follow-on paper published in 2020, they revealed an intriguing paradox: C4A also conferred protection against autoimmune diseases such as systemic lupus erythematosus and Sjögren's syndrome. The protective effect was especially pronounced in men, who benefited twice as much as women. Some important lessons emerge from this in the context of hypothetical human genome rewriting.

The use of a single protein to perform two different functions – in this case, innate immune response and synaptic pruning – reflects evolution's pragmatic but suboptimal moonlighting strategy, and is the root cause of C4's contribution to schizophrenia. Had evolution selected different proteins to perform these unrelated functions, they would not have become intertwined, and this component of the risk of developing schizophrenia might have been avoided. It makes no engineering sense to repurpose the same protein for such divergent functions. A gene predisposing to one disease may, paradoxically, protect against another, underscoring the unpredictable effects of genetic optimisation. Such linkages can result in the propagation of suboptimality in biological systems. Optimisation in one domain to reduce the risk of developing a specific disease may introduce suboptimalities in others that predispose to the development of different conditions. Males and females may also differ in the portfolio of risks and benefits conferred by particular sequences.

These complexities highlight the immense challenges involved in rewriting and refactoring complex genomes. Any attempts to rewrite the human genome to extend healthy longevity must thoughtfully balance the associated risks and benefits. One of these risks is the fact that the human genome may not be fully computable, even in principle. An apparently successful redesign may skew the overall risk profile, reducing the likelihood of some diseases while increasing susceptibility to others. The introduction of new vulnerabilities following a specific intervention is almost inevitable. However, through careful optimisation, it should often be possible to define design solutions that minimise the probability of introducing new jeopardies. It may also become feasible to untangle moonlighting functions by engineering genomes so that each function is performed by distinct artificial proteins, rather than relying on a single multifunctional one.

In attempting to achieve specific, favourable health outcomes, the natural world may serve as a source of inspiration. Should we, for example, wish to protect human cells from developing cancer, we might consult the strategies of species that have already evolved strategies to do so.

In 2016, Vincent Lynch and Michael Sulak demonstrated that the genomes of elephants contain twenty copies of the tumour suppressor gene *TP53*, which contributes to their remarkable resistance to cancer. *TP53* (commonly referred to as *p53*) is the gene most frequently mutated in human cancers. When functioning correctly, it suppresses tumour development by responding to cellular stress, inhibiting cell division, and invoking programmed cell death through a mechanism called apoptosis.

It is no coincidence that elephants have multiple copies of *p53*, as their large body size elevates their risk of developing cancer. An increased number of *p53* genes may have been an evolutionary prerequisite for achieving such size – a simple fix for a major biological challenge. But, as is often the case, solving one problem may create another. Mice provided with an extra copy of *p53* develop significantly fewer tumours following exposure to carcinogens, but acquire an unwelcome trade-off – they age prematurely. If we attempted to protect ourselves from cancer by adopting the elephant's elegant strategy, we might encounter a similar trade-off. In some instances, however, addressing a health issue can be more straightforward.

John Tacket was an exceptionally rare teenager. At the age of just fifteen, he had already exceeded his predicted life expectancy. One year later, he was dead. He suffered from Hutchinson–Gilford Progeria Syndrome (HGPS), a disease that causes accelerated ageing and is driven by a single mutation in a gene called *LMNA*. This results in the production of a truncated lamin A protein, which normally functions to maintain the integrity of the inner nuclear membrane. Although healthy at birth, symptoms began to emerge at around one year of age. The disease is characterised by severe atherosclerosis – typically the cause of death – as well as osteoporosis, early growth cessation, degeneration of skin and muscle, hearing loss, and baldness. Affected individuals age so rapidly that, by the age of ten, they acquire the appearance of extreme old age, typically resembling octogenarians.

In 2021, David Liu and Luke Koblan showed that a precise form of gene editing that targets a single-nucleotide base, known as base editing, could correct the mutation in mice. This raised the hope of an eventual cure. By seven months, untreated mice had grey fur, were hunched over, and barely mobile. In contrast, mice treated shortly after birth had healthy coats and moved vigorously. The disease does not, however, recapitulate all the features of natural ageing, such as neurodegeneration or increased cancer risk, and its direct relevance to natural ageing therefore remains uncertain. Nevertheless, it suggests that many aspects of natural ageing may have a strong genetic component, which is potentially modifiable through genome redesign and rewriting.

A perusal of the lifespans of various species, and of the impact that interventions may have on them, provides further evidence that, rather than being fixed and immutable, lifespans have a significant genetic component. The mushroom-forming fungus *Armillaria bulbosa*, for instance, is especially adept in this domain. It may live for several thousand years and is one of the longest-living organisms on Earth. It achieves this, in part, by maintaining an exceptionally low mutation rate. In contrast, other species of fungi, such as *Podospora anserina*, live for just a few weeks. Studying the genomes of closely related species with dramatically different lifespans may help uncover the principles that could extend the human lifespan and healthspan through genomic modification.

A striking example of this is found in some Pacific Ocean rockfishes. The short-lived species *Sebastes minor* lives for just eleven years, while its closely related cousin, the rougheye rockfish *Sebastes aleutianus*, can live for more than 200 years. In 2021, Peter Sudmant and Sree Kolora at the University of California, Berkeley, compared the genomes of 88 different species of long- and short-lived rockfish. They found that longer-lived species had an increased number of genes associated with DNA repair and maintenance. They also had more genes coding for nutrient-sensing pathways. Remarkably, the same DNA repair genes enriched in long-lived species of rockfish were also found in the bowhead whale *Balaena mysticetus*, whose lifespan may likewise exceed 200 years.

In contrast, these genes are mutated in the short-lived turquoise killifish *Nothobranchius furzeri*, which inhabits the transient pools that form during the rainy season in Mozambique and Zimbabwe, and lives for less than a year. Long-lived rockfish species also have more copies of genes involved in reducing inflammation, while short-lived rockfish species are generally smaller and live in shallower waters, making them more vulnerable to predation and competition. Given their short life expectancy due to mortality from environmental causes, their genes would have little opportunity to be selected for increased longevity.

Turtles also have exceptionally long lifespans, sometimes living for more than 150 years. In a 2022 study of tortoises and turtles living in zoos and aquariums, Fernando Colchero and Rita de Silva at the University of Southern Denmark found little evidence to suggest that many of these species age significantly, if at all, or that their mortality increased with age. This suggests that ageing – at least in some species – may not be inevitable. The bristlecone pine *Pinus longaeva*, the longest-living tree species on Earth, can live for several thousand years. The oldest known individual bristlecone tree, known as 'Methuselah' and located in the White Mountains of California, is more than 4,800 years old.

Lifestyle and environmental factors also have a significant impact on longevity. In 1678, the Greek Orthodox bishop Joseph Georgirenes published a book in English, *A description of the present state of Samos, Nicaria, Patmos and Mount Athos*, which provided an account of these islands, including their geography, religious practices and culture. He noted the exceptional longevity of the inhabitants of Nicaria (now known as Icaria) and its abundance of centenarians. This small Greek island, located in the Aegean Sea, has one of the world's longest-lived human populations. Georgirenes attributed the islanders' longevity to a combination of community spirit, a simple diet and favourable genetics, observing that the small island was 'the poorest and yet the happiest of the whole Aegean Sea'.

The oldest living human on record was Jeanne Calment, who was born in 1875 and lived to the age of 122 years and 164 days. In 2018, Kenneth Watcher and Elisabetta Barbi at the University of California, Berkeley, presented evidence suggesting that the human lifespan may have no natural boundaries. Examining 4,000 'super-elderly' individuals in Italy aged 105 or older, they demonstrated that mortality appears to level off after age 105, creating a so-called 'mortality plateau'. However, this conclusion has been challenged on the basis that the design of the human body contains intrinsic limitations. Neurons, for example, cannot replicate or renew themselves, thereby setting intrinsic limits on the human lifespan.

While the genomic blueprints and lifestyles of long-lived species like the Pacific Ocean rockfish and turtles embody hidden rules and evolutionary Earth-secrets that offer insights into human health, it is clear that human beings must now become their own model organism. As the biologist Sydney Brenner stated, 'we must look at humanity's genome, not the human genome. We must study ordinary people – with no hypothesis – in large numbers.'

As databases of human genome sequence information reach unprecedented sizes, we are entering an era characterised by an extraordinary degree of fact generation and AI-mediated knowledge extraction, as neural networks and deep learning algorithms interrogate vast datasets. A crucial next step will be persuading institutions and nation states to share their anonymised genomic databases and pool them into a single global resource. However, the data will only be as useful as the quality of the accompanying healthcare information.

Medical records are notoriously inconsistent, and efforts will be required to increase their quality and accuracy.

We will need to take the science of documenting human characteristics more seriously. In 2018, Dan Rader and Garret FitzGerald at the University of Pennsylvania called for a new kind of 'human phenomic science', characterised by the 'deep phenotyping' of individuals. But it is not just genetic information that matters. In 2024, G. Cavalli and V. Parreno at the University of Montpellier in France showed that a brief and reversible inhibition of a gene-silencing mechanism – an epigenetic process – could result in irreversible tumour formation in the fruit fly, *Drosophila melanogaster*, without introducing any permanent changes into the DNA.

We will also need to complete the cartography of human cellular organisation, mapping behaviour at the systems level. This includes the wiring diagrams both within and between cells, encompassing, for example, the estimated 86 billion neurones in the human brain, and an equivalent number of glial cells. The feasibility of this was demonstrated in 1986 with the mapping of the 302-neuronal cell network of the *Caenorhabditis elegans* (roundworm) brain – the so-called 'mind of a worm'.

In 2023, the BRAIN Initiative Cell Census Network published a cellular atlas of the entire mouse brain. These types of studies may eventually reveal the molecular and cellular basis of consciousness, and suggest how the human brain constructs complex phenomena such as morality, goodwill and creativity. Fortunately, as Paul Berg and Janet Mertz observed, 'nature's secrets are not beyond our capabilities'.

In his 2002 Nobel Prize lecture 'Nature's Gift to Science', Sydney Brenner remarked: 'There are many aspects of humanity that we still need to understand for which there are no useful models. Perhaps we should pretend that morality is known only to the gods, and that if we treat humans as model organisms for the gods, then in studying ourselves we may come to understand the gods as well.' If there is a God, or a Creator, then humankind may be its model organism. The detailed study of our own species may be the only way of comprehending what exists beyond us.

12
A Manifesto for Life

> To deride the hope of progress is the ultimate fatuity, the last word in the poverty of spirit and meanness of mind.
>
> – PETER MEDAWAR, 1972

The ongoing advances in our ability to design and synthesise the genomes of living organisms from first principles are unprecedented in human history. They are informed by the convergence of emerging insights into the generative rules of biology – derived through the application of AI to vast biological databases – with technologies that allow DNA to be constructed at scale. Together, these nascent enterprises hold the potential to fundamentally alter the nature of life on Earth and to unlock the potential for entirely new species. Such advances offer the promise of significant benefits to society, including the transformation of healthcare and the generation of biologically inspired materials and devices. They also convey great risk.

The eventual outcome of such activities – the attainment of Artificial Biological Intelligence – will provide humankind with the ability to predict the nature of an organism from its genome sequence, and to define the composition of genomes necessary to realise specific biological outcomes. This design capability, based upon a mastery of the generative grammar of genomes, will be accompanied by a constructive capability that enables the efficient synthesis of artificial genomes of any length or sequence, and their actualisation in living cells.

It is the biological equivalent of Artificial General Intelligence. ABI represents the attainment of a comprehension of the generative grammar of biology that approximates to that of nature. Fluency in this biological language will allow the genomes of any possible species to be designed and constructed at will. ABI will also allow for the specific modification of existing species in a pre-determined manner. Its attainment must become the central objective and cornerstone of the biological sciences. The eventual elucidation of the generative rules of biology underlying ABI will transform biology into a predictive engineering material. Just as AGI began with video arcade games, progressed to games such as checkers, chess, Go, and then to chatbots and the prediction of protein structures at an atomic level of resolution, so the generative competence of ABI will also be acquired incrementally.

At first, we will learn how to 'speak' the languages of viruses and bacteria. Once we have mastered the ability to converse with simple biological entities, we will progress to 'talking' to more complex, multicellular forms of life. Eventually, we will learn the languages of all species – including that of the human genome itself. When the universal generative grammar of biology has been fully comprehended, the possibilities for authorship will be limitless.

Given ABI's profound significance, there is a compelling need to consider how its anticipated generative capability may be deployed in a safe, responsible, inclusive, and ethical manner. It is essential that it is used in a way that benefits all humankind equally, while safeguarding the natural species and ecosystems shaped by life's evolutionary history. While discussions of the potential consequences of biology's future capabilities are often postponed, we must prioritise the debate and initiate it in a timely manner.

What follows is a preliminary attempt to sketch out an outline vision of life's potential future, and to detail the exploration of the – as yet poorly characterised – biological landscapes that ABI will reveal. It is not intended as a definitive manifesto, but rather as a tentative starting point for an ongoing discussion. The definitive manifesto will not be drafted by any individual. It will emerge instead from the aligned contributions of representative members of humanity. Referencing diverse and contrasting perspectives on a subject of such fundamental importance to humankind's future is both expected and necessary.

The manifesto embraces the principles of equality and diversity, as well as religious, political, and ideological tolerance. It also acknowledges the intrinsic value and legitimacy of the many different ways of being human. These inviolable principles form both its foundations and its guiding light. It addresses the need to preserve the integrity of the natural world in its broadest sense, as well as the non-negotiable aspects of authentic human nature as we currently experience it.

A framework for navigating the developments in genome design and synthesis within the context of human healthcare is provided, along with a discussion of how the emerging ability to programme biology may be used to create living structures that exist beyond nature. By necessity, it also touches on some of the broader issues these technological advances will invariably invoke, and offers guidance on the types of interventions that may be considered reasonable.

The eventual consensus manifesto will emerge following extensive debate and alignment among multidisciplinary experts, governmental agencies, and nation states. This must also, necessarily, include extensive public engagement. We will need to determine what kind of world we wish to live in and align, as a species, to define a coherent vision for the future of humanity and of natural and artificial species. There is a compelling existential need to chart a coherent roadmap for the responsible navigation of life's future. This will, among other things, involve defining what it means to be human.

In his 1980 Nobel Prize lecture 'Dissections and Reconstructions of Genes and Chromosomes', the biochemist Paul Berg questioned whether 'certain enquiries at the edge of our knowledge and our ignorance should cease for fear of what we could discover or create'. He concluded by referencing Peter Medawar's book *The Hope of Progress* (1972), in which Medawar denounced the cynicism of rejecting progress, stating: 'To deride the hope of progress is the ultimate fatuity, the last word in the poverty of spirit, and the meanness of mind.' Despite the risks, we cannot afford to ignore the unprecedented opportunities before us. On the contrary, we must actively and responsively explore these possibilities to determine how they might benefit humankind and other species, while drawing awareness to their pitfalls and potential for misuse.

In short, we are compelled to proceed – and to search for new ways to improve the human condition. This includes the elimination of disease; securing the basis for a second green revolution to establish global food security; extending healthy longevity; reducing suffering; and safeguarding the natural world. The desire to explore alternative worlds and to expand the possibilities of this one is an intrinsic part of the human spirit. It has defined the history of our species. To deny the possibility of such opportunities would be to undermine the fabric of our humanity. Yet we should undertake this journey with great circumspection.

The most pressing issue is whether we are, as a species, capable of uniting to prioritise nature and ensure its continued welfare over short-term economic gain and expediency. We are its privileged custodians, and must redefine our transactional and utilitarian relationship with the natural world into one that is more nuanced, respectful, and enduring. While exploring the potential of generative biology, we must, at each step, take significant care to preserve, protect and respect natural species. No issue is more urgent. It must form the cornerstone of any discussion concerning life's synthetic future.

In reducing nature's creations to the products of rules-based generative principles, we must not lose our appreciation of their intrinsic wonder. Gaining insights into the mechanisms underlying nature's complexity does not diminish its mystery. While the molecular logic of living organisms presents formidable challenges to our ability to engineer biology, it remains as valuable and aesthetically compelling as Manet's *Le Déjeuner sur l'herbe*, or da Vinci's *Mona Lisa*. The following is an outline of the principal issues that must form the essential foundations of any manifesto that attempts to address life's future.

I
The Conservation of Nature

Humanity must urgently re-examine its relationship with nature. Life's extraordinary creations are the product of approximately four billion years of evolution – both our inheritance and a precious resource. The idiosyncrasies and diversity of living things are nature's generous gift to humanity. Nature must be protected, not just for its utility, but for its own sake.

The biologist J. B. S. Haldane remarked in his 1947 book *What is Life?* that 'The Creator would appear as endowed with a passion for stars, on the one hand, and for beetles on the other.' With more than 400,000 known species, beetles have a staggering diversity. Would it matter if just one were to disappear? From a heritage perspective, the answer is an unequivocal yes. The collection of Earth's species comprises a unique masterpiece – each an essential detail of the whole. Moreover, every species, no matter how apparently irrelevant or inconspicuous, plays a unique and often unpredictable role in the delicate balance of ecosystems.

Should such arguments fail to persuade, there is a more pragmatic one. The genomes of Earth's species represent an unparalleled archive of biological ingenuity – a living library that has been refined and curated across evolutionary history. To squander this resource would be an act of extreme recklessness and disregard. The loss of even the most seemingly insignificant species should concern us. Each holds biological Earth-secrets – insights that could enhance our understanding of the workings of life itself.

Biological diversity has proved invaluable in advancing scientific progress, offering ready-made solutions to a broad range of technological and medical challenges. But the species that share the world with us are more than just a repository of knowledge. They are an essential part of our humanity – a mirror of human nature. They help us to trace our evolutionary past and, in so doing, teach us who we are. Yet we have placed ourselves above nature, treating it with indifference and disdain. To redress this imbalance, we must learn once again to value the natural world and acknowledge its importance for our survival and well-being.

We continue to discover potentially habitable exoplanets orbiting stars beyond our solar system, but we may have overestimated the likelihood of there being life elsewhere in the universe. The emergence of life – especially intelligent life – may be a highly improbable event, perhaps even a singular one. In light of this, every Earth species becomes significant.

Human activity has profoundly disrupted Earth's climate, which has been relatively stable for the past 12,000 years, since the end of the last Ice Age. This has led to desertification, the destruction of natural habitats, ecosystem degradation, and the extinction of countless species. While extinction is an expected part of evolutionary processes, the current rate of species loss is unprecedented. We are presiding over an ecological crisis on a global scale, and of our own making. The destruction of the natural world and unravelling of life's fabric must stop and, where possible, be reversed.

The diversity of habitats and natural species must be preserved for future generations. They are sacred. No species exists in isolation. All life on Earth is part of a complex, entangled system of ecological networks. Humankind has a responsibility to future generations to operate within safe and sustainable boundaries to maintain the planet's stability and resilience. We must protect Earth from the unjust planetary change that is already subjecting hundreds of millions of individuals to extreme climate conditions, while simultaneously undermining global biodiversity. We must take our stewardship of Earth seriously.

Currently only 19 per cent of Earth remains true wilderness. Urgent action is required to protect these areas, preserve habitats, safeguard biodiversity, curb pollution, halt and reverse climate change, secure migration corridors, adopt sustainable technologies for energy production, prevent further deforestation, and reforest the planet. These actions will reduce greenhouse gas emissions such as carbon dioxide and methane, protect ecosystems, and help restore predictable seasonal patterns. They should also slow the rise in sea level caused by the melting of the Antarctic Ice Sheet – the largest reservoir of land ice on Earth. With the global human population set to reach eleven billion by the century's end, we must stabilise it at a more sustainable level and embark on a global rewilding effort. The suggestion by E. O. Wilson in his 2016 book *Half-Earth*, that we should preserve half the world as wilderness, is reasonable and potentially achievable.

We are the custodians of human nature's future, and with that comes the responsibility of preserving its defining features. These include: free will, individual liberty, empathy, freedom of speech, privacy the capacity for morality, the right to dissent, self-determination and cognitive and intellectual liberty. These, in turn, encompass the right to control our brain chemistry and protect our innermost data – our emotions and thoughts. Without robust political structures that respect and uphold such fundamental human rights, the essence of authentic human nature, as we know it, is at risk of being undermined – perhaps irreversibly.

In light of the emerging potential to erase such essential facets of the human construction, there has never been a moment in history where the perpetuation of political structures aligned with the preservation of such inalienable rights, has been of greater importance. The ability to eliminate the possibility for dissent and permanently obliterate free will may one day become a technologically feasible prospect.

Beyond political and ideational freedoms, the integrity, sanctity, and authenticity of the fabric of human nature itself must be safeguarded. No attempts should be made to interfere with its essential, non-negotiable elements, including memory, the right to construct authentic experiences, unconstrained cognition, authentic emotions, the integrity of the subconscious mind, irrationality, individuality, autonomy, moral intuition, justice, instinct, and the ability to distinguish good from evil. Altruism, empathy, reciprocity, the capacity for unconditional love, forgiveness, kindness, humour, generosity, selflessness, and compassion must also remain inviolate.

Even the so-called fallibilities of human nature – such as jealousy, inconsistency, impulsivity and avarice – should not be perceived as problems to be solved, so much as part of the cement that holds our necessarily flawed and imperfect nature together. It is precisely these types of paradoxical constructions that make us human. To deny them would be to diminish our agency. We might call this disparate collection of characteristics the human 'soul', without needing to invoke a metaphysical agency. The human brain is an ancient machine that evolved to keep us safe in a world that no longer exists. Yet its peculiarities and contradictions endow it with an intrinsic authenticity and beauty. Our imperfections should be cherished.

Humanity is defined by diversity – not only of biology, but of thought, culture, and societal structure. These differences should be celebrated and embraced. Variation in gender, class, aesthetics, ideology, sexual orientation, intellect, economics, and beliefs is integral to human civilisation. Even at the genetic level, there is no singular, Platonic ideal of a human genome. While we can define a pangenome – a reference map of common genetic variation – every individual deviates from that consensus in their own unique way. These differences are what make humans such a resilient, adaptable, and culturally divergent species. If human nature is to be preserved, we must delineate the landscape of tolerated genetic variation that defines its boundaries and demarcates its natural operations. Without such definitions, we risk inadvertently altering what it means to be human while being unaware that we have done so.

The rise of artificial intelligence, and in particular deep learning applied to vast, annotated databases, will help reveal which parts of the human genome shape individual and social behaviour. But such knowledge should not be used to engineer human nature through either direct or indirect interventions. As Artificial General Intelligence emerges – an aspirational human-like level of machine intelligence with an applicability that generalises across multiple domains – followed, potentially, by Artificial Superintelligence, which may surpass most natural human capabilities, we will be tempted to rewrite human nature, either in part or in its entirety. This must be vigorously resisted.

Human reproduction should remain a natural process, except in cases that necessitate intervention – for example, infertility – in which case the means to do this should be readily available. Synthetic human genomes, while valuable for medicine and research – especially for engineered cell therapies – should not form the basis of parentless reproduction. Nor should we attempt to 'switch off' human evolution by disabling mutational mechanisms or artificially stabilising the genome. Human nature must retain its ability to evolve, as it always has.

A new kind of science may yet emerge – one that we might call *culturomics* – which explores how genomes enable and constrain the repertoire of cultural possibilities available to a species. This may never predict the existence of specific cultural artefacts – a game of football, wristwatches, coffee drinking, tiddlywinks, a symphony orchestra, or the origin of democracy in ancient Greece – but it could reveal the underlying genetic architectures that determine the types of possible human cultures that may be generated. No attempt should be made to alter the intrinsic cultural potential of the natural human genome.

II
The Information of Species

The various components of information that define a species – rooted in its genome, yet nested hierarchically above it – may be thought of collectively as its informiome. This conceptual framework extends beyond genetic sequences to encompass information distributed across multiple levels and substrates. In this view, species function as information superclusters: anchored in their genomes, yet extending far beyond them. This informational penumbra, built upon genetic foundations, incorporates development, epigenetics, learning and culture, along with the complex ecosystems in which organisms are embedded. It also includes the viral, microbial, fungal and parasitic genomes that coinhabit individual organisms. In this sense, the informiome is a kind of informational 'metagenome'.

While the genome serves as a digital code script capturing a species' foundational genetic information, not all biological information is encoded. The self-assembly of protein complexes, the folding of individual proteins, the assembly of viral particles, and other emergent phenomena arise from the agency of self-organising processes. Such properties are not directly specified by the genome, emerging instead from the behaviour of its encoded components. A predictive science of such emergent, unspecified behaviours remains a significant challenge.

Despite their diversity, all natural species rely on a remarkably narrow biochemical repertoire to encode and express biological information. With the exception of certain viruses that use RNA genomes, the genetic information of all known organisms is encoded using the same four DNA nucleotides. Likewise, the twenty naturally occurring amino acids serve as the fundamental building blocks of proteins across the tree of life. The genetic code – the rules that translate nucleotide sequences into the amino acid sequences of proteins – is near universal, with minor deviations in mitochondria, ciliates and certain fungi and protozoa. Yet, living systems could, in theory, be constructed from unnatural biochemical 'software' and 'hardware' – chemical systems that nature has never previously explored. Each of these alternative chemistries generates its own combinatorial information space, analogous to the sequence space of DNA.

Both the genetic material and the genetic code connecting a genome's code script to the molecular machinery of life can be altered artificially. DNA could, for example, be replaced by xeno-nucleic acids – synthetic nucleic acids with chemically distinct backbones or bases. Examples include HNA (hexitol nucleic acid), PNA (peptide nucleic acid), TNA (threose nucleic acid) and GNA (glycol nucleic acid). Each of these alternative genetic systems would require its own specialised enzymes for replication and repair, and each would constitute a novel biological language with its own generative grammar.

Beyond the genome's material composition, the genetic code itself can be rewritten through the design and synthesis of recoded genomes, enabling the incorporation of non-natural amino acids – also referred to as non-canonical amino acids or unnatural amino acids. These include O-methyltyrosine and L-homopropargylglycine (HPG), which depending on their structure, may resemble or diverge significantly from natural amino acids. The introduction of unnatural amino acids into biological systems expands the functional repertoire of natural proteins, opening up new possibilities for synthetic biology.

Amino acids themselves could, in theory, be replaced by entirely novel polymers constructed from non-canonical monomers. These might include α, α-disubstituted amino acids, macrocyclic peptides and depsipeptides. Whether such alternative building blocks could support life as we know it remains an open question.

The term Artificial Biological Superintelligence (ABS) captures the potential to define different generative grammars of life using alternative chemistries that go beyond those available to natural biological systems. These might include distinct types of genetic materials, or non-canonical biochemical components. There is a sense in which the mathematical space containing the potential for all biological structures, exists independently of the biological languages available to articulate them. While some of these 'logical forms' are readily expressed using natural biological technologies, others are not.

To the extent that all biological structures must inevitably be imperfect representations of the primary mathematical structures describing them, there is a sense in which all natural biological structures are approximations or 'mispronunciations'. It must also be the case that there are many potential biological structures that natural biology must pass over in silence. The particular language of biology utilised by a naturally evolved or artificially generated system therefore sets an intrinsic limit on what can be expressed or realised by it, much in the same way that the use of a specific spoken language constrains the manner of articulation of a particular thought.

III
A Paradigm Change in the Biological Sciences

We are entering a golden age of molecular genetics, poised to form the basis of an unprecedented, biologically inspired industrial revolution in science and medicine. Biology is evolving from a descriptive science into a predictive engineering discipline that will impact almost every human endeavour. The biological sciences have, for centuries, used a reverse-engineering paradigm to decipher nature's creations. Now, informed by artificial intelligence and large-scale genome synthesis, biology is transitioning into a forward-engineering, bottom-up, constructional enterprise. This marks a fundamental paradigmatic shift. Instead of breaking biological systems down to determine the function of individual components, we will build them to understand them.

Artificial intelligence, in combination with synthetic genomics, allows for the interrogation of vast biological datasets, the generation of hypotheses and for iteratively designing, building and testing them – linking specific sequences to defined outcomes. This process will help reveal biology's generative rules – its hidden logic – enabling the long-sought-after mathematisation of life.

In this new constructional paradigm, natural species, shaped by Darwinian evolution, will be joined by artificial species designed according to predefined human specifications. Whole-genome design and synthesis go far beyond gene editing. Whereas gene editing modifies existing genetic texts, genome writing dispenses with inherited templates altogether. It liberates life from heredity and descent, making it timeless and allowing genomes to be reimagined from scratch.

For the first time ever, species generation will no longer be dictated solely by natural selection. Rather than remaining passive observers of nature's handiwork, we are instead becoming authors of life. The creative potential of generative biology technologies is considerable. They will transform agriculture, industry and – perhaps most significantly – medicine. We are transitioning from a Darwinian world into a post-Darwinian landscape: a limitless synthetic frontier. Like the pioneering hominins who first ventured out of Africa, we are embarking on our own migration – into the uncharted and borderless territory of engineered life. A landscape that we are only just beginning to comprehend.

For much of the twentieth century, the term 'gene' was equated with DNA sequences encoding proteins. Large stretches of DNA in the human genome with no apparent function were dismissed as 'junk'. We now know that much of this so-called junk DNA comprises non-coding elements encoding regulatory RNA molecules, which exert a profound influence on the biology of species. These RNA-based components of biological machines are every bit as important as protein-encoding genes, and control and fine-tune their behaviour.

To fully grasp the molecular logic of complex genomes, we must move beyond the protein-centric view of genomes and systematically explore their non-coding 'dark matter'. Genome synthesis will be key to this effort, enabling the iterative rewriting and testing of specific genomic regions to uncover the rules governing regulatory sequences. We are entering an 'RNA renaissance', in which the role of non-coding RNA in health and disease is coming into sharp focus. This new focus on RNA will be especially critical for understanding human pathology.

Though biological systems differ fundamentally from digital computers, the 'cell-as-computer' and 'cell-as-machine' metaphors retain considerable practical utility. The notion of the genome as an operating system comprising a set of curated programmes, is especially evident when a transplanted genome transforms the cell of one species into that of another. Yet this analogy has its limits. Many components of biological systems, for example, operate in a continuous, analogue-like manner. The logic of biology is therefore 'fuzzy'. Furthermore, rather than being programmed, much of its order emerges spontaneously as a result of self-organising processes. Protein folding and the assembly of viral capsids – where complex structures emerge in the absence of explicit encoded instructions – are striking examples.

Moreover, most biological components do not behave like engineered parts that perform single, well-defined functions. Instead, they often 'moonlight' – performing multiple functions, depending on their context. They also form the parts of complex, highly interconnected molecular networks. To transform biology into a truly predictive engineering discipline, we will need to deconvolute this complexity. By refactoring biological systems – simplifying and modularising their components – we may eventually render living systems as engineerable and programmable as silicon-based computing systems. The requirement for network operations may, however, set a functional limit to the extent to which this can be achieved.

Biology is on the cusp of becoming fully programmable – the engineering material of choice for numerous applications. Engineering biology – variously known as synthetic genomics, generative biology, generative genomics or genome writing – is now converging with AI-augmented genome design and has the potential to address many of the world's most pressing challenges. It promises to transform human existence and the living world in ways barely imaginable. The benefits could be immense. It could also cause incalculable harm.

The ability to design and synthesise genomes with predictive precision marks the closure of a historic biological information loop.

After billions of years of evolution by natural selection, life is now at the earliest stages of decoupling from heredity. It is becoming free to reinvent itself under human direction. That we have developed techniques capable of this kind of bottom-up design may be seen as a natural corrective mechanism to address the exponential changes generated by human culture. In short, the incremental processes of biology can no longer keep pace with culture.

The velocity of cultural change driven by modern technological innovation greatly exceeds anything previously observed in human history. From the discovery of agriculture through to the invention of the printing press, internal combustion engine, computers, AI, and synthetic biology, humanity has launched itself into a state of hyper-accelerated transformation. We might call this phenomenon 'hyperculture' – a mode of human existence in which environmental, technological, cultural, and societal changes occur at an extraordinarily high (and exponentially increasing) frequency. The ability to design and write genomes may offer a way to reconcile biology with hyperculture – providing life with the means to update ancient biological systems designed to operate in a world that no longer exists.

Although still in its infancy, the science of genome design must overcome formidable technological hurdles to realise the promise of strong ABI: a fusion of Artificial Biological Intelligence capable of genuine genomic authorship, combined with the ability to construct and activate the corresponding genome sequence.

It will also require the ability to build DNA of any complexity, rapidly, efficiently, accurately, at scale, and at minimal cost. If the capacity to write genomes progresses along the same trajectory as DNA sequencing, exponential improvements are likely. Once sufficiently advanced, genome writing will enable the on-demand production of prespecified genome sequences, and even whole libraries of sequences of a particular type. Genomes themselves will be printed out routinely – like documents on a laser printer.

For now, however, we must remain imitators. Current genome synthesis efforts – even at their most innovative, for example, the creation of a novel seventeenth neochromosome by the Sc2.0 synthetic yeast project – remain principally plagiaristic. They resynthesise natural genome designs rather than inventing truly novel ones. Such engineered genomes are still constrained by the pre-existing architectures and mandates of nature. But we are moving steadily towards strong ABI, where the genuine creative authorship of unnatural genomes will become possible. The pathway to this future will be paved incrementally by increasingly capable forms of weak ABI, with progress depending on the choice of model systems, AI architectures, computing hardware and training datasets.

Bacteriophages – viruses that infect bacteria – have historically proven foundational to many of the key breakthroughs in molecular biology. They are, therefore, a logical starting point for weak ABI, providing a tractable testing ground for artificial genome design. Mastering the language of viruses will be highly consequential. It is unsurprising that the first fully artificial biological entity to be designed, physically constructed, and tested – an AI-generated Φ-X174 bacteriophage called Evo-Φ2147 – was a virus. Based on this landmark achievement, future success in the emerging field of 'generative genomics' will be measured not by imitating existing genomes, but by crafting entirely new ones, in order to generate biological entities that have never previously existed in nature.

IV
The Generative Rules of Biology

While we have made remarkable progress in reading the alphabetic nucleotide letters of life's language, our ability to comprehend the deeper meaning of biology's utterances remains rudimentary. Beyond translating nucleotide sequences into amino acids – the most rudimentary level of meaning – we know little about the grammatical rules and design principles governing life's complexity. To acquire the fluency in this language necessary to author new species, we will need to decipher its hidden grammatical structure. AI, in the guise of large language models, will play a central role in this. A key part of the puzzle lies in understanding how genomes fold within cells and how this folding programmatically influences gene expression.

The language of life, much like spoken and written languages, accommodates a rich variety of alphabets, vocabularies, grammars, syntactic structures and dialects. While there are areas of evolutionary overlap, the grammar, syntax and vocabulary of bacterial genomes, for example, differ significantly from those of flies and humans. Despite the presence of conserved construction pathways, you could not build a human using the grammar of a nematode. Mastering these languages on a species-by-species basis will ultimately allow us to programme all forms of life. Deciphering the grammar of the human genome will have profound implications for human health, with the potential to extend healthspan and lifespan in ways never before imagined.

Unlocking biology's 'codes within codes' will require the interrogation of extensive, multimodal datasets at an unprecedented scale. AI will be indispensable in navigating this complexity. Yet, unlike the laws of physics – elegantly described by mathematical formulae – the inner workings of AI remain poorly understood. As a result, while we may soon be able to apply biology's generative rules with precision, attaining a true understanding of them may prove challenging.

In the future, we will construct an exhaustive chart of genome sequence space – a biological atlas detailing all possible extinct, existing, and potential future natural species. This conceptual inventory of theoretically realisable and unrealisable life forms – constrained by the limits of natural biochemistry – would serve both as a compendium of potential species and as a guide for navigating genome sequence space. Such a map would not only delineate the boundaries of biological possibility, but also highlight genomes with potential utility for humankind. Since genome sequence space has a timeless mathematical existence, this project represents less the creation of new life than the discovery of virtual organisms – an exploration of a computationally defined, infinite library of possibilities. In a mathematical sense, these species already exist within this vast archive of biological potential. Early, low-resolution maps will help identify promising regions of genome sequence space, replete with biologically plausible organisms, with successive refinements leading to increasingly higher levels of resolution. Ultimately, this will result in a comprehensive atlas of organismal possibility – an exhaustive 'natural' species map.

Beyond natural biochemistry, we may also map the genomic potential of alternative biological universes – those governed by informational systems beyond DNA and RNA, or composed of recoded species built from non-canonical amino acids and other 'unnatural' molecular architectures. Most of these potential species will never exist, or cannot exist, in the physical world, but they all, regardless, possess a concrete mathematical reality, forming a kind of Platonic 'species real estate' – a virtual landscape that can be charted and explored as if it were real.

The programmable design of synthetic species – including genomically rewritten microorganisms – holds immense potential for biomanufacturing. In the future, biologically inspired materials and devices will integrate into every aspect of our daily life. Smartphones may be living; we may grow houses rather than build them; and our clothing items may converse with us and hold opinions. We may archive our information in DNA storage media, write documents on biocomputers, and package our food using sustainable biofilms. Synthetic species could be harnessed to produce biofuels, medicines, biosensors, drought-resistant crops and countless other innovations. By replacing plastics and other environmentally harmful materials, biomaterials derived from synthetic species could dramatically reduce global pollution and contribute to the stabilisation – and potential reversal – of climate change.

But synthetic species must not be allowed to endanger natural organisms or destabilise ecosystems. Their deployment should be restricted to controlled environments and applications that do not require release into the wild. The criteria for their environmental release must be stringent, demanding a comprehensive understanding of the ecosystems they may affect. To support this, we will need a predictive science capable of modelling interactions within and between ecosystems with great precision. One can imagine a new science – 'ecogenomics' – dedicated to modelling the impact of synthetic species on natural ecosystems. Mastering this will be critical to life's future and to preserving the unity of life on Earth.

In a world where species can be artificially generated using AI-informed genome design and genome synthesis, the concept of a species – rooted in ancestry and reproductive barriers that restrict gene flow – may require revision. If a species can be created by synthesis, the classical definition becomes obsolete. Yet it remains clear that living organisms exhibit distinct anatomical, biochemical and behavioural characteristics. We might therefore choose to introduce a new term – 'morphora' – to replace *species*, better reflecting the reality of AI-designed and synthetically generated life.

Individual genes and their regulatory sequences can be redesigned and constructed to optimise their function. AI is ideally suited to this task, facilitating the creation of novel protein and RNA structures at an atomic level of resolution that extend beyond the natural repertoires discovered by evolution. The ability to design and construct new genes and libraries through the convergence of AI-guided design technologies and versatile construction technologies carries significant biosecurity risks. It could be used to create proteins with pathogenic potential that evade detection through having only minimal sequence identity with their natural counterparts.

The systematic reprogramming of the human microbiome presents another frontier with profound implications for healthcare. To achieve this, we must first construct comprehensive inventories detailing the collection of species from which individual microbiomes and their genomic compositions are formed. The totality of the genes of the microbiome should be regarded as an extension of the human genome, requiring continuous monitoring and curation. This includes not just bacteria, but viruses (virome), fungi (mycobiome) and parasites (which we might call the parasitome).

A more detailed understanding of microbiome structure could lead to novel interventions for disease prevention and treatment. Certain bacteria, such as *Helicobacter pylori*, are known to drive disease processes and may even influence the effectiveness of immunotherapies. In a study of an anti-PD1 monoclonal antibody immunotherapy that reactivates immune cells, melanoma patients with a high abundance and diversity of bacteria from the family *Ruminococcaceae* were more likely to respond to treatment. This provides a potential basis for future patient selection and for the systematic engineering of the microbiome to enhance clinical sensitivity to immunotherapy.

It is necessary to address the question of whether synthetic species – artificially designed and constructed by humans – possess the same legitimacy and authenticity as natural species, which have been crafted through billions of years of evolution by natural selection. Although built using the same biochemical principles as natural life, synthetic organisms have not co-evolved with other species. Nor have they endured the relentless selection pressures of evolutionary time, or adapted in concert with other species and ecological networks. Synthetic species, while undeniably biological, exist beyond the natural web of life rather than within it.

For this reason, naturally evolved organisms should, for now, be regarded as more authentic and legitimate than their artificial counterparts. The distinction between natural and artificial will, however, become increasingly blurred. We know that evolved life is highly historically contingent, and that the emergence of individual species, including humans, is by no means inevitable. On that basis, naturally evolved life has no intrinsic monopoly over claims for authenticity and legitimacy.

If de-extinction activities are to advance at all, efforts should be directed at prioritising species that have disappeared within the past few decades, where there remains a chance that suitable ecological niches may still be available to accommodate them. Extreme caution is warranted before releasing ancient de-extincted species into the wild. The world they once inhabited no longer exists and their reintroduction could drive other species to extinction. The natural world is in constant flux, making history an unreliable reference point for ecological restoration. We should not treat the distant past as a baseline to which we should return.

De-extinction is an imperfect and logically flawed process. We will never, for example, be able to resurrect the historical dodo or woolly mammoth. The information defining them transcends genome sequences. Though such de-extincted creatures may superficially resemble their historic counterparts, they will not be exact historical facsimiles. Even small genetic variations – let alone differences in their microbiomes – could alter their behaviour in unpredictable ways. The cultural aspects of these organisms are likely impossible to reconstruct. The song of the dodo has been lost for ever. The thorny issue of the potential de-extinction of ancient species of human, such as Neanderthals or Denisovans, will be an area of intense future debate. The full genome sequences of Neanderthals or Denisovans are now available, but the potential de-extinction of ancient human species should – at least for the time being – remain firmly off-limits. The implications of de-extincting ancient human species go far beyond science, and raise substantive moral, ethical, societal, and philosophical issues.

V
Biosecurity and Safety

The unintended release of highly engineered natural biological entities, or entirely synthetic species – such as artificial viruses with no natural counterparts – poses a profound threat to global ecosystems. Even a seemingly benign engineered bacteriophage could have unpredictable and potentially catastrophic consequences. For instance, any significant disruption to the natural population of marine cyanobacteria (blue-green algae) could destabilise the oceanic carbon cycle and accelerate climate change. These microbes play a critical role in regulating atmospheric carbon dioxide and form the base of marine food webs. Their collapse could trigger mass extinctions, disrupt fisheries, and threaten global food security.

Given these risks, synthetic organisms, whether intended for deliberate environmental release or not, must undergo rigorous ecological impact testing. Coordination with regulatory bodies is essential, and strict quarantine protocols should be put in place to prevent their unintended dispersal. They should also incorporate multiple redundant fail-safe mechanisms, including built-in vulnerabilities and safety switches that enable their selective destruction in the event that unexpected ecological, ethical or biosafety concerns arise. Additionally, all synthetic species should be watermarked for traceability. To limit the effects of natural evolution, which may unpredictably alter genomes, synthetic organisms should also be engineered with genetic 'firewalls'. The most effective way to establish and maintain containment is to recode their genomes. This would generate a biological 'language barrier' between natural and artificial species and prevent their genetic information from intermingling. The use of alternative fail-safe genetic systems – such as quadruplet codons – could slow functional mutation rates and enhance long-term genomic stability.

The genomes of wild species should be systematically monitored on a global scale to proactively seek out the undisclosed generation and release of synthetic organisms outside of an approved international framework. This could be achieved by initiating an ongoing global biosurveillance programme, which through the use of next-generation sequencing, would monitor metagenomic DNA sequence data obtained from representative sources across a wide variety of environments. Extremely rapid new sequencing methods like the Roche SBX system would be ideally suited for this purpose. The use of genome language models like Evo 2 have demonstrated that the sequence to function relationship has been broken. It is no longer possible to directly look at a sequence and conclude that the biological entity it computes is pathological or benign.

The sequencing of metagenomic samples will therefore need to be accompanied by attempts to interrogate any unnatural sequences identified using genome language models, in order to determine whether apparently benign sequences are able to compute predicted pathogenic phenotypes. These may include the phenotypes of biological entities that have never previously existed in nature and for which there is no precedent. Simultaneous with such efforts, it may be necessary to construct any such identified artificial metagenomic genomes to directly determine the nature of the biological entity that they compute. In the case of totally new biological entities, there may be no shorter way to infer their phenotype than to construct the biological entity, thereby 'running' its biological programme.

The dual-use nature of genome design and synthesis technologies demands robust security measures as, for example, detailed in the 2024 'Responsible AI x Biodesign: Community Values, Guiding Principles, and Commitments for the Responsible Development of AI and Protein Design.' The 2025 Hoover Institution 'Biosecurity Really' report argues that 'the dangers lurking within the possibilities of biology are worse than those in nature'. Governments must prepare for the possibility of bioterrorism or biowarfare involving synthetic pathogens – whether deployed by nation states or rogue actors. Rapid response capabilities must be established to address both the accidental and deliberate release of synthetic pathogens. As genome synthesis becomes more accessible – decentralised, inexpensive, and democratised – so too does the potential for misuse. A global screening system must be established to monitor the production and distribution of synthetic DNA sequences using internationally approved protocols to detect the presence of potentially hazardous designs, including engineered pathogens that could threaten public health or biodiversity.

Surveillance should extend both to the environment and to synthetic DNA orders placed through central facilities. All requested sequences must be screened against security databases, and the identities of individuals and organisations placing orders verified. Predictive modelling should be used to assess the likely impact of introducing a new sequence into an organism, or the potential of a synthesised DNA fragment to generate a novel, high-risk species. An international regulatory agency should be established to ensure rigorous oversight of synthetic DNA generation, distribution, and deployment.

As benchtop DNA synthesisers capable of assembling entire genomes eventually become widespread, it will become increasingly difficult to track what is being synthesised, and by whom. Vendors and customers using such devices must be carefully vetted, and strict safeguards established to prevent their unauthorised resale or misuse. The emergence of AI-driven protein design at the atomic level, especially using transformer-based models leveraging extensive computational power, adds another layer of complexity. These technologies enable the creation of synthetic protein surrogates that have significantly reduced sequence resemblance to known pathogens, making them more difficult to detect using standard biosecurity screening methods. For example, one of the AI-generated Φ-X174 bacteriophages constructed and tested using Evo 2 in conjunction with supervised training, called Evo-Φ2147, had just 93 per cent sequence identity to the natural parental virus. As the relationship between the structure of biological entities and the genomic sequences computing them becomes increasingly opaque, conventional screening will struggle to determine that artificially designed biological entities with substantial divergence at the nucleotide sequence level may encode near-identical biological entities and functions. The situation has been considerably exacerbated by the development of Brian Hie's Evo-enabled semantic design technology, which will enable the potential mining of pathogenic function in areas of unnatural DNA sequence space that may have no correspondence whatsoever to the parental sequence, or to any pre-existing natural sequence. To counter this, advanced computational tools and AI-driven surveillance systems should be deployed to proactively identify emerging biosecurity threats. Centralising genome-scale DNA synthesis capabilities may offer the safest way forward, reducing the risk of unmonitored and potentially malicious use.

VI
Synthetic Human Genomes

A global moratorium should be placed on any attempts to introduce synthetic human genomes into germ cells – sperm and eggs – with the intent of creating parentless, living humans. Rewriting human genomes for this purpose is unlikely to ever be safe, ethical, or desirable. The intrinsic complexity of the human genome, especially in the broader context of its informiome, renders the full consequences of such potential modifications not only uncertain, but likely non-computable. A single microorganism species, more or less, may influence gene expression in unexpected ways. Even apparently minor alterations may have far-reaching effects.

Beyond the correction of simple monogenic diseases, the risks associated with genome modification rapidly increase, and the outcomes become progressively more uncertain. More extensive modifications will inevitably introduce some degree of indeterminacy, further exacerbated by the lack of suitable model systems in which to explore their effects. This may ultimately place fundamental limits on the extent to which human genomes can, or should be safely and ethically rewritten. Nonetheless, as our predictive capabilities improve, the concept of localised or regional genomic rewrites, aimed at achieving specific healthcare outcomes, may become more acceptable. This is especially the case for somatic (non-heritable) modifications.

Limited germline modifications targeting well-characterised monogenic (involving a single gene) diseases may, in time, and in specific cases, be considered acceptable for therapeutic purposes. However, the threshold for addressing oligogenic (involving a few genes) or polygenic (involving many genes, each with small effects) conditions remains very high. The risk of unintended genome-wide effects is considerable. The complexity of such modifications, and the uncertainty they invoke, may prove insurmountable for the foreseeable future. The moratorium should therefore also extend to the therapeutic introduction of complex modifications into the human genome – whether at the non-inherited somatic, or at the inherited germline level – until the science of genome engineering becomes sufficiently predictive. In contrast, the targeted redesign of individual protein- and RNA-encoding genes in somatic cells, when there is sufficient evidence of benefit, should remain a viable area of clinical exploration.

The ability to synthesise human genomes holds enormous promise for next-generation cell therapies targeting cancer, autoimmune diseases, and a range of other diseases. Synthetic human genomes tailored for cell therapy should represent the full spectrum of human diversity – ideally in the form of streamlined homozygous pangenomes. Synthetic genomes could also revolutionise organoid technology, enabling the creation of miniature, functional replicas of human organs.

The synthesis of large sections of human genomes will play a pivotal role in mapping out functional regions within the genome identified through genome-wide association studies. This will help accelerate the discovery of therapeutic targets. Writing out alternative versions of genomic loci will advance our understanding of non-coding DNA – systematically distinguishing functionally relevant regions from irrelevant sequences.

Engineering a minimal human genome for use in cell therapy would provide a compelling rationale for a human genome-writing initiative. Redundant genes – conferring functions irrelevant to single cell function, such as those associated with multicellularity, immunity, or specialised structures like muscle dystrophin – could be selectively eliminated. Initially, only essential non-coding DNA would be retained, with further reductions guided by experimental evidence. Such minimal genomes would be faster, cheaper and more efficient to synthesise than their natural counterparts, while at the same time providing profound insights into genome function. Their operating systems would underpin a new class of cell therapies with curated and regulatable therapeutic content.

It is in this context notable that any individual human cell type – from lymphocytes to stem cells – transcribes less than approximately 3 per cent of the genome's protein- and RNA-encoding information. In some cases this it may be as little as 0.02%. As such for any particular cell at any particular time, only a fraction of the genome's information is utilised. This suggests that the extent of any potential human genome minimisation may be considerable.

Cells with streamlined genomes would have many applications. Synthetic human genomes could, for example, be designed to facilitate cellular recoding, allowing cells to incorporate non-canonical amino acids or other synthetic monomers – opening the door to the construction of novel therapeutics, advanced biomaterials, and new types of cell therapies. The size of the human genome could be further reduced through the use of AI-guided protein design to miniaturise every protein-coding gene while retaining their function.

One intriguing prospect is to explore 'unused' cell types – cellular identities that evolution never discovered, but which have the potential to introduce new properties and behaviours into living systems. These hypothetical cells would be defined by distinct gene expression patterns and offer the potential to explore an uncharted space of cellular biological possibility. An infinite cell-type space, comprising all possible cell types, can be imagined – each with its own functionalities and emergent phenomena. Exploring this latent biological possibility, and how different combinations of such cells perform computations and interact, will open up a new frontier of biology and medicine.

VII
Editing Human Genomes

Whole-genome synthesis is unlikely to ever be used therapeutically in humans, except in the context of cell therapy, where modified cells are infused into patients and persist primarily in the bloodstream. However, it is conceivable that generative genomics and accompanying genome synthesis methods could one day be applied to replace specific regions of the human genome for therapeutic purposes. The introduction of such large DNA segments, in a manner analogous to pasting paragraphs of text into a document, might be delivered using technologies such as integrases and recombinases – enzymes capable of inserting sizeable DNA fragments into genomes – or retrotransposons, which are mobile genetic elements that copy and paste themselves, along with sizeable DNA payloads, into host genomes. The payload capacity of both these delivery systems could be expanded through design and engineering.

At present, gene editing remains the most practical method for modifying human genomes for therapeutic purposes. Whenever possible, monogenic (single gene) diseases should be addressed using somatic (non-heritable) gene editing technologies. A global moratorium should remain in place for all forms of germline (heritable) gene editing. Gene editing methods that avoid cutting DNA, such as epigenetic editing, are generally preferable, as they are likely to be safer. Epigenetic editing works by activating or silencing genes without altering their underlying sequence. However, this approach is ineffective in situations where a faulty gene needs to be replaced.

All genome editing techniques must undergo rigorous assessment for off-target effects, using at least two independent, highly sensitive detection methods. Off-target effects – unintended changes occurring in regions located beyond the area being edited – represent a serious risk. When therapeutic genes are introduced into genomes, they should, whenever possible, be integrated at precisely defined genomic locations, rather than inserted randomly. Lifetime safety monitoring of patients receiving such treatments is essential.

Genetic conditions caused by a small number of genes (oligogenic) and complex disorders influenced by many genes with small effects (polygenic) might, in principle, be addressed through multiplex editing – the simultaneous modification of multiple genes. However, multiplex editing significantly increases the risk of off-target effects and should be performed only in a somatic, non-inheritable context.

Efforts to address complex diseases through genome editing should proceed only when the underlying biology is well understood. A moratorium should remain in place on all forms of polygenic editing, given the complexity and unpredictability of its outcomes. Oligogenic editing may be acceptable if the number of target genes is small, their biological roles are well characterised and safety concerns are minimal.

For early-onset monogenic diseases, once the relevant technologies have matured and undergone extensive safety assessments, *in utero* somatic editing could be considered to prevent the irreversible damage caused by disease progression early on in life. Such interventions should be limited to severe, early-onset medical conditions, such as cystic fibrosis or spinal muscular atrophy, and never performed for non-medical purposes. The definition of 'medical' should align with internationally accepted principles, and lifetime monitoring of all treated individuals is mandatory.

In utero somatic editing for early-onset diseases should take precedence over germline modifications, at least until gene therapy becomes more precise, and the associated social, ethical and societal issues have been adequately addressed. Economic considerations will likely influence future policy, as the cost of treating successive generations of individuals with monogenic diseases – especially in regions of high prevalence – will impose a growing financial burden on healthcare systems.

Following extensive interdisciplinary debate, ethical evaluation and international consensus, germline editing may, in select cases, one day be considered acceptable. This would likely only be relevant to severe, life-threatening diseases caused by single genes, and only where stringent conditions have been met. Even then, the underlying biology must be extremely well understood to prevent unintended consequences. The loss of the protective effects of sickle cell trait may, for example, lead to an increased susceptibility to malaria. Such interventions would, furthermore, only be considered reasonable if no safer, feasible, and affordable alternative were available. They should be high-precision to minimise 'off-target' effects, and commenced only after obtaining extensive and culturally appropriate informed consent. Long-term safety must be assured and intergenerational safety monitoring provided. Rigorous oversight would be required to ensure scientific merit, a favourable risk-benefit profile, and ethical integrity.

Any permitted germline intervention must include a lifetime commitment to long-term safety monitoring, including intergenerational follow-up. A permanent moratorium should remain in place for germline editing aimed at modifying polygenic traits or enhancing human characteristics. Germline editing must never be performed for non-medical purposes.

In the future, national or global germline elimination programmes targeting severe monogenic diseases may, following appropriate debate and international alignment, be considered. Conditions such as sickle cell disease and beta thalassemia, for instance, could become candidates for consideration, particularly in regions where their prevalence renders repeated somatic editing economically or logistically unsustainable.

VIII
The Goals of Medicine and Nature of Disease

The fundamental goals of clinical medicine remain unchanged: to prevent, cure, and ameliorate human disease. These efforts are driven by the imperative to reduce suffering and extend healthy longevity. However, the ability to cure disease at the genomic level will inevitably encounter ethical and practical boundaries.

While monogenic diseases are relatively straightforward to address, efforts to rewrite multiple genes, large genomic regions, or entire genomes in order to treat complex conditions risk introducing unpredictable changes into human biology. As the limitations of gene-focused therapies become apparent, network-based treatments – targeting the dynamics and self-organising behaviours of metabolic systems – are likely to assume an increasingly important role in future medical interventions.

The distinction between *illness* – the subjective experience of poor health – and *disease* – a formally recognised pathological condition – remains critical. The definition of disease should be established by internationally recognised medical and scientific authorities, rather than in an ad hoc, politicised, or ideologically driven manner.

To evaluate the effects of genomic interventions, extensively humanised mice – engineered to replace mouse genes and regulatory elements with human equivalents – should likely serve as the standard model organism. The extent of humanisation must be carefully considered and debated. Humanised versions of other rapidly reproducing model organisms, such as the nematode *Caenorhabditis elegans*, may also yield valuable insights. Primates should not be humanised.

The use of pre-implantation genetic diagnosis to select for preferred characteristics should be prohibited, except in specific cases such as the use of sex selection to prevent sex-linked genetic disorders.

IX
Governance, Bioethics and Public Discourse

A comprehensive, inclusive, and apolitical international framework must be established to provide robust ethical oversight of synthetic genomics. This should foster an ongoing dialogue among willing nation states and a broad spectrum of stakeholders, including policy-makers, regulatory experts, bioethicists, and environmental advocates. While corporate sponsorship may help fund these efforts, decision-making must remain free from for-profit influence.

This regulatory framework should work in concert with existing international organisations and initiatives. An appointed global body should be tasked with crafting and endorsing a universal code of good practice accompanied by dynamic, regularly updated guidelines. These should reflect consensus views on the collective responsibilities of participating nations. The regulatory body will be responsible for monitoring synthetic biology activities worldwide to ensure that research and applications proceed in a responsible, transparent, and accountable manner.

Human genetic material should never be modified without first obtaining comprehensive and culturally appropriate informed consent. Such consent must be given freely, and without coercion, undue pressure, or financial inducement. Genetic modification must never be used as a form of punishment or deployed in warfare. Experimental interventions in human subjects should occur exclusively within voluntary, ethically governed clinical trials, conducted with good intent and designed to safeguard the health and well-being of participants.

It is, however, acceptable to synthesise a human genome – including minimal or refactored versions – solely for the generation of cell lines intended for research, or industrial or therapeutic use in the context of engineered cell therapy.

Governance in synthetic genomics must carefully balance risk mitigation with the need to foster innovation. Embracing the principle of regulatory parsimony – intervening no more than necessary – will facilitate a flexible yet vigilant oversight model. This approach should promote progress while also upholding safety, justice, intellectual freedom, responsibility, inclusivity, diversity, and fairness, in a way that maximises the benefit to society. Transparency in policymaking is paramount to fostering trust, and must be reinforced through ongoing public engagement.

A global citizens' assembly, drawing on the perspectives of diverse religions, cultures, socioeconomic backgrounds, ages, and genders, should be central to this dialogue. It must operate independently and free from political interference and external lobbying. This assembly would help address the interconnected global challenges posed by synthetic genomics, including climate change, biosafety and the erosion of global diversity. These issues cannot be addressed by individual nation states acting in isolation. Those that fail to uphold international agreements should be met with coordinated diplomatic and economic pressure to align with globally acceptable ethical standards.

Public awareness of synthetic genomics remains limited. To maintain public trust and social legitimacy, transparent and responsible science communication is imperative. A sustained and inclusive dialogue must be established, providing the public with the knowledge necessary to evaluate its risks and potential benefits, and fostering a collective sense of responsibility. This will empower the public to participate in an inclusive, pluralistic and far-reaching debate referencing moral, ethical, and societal implications.

Nothing in this manifesto is intended to conflict with religious beliefs or creation narratives. Science and religion need not be adversaries: one investigates the world through observation and experiment, while the other explores questions of meaning and purpose. The origin of a universe as vast and complex as our own remains a mystery that currently lies beyond the reach of science and philosophy. As Galileo is said to have stated, the 'book of nature' is written in the language of mathematics – a truth that can coexist with a diversity of views on the origin of those mathematical laws.

The transformative potential of AI-informed genome design and synthetic genomics will underpin a sustainable global bioeconomy. These technologies – poised to redefine almost every facet of human endeavour – demand substantive public sector and governmental investment. No nation state can afford to fall behind in progressing such innovations, which promise not only to reshape the global economy but to reimagine the fabric of our daily life.

Modern cities, for example, should be constructed in a way that integrates and accommodates nature rather than excludes it. Urban planning must incorporate habitats for wildlife, and protect and establish migration corridors. Engineered living materials – comprising living matter embedded within intelligent, responsive matrices – should be integrated into our houses and buildings. We should construct cities using smart materials that incorporate biological sensors and actuators that are able to respond to the environment and human needs. We might one day grow our homes rather than build them.

Basic resources – including clean water, food security, reliable electricity, healthcare, and medicines – must be universally accessible. Securing global health and well-being for all is achievable, but will require an aligned, focused and coordinated global vision that transcends national and geographical boundaries.

Personal genomic data must be encrypted and safeguarded against misuse. No government, organisation, or private entity should be permitted to exploit genetic data to discriminate against individuals or undermine individual rights. The principle of equality among all human individuals must inform and guide every technological and policy decision. Equal access to technologies that enable genetic modifications and to the benefits they produce, is a matter of distributive justice.

The ownership of the artificial genomes of unrealised or extinct species for which no physical DNA material exists, remains an open question. Their future legal status will depend on whether they are considered part of nature or the product of artificial design. If determined to be artificial – and therefore patentable – we may soon witness a competitive race to secure intellectual property within this emerging domain of 'species real estate', unlocked by AI's capacity to map the virtual expanse of DNA sequence space.

X
The Rights of Natural and Artifical Intelligence

Artificial intelligence may soon rival – or perhaps even surpass – natural intelligence in terms of creativity, intellectual capacity, and raw computational power. One cannot help but wonder whether natural intelligences like our own – or forms far beyond our current capabilities, whether natural, artificial, or hybrid 'polyintelligences' – already exist elsewhere in the universe. In light of this possibility, and of the exponential growth of AI, the development of biologically-based superintelligence may emerge as humanity's only plausible countermeasure to maintain its continued relevance. Yet it seems unlikely that we will achieve biological superintelligence simply by modifying our natural molecular architectures. We will, at some point, also need to consider the rights of machine intelligence, and hybrids thereof, incorporating biological structures.

Every human being is born with an inviolable right to physical, emotional, and intellectual freedom – a right that must be upheld without compromise and in perpetuity. All humans must be afforded equal dignity and rights, regardless of their socioeconomic status, ideology, or morphology. These rights must extend to all extinct species of humans and to every future, and as yet unrealised type of human nature. No intervention in the human genome – whether somatic or germline – should ever violate these inalienable rights.

The ethical dilemmas posed by humanised species are equally complex. Consider, for example, extensively humanised mice. While humanised organisms promise to be invaluable for studying human disease and validating hypotheses derived from human genomic data, the extent to which they should be humanised remains unresolved. We do not yet understand which regions of the human genome capture our morality, aesthetics, free will, and conscious awareness. It would, furthermore, be unethical – and for some complex traits, likely impossible – to perform the experiments necessary to explore these questions.

Nonetheless, partially humanised mice could serve as effective model systems. In many cases, it should be possible to demonstrate our understanding of the genetic basis of a disease or trait by synthesising the relevant genetic sequences and inserting them into a humanised mouse, to see whether the feature of interest can be recapitulated. Ultimately, the rate of progress in synthetic genomics as it applies to human health, will be limited by the availability of appropriate model systems.

History abounds with philosophies and ideologies that reflect the darkest and most unsavoury facets of human nature. We should therefore adopt a cautious, yet optimistic view of life's future possibilities. Genome writing holds transformative potential and may convey great benefits to humankind, provided its agency is distributed equitably and ethically, and used with integrity. This manifesto has deliberately avoided referencing the various abhorrent attempts across history to undermine and dissolve the foundational ethical structures of humanity. These have included the propagation and implementation of pseudo-scientific ideologies based on spurious representations of genetic principles and, more recently, the reckless and irresponsible use of gene editing.

Such aberrations of fundamental decency have no place in a manifesto designed for justice and the equal benefit of all humankind. Genome modification – whether somatic or germline – if performed safely, ethically, responsibly, and inclusively, in a manner that is free from politics and ideology and focused exclusively on individual health and the well-being of all, has the potential to bring great benefit to society.

XI
A Roadmap for the Future of Species

We must begin by proactively cataloguing all life on Earth, constructing a comprehensive archive of the complete collection of all natural genome sequences that captures the full spectrum of the planet's diversity. Every species on Earth, both existing and, where possible, extinct, should be represented. The repository – an expansive 'species ark' accommodating the complete repertoire of genome sequences of all of Earth's biodiversity – should not be limited to one-dimensional genetic sequence alone. It should encompass as many other aspects of a species' informiome as possible, including three-dimensional genome structures, epigenetic landscapes, behavioural traits, microbiomes, lifecycles, interspecies relationships, culture (where relevant), and ideally a representative pangenome for each species. Where possible, physical specimens should accompany this information. This biological species archive will function not only as an essential reference for future science and conservation, but as a time capsule.

Active steps must be taken to compile an inventory of every living species. The diversity that we have catalogued so far likely represents just a fraction of nature's repository. It is estimated that as many as 85 per cent of species remain undiscovered, each holding its own hidden lessons and insights. Collectively, Earth's species comprise the pages of a guidebook to life.

Natural species are our most precious asset. We must act now to save our wildernesses and 'nature capital' for future generations. We cannot protect what we have not yet discovered, which is why there is an urgent need to document all living species and to monitor their health, genetic diversity, distribution, prevalence, migratory behaviour, and vulnerabilities.

The achievements of medical science to date represent a triumph of the human intellect. Yet if we continue to unquestioningly pursue its implicit goal of eradicating all human disease, we will eventually encounter a philosophical landscape of unprecedented complexity. Is ageing, for example, a disease to be cured, or a defining feature of human nature? Might the pursuit of indefinite life itself become a paradoxical affliction? We must define what success in medicine looks like, and foster a shared vision for it.

The roadmap for humankind's future must balance the preservation of natural species with the responsible exploration of artificial ones. History unfolds unpredictably, and the powerful technologies we are developing – while conveying an unprecedented agency to create new worlds – may overstep our utopian aspirations. Ongoing governance, caution, and vigilance are essential.

The more engineerable a synthetic species becomes, the more brittle it is likely to be in the face of environmental variability. Highly engineered organisms will likely lack the resilience and evolvability that characterise natural life. An inverse relationship exists between engineerability and natural evolvability. Awareness of this trade-off is essential to prevent the collapse of synthetic species in unpredictable ecological circumstances.

Human nature has always been fluid – a product of contingency and Darwinian evolution. To believe that human nature in its current form represents an apotheosis or endpoint of a historical journey would be inconsistent with our understanding of the process of evolution by natural selection. In the future, we may speak of multiple 'human natures' rather than of a singular 'human nature'. We should not be surprised to encounter species with intelligences, moral sensibilities, and ethical systems that greatly surpass our own. Such encounters may lead us to re-evaluate our legacy as a species. Human nature in its current form, with all its contradictions and irrationalities, is nevertheless a uniquely precious phenomenon. It is incumbent upon us to safeguard its integrity.

The advent of Artificial Biological Intelligence – even in its current rudimentary form – will enable us to gain full command of the language of DNA, with an eventual fluency similar to a natural language speaker. This mastery promises to deliver transformative benefits for human health. It will also offer innovative solutions to a host of global challenges, such as food security. A second green revolution, driven by the rational design of plant genomes, could help resolve inequities in global food distribution and ensure that population growth does not outstrip agricultural productivity.

As a result of the emerging ability of generative AI to design proteins with atomic-scale precision, molecular biology is transitioning to a new age of 'atomic level' design. Simultaneously, generative genomics – the use of genome language models to infer deep biological rules and the relationships between distant nucleotides to design DNA at the level of whole genomes – along with new technologies for building DNA at scale, is ushering in a new age in which the genomes of species will eventually be authored like books. The use of AI to design, construct and validate the artificial genome of the first ever viable biological entity that has never previously existed in nature, is a harbinger of a future in which entirely new species – unconstrained and unfettered by Darwinian evolution – may be scripted with impunity.

The future of species, both natural and artificial, promises to be bright, provided we exercise our new powers ethically and responsibly. We must preserve species and their habitats, and honour the mystery of nature, while being open to the possibility of cautiously and thoughtfully engineering aspects of it for the common good and benefit of society.

It is possible to imagine a time when almost every human need will be furnished through sustainable biological innovation. By gaining control over life itself, we will be able to nurture it, and above all, ensure that the human spirit remains free, and that our nature remains intact, paradoxical, and imperfect.

Acknowledgements

I would like to thank my mother, Sheila Banks, for her endless enthusiasm and encouragement throughout my life. You were always there when I needed you, and sadly are now no longer with us to see the completed book – the product of your nurturing, intellectual stimulation, and thoughtfulness. You believed in me, and I in you. Thank you for naturally engineering me, and for your sparkling and uncompromising intellect, brilliant mind, sense of humour, and devotion.

My father, Gerald Woolfson, has always taken a great interest in my work and has diligently helped me at every stage, and in every way that he possibly could. I am deeply grateful to him for reviewing the book in its entirety, assisting with the bibliography, and for helping to rationalise the epigraphs. When I was a child, you told me that the human being is the most complex machine in the world. I have never forgotten that. It has guided my imagination throughout my life. You were right.

Margherita supported me throughout the writing process and has been my constant companion and inspiration in San Francisco, New York, Primrose Hill, and Vicenza. I have benefited greatly from my discussions with her. She has informed my writing style, stimulated my imagination with trips to various locations, and provided me with an environment conducive to writing. She also reviewed the book in its entirety, reminded me to keep it simple, and made detailed and thoughtful comments throughout. Like the encrypted instructions of genetic sequences, her spirit accompanies my own throughout the book.

Further thanks to my grandfather Ephraim Charlaff, for infusing me, by proxy, with the tenets of mathematical Platonism and the notion of a timeless mathematical reality of all possible things, even though I was too young to know you properly. To my grandmother Rachel Charlaff, for buying me, when I was a child, the delightful chocolate animals from the John Barnes department store on the Finchley Road, where the blind man used to play his accordion on the street. To my grandfather Joseph Samuel Woolfson, for being an entrepreneur and pioneer who was brave enough to strike out into an unknown world.

To Michael Alcock, a gentle and humorous man who showed me great kindness in becoming my literary agent, finding Bloomsbury to publish the book, and accompanying me on the publishing journey at every point. To Charlotte Seymour at Johnson & Alcock for her support and guidance. To my commissioning editor, Ariel Pakier, and the Bloomsbury team, including Shanika Hyslop, for seeing value in my book proposal, supporting it, and guiding it through to publication. Thanks also to the foreign rights team at Bloomsbury, including Claire Kennedy, Stephanie Purcell, and Hannah Stokes. Juliet Brooke made astute and enormously helpful editorial comments on the early drafts of the book, which helped me to reconfigure it in a more suitable manner.

Mike Jones greatly improved the manuscript with his thoughtful structural editing. My managing editor, Lauren Whybrow, astutely managed the late stages of production, including the copyediting, which was efficiently performed by Katherine Fry. Thanks also to Amy Whitaker, Fabrice Wilmann, and Francisco Vilhena for their diligent work on correcting and coordinating the proofs. Thanks to Ian Marshall for believing in the book from the outset and for nurturing it through its final stages. Thanks also to my US editors Andrew Kinney and Janice Audet at the MIT Press in Boston, and to my UK and US publicists Ruth Killick and Cassandra (Cassie) Birk.

To my mentor César Milstein, who took a chance on me when I became his PhD student in the division of Protein and Nucleic Acid Chemistry at the MRC Laboratory of Molecular Biology in Cambridge, and who taught me to pay attention to detail. To Michael Neuberger for his intellectual brilliance, kindness, and generosity of spirit. To Barry Keverne, who gave me my first academic opportunity in the Sub-Department of Animal Behaviour in the Department of Zoology at Cambridge University, mentored me, believed in me, and whose curious and thoughtful nature guided me to new pathways. To Peter Salmon, for supporting me early on in my intellectual development, seeing potential in me, and for helping me progress my academic career. To Benny Chain, for his great generosity, kindness, intellectual curiosity, and belief in my ideas at the earliest stage of my academic career. To Henry Plotkin, for introducing me to evolutionary epistemology, and to Avrion Mitchison, for his eccentricity, belief in me, and for teaching me that you have to 'earn your ticket to theorise'.

To Margaret Johnson, whose kindness in inviting me to deliver one of the plenary lectures at the 500th anniversary of the Royal College of Physicians in 2018 provided the inspiration for this book. To Roger Kornberg, for his tremendous generosity, support and kindness, and with great thanks for reviewing the biographical details relevant to his father, Arthur Kornberg. To Keith Peters, who supported me early on and encouraged me, and to David Oliveira for his kindness in nurturing me. To Herman Waldmann, for being there for me, helping me to make the right decisions, and for his great loyalty over the years. To Greg Winter, for his support in my various activities.

To Chris Bulstrode, a free and unconventional spirit, who inspired me to write my first book when I was a medical student at the University of Oxford, attached to the Accident & Emergency Department at the John Radcliffe Hospital, and to whom I will always be immensely grateful. Thanks to Jef Boeke and Patrick Cai for reviewing the discussion of the synthetic yeast genome, to Drew Endy for reviewing the chapter on genome refactoring and to Jeff Sheedy for his comments on the manuscript. Thanks to Kaihang Wang for his thoughtful and helpful comments on my manifesto for life and for his input, along with that of Noah Robinson, into the section on the Sidewinder DNA construction technology. To Gottfried Fischer for his comments on the subtitle of the book and for his companionship in Cambridge.

Thanks to Lynne Woolfson for her helpful discussion in Sicily relevant to the subtitle and Introduction, and for providing a helpful and supportive sounding board in the later stages of the book's production. I am grateful to Alexander Woolfson for his insightful thoughts and observations on the text. Thanks to Andrew Hessel for reviewing the section on HGP-write, and for being such an enthusiastic and energetic advocate of synthetic biology and an inspiration. To Claire Sefton, for reviewing the manuscript in its entirety and for her support, thoughtfulness, and kindness. Thanks to Alex Salam, for reading the manuscript in full and for his helpful comments. Thanks to Sajob Ramakurup Gopinathan for our discussion on the plane from Doha to Trivandrum, on the importance of the subconscious mind in the context of preserving authentic human nature.

Thanks to Ralf Dahm for his invaluable communications on Friedrich Miescher, and to Kersten Hall for providing me with a translation from the German of Suter's 1944 reminiscences of Miescher. To Aldo, Clive, and Boz for providing ongoing companionship and inspiration, and to Percy for his help with editing the text. To Joanna Amberger at the Online Mendelian Inheritance in Man (OMIM) database for her kind and diligent assistance. To Nick Mackintosh, for his great support, advocacy and kindness. To Miriam Rothschild, with her childlike skip, owls, moths, and dragonfly parties, and unshakeable belief in my ability and future prospects. To George, for our various discussions at the bar at Cotogna in San Francisco.

And finally, to Caffè Trieste, on the corner of Vallejo and Grant in San Francisco and its staff including Kim, Jason, Nathan, William, Nicky, Paul, Yates, Matt, Victor, and Wolfie, and its customers – my inadvertent companions – such as Carl (and the flying car he has constructed), the poet Owen, Linda, Mona, Lani, Michelle, Bugsy the dog, Herman and Nickie, Mayumi, Hollywood, the lingering spirit of Francis Ford Coppola, and the ghosts of Jack Hirschman, Allen Ginsberg, Lawrence Ferlinghetti, and Jack Kerouac, who provided me both with inspiration and a suitable environment for writing this book. Thanks also to the assorted collection of North Beach characters, including Jack the Hat, Rooster, Scott, Beth and Dave, Ysidro, Damaq, Samuel, Stella, and Jessica, who in their own unique ways provided a backdrop for my thoughts. I treasure the kindness of all of you, and remain for ever indebted.

Bibliography

EPIGRAPHS

Blake W. The Fly. In *The Complete Poems*. Penguin Classics (1978).

Borges JL. La Biblioteca de Babel. In *El Jardín de senderos que se bifurcan*. Editorial Sur (1941).

Brenner S. *Sydney Brenner's 10-ON-10: The Chronicles of Evolution* ed. Shuzhen Sim and Benjamin Seet. Wildtype Books (2018).

Chargaff E. Preface to a Grammar of Biology: A hundred years of nucleic acid research. *Science*. May 1971; 172(3984):637–42. doi: 10.1126/science.172.3984.637. PMID: 4929532.

Dennett DC. *Darwin's Dangerous Idea: Evolution and the meanings of Life*. Allen Lane (1995).

His W et al. (eds). *Die Histochemischen und Physiologischen Arbeiten von Friedrich Miescher: Aus dem wissenschaftlichen Briefwechsel von F. Miescher*. Vol. 1: 116–17. FCW Vogel (1869).

PREFACE

Byron, G, G. *The Curse of Minerva*. London: Printed by T. Davison, 1812.

Darwin, C. R. and A. R. Wallace. 1858. On the tendency of species to form varieties; and on the perpetuation of varieties and species by natural means of selection. [Read 1 July.] *Journal of the Proceedings of the Linnean Society of London*. Zoology 3 (20 August): 45-62.

Kavanagh, K. World's first AI-designed viruses a step towards AI-generated life. *Nature* 646, 16 (2025) doi: https://doi.org/10.1038/d41586-025-03055-y.

King SH, Driscoll CL, Li DB, Guo D, Merchant AT, Brixi G, Wilkinson ME, Hie BL. Generative design of novel bacteriophages with genome language models bioRxiv 2025.09.12.675911; doi: https://doi.org/10.1101/2025.09.12.675911.

McCarty, N. 'AI-Designed Phages.' Asimov Press (2025). https://doi.org/10.62211/21er-45fg.

Partridge, D. The famous Linnean Society meeting: from old errors to new insights, *Biological Journal of the Linnean Society*, Volume 137, Issue 3, November 2022, Pages 556–567, https://doi.org/10.1093/biolinnean/blac108.

Woolfson A. ABI and generative biology: A new paradigm for gene therapy, genome engineering, and engineered cell therapy. *Molecular Therapy*. 2025 May 7; 33(5):1881–5. doi: 10.1016/j.ymthe.2025.02.021. Epub 2025 Mar 21. PMID: 40120587.

INTRODUCTION

Al-Hashimi HM. Turing, von Neumann, and the computational architecture of biological machines. *Proceedings of the National Academy of Sciences*. June 2023; 120(25):e2220022120. doi: 10.1073/pnas.2220022120. PMID: 37307461; PMCID: PMC10288622.

Brenner S. The Human Gene Kit. Unpublished lecture presented at the Addenbrooke's Hospital School for Clincial Medicine (1998).

Brenner S. Turing centenary: Life's code script. *Nature*. February 2012; 482(7386):461. doi: 10.1038/482461a. PMID: 22358811.

Brenner S. Nature's gift to science (Nobel lecture). *ChemBioChem*. August 2003; 4(8):683–7. doi: 10.1002/cbic.200300625. PMID: 12898617.

Brenner S. The genetics of *Caenorhabditis elegans*. *Genetics*. 1974 May; 77(1):71–94. doi: 10.1093/genetics/77.1.71. PMID: 4366476. PMCID: PMC1213120.

Brenner S. Sequences and consequences. *Philosophical Transactions of the Royal Society of London B, Biological* Sciences. 2010 Jan 12; 365(1537):207-12. doi: 10.1098/rstb.2009.0221. PMID: 20008397; PMCID: PMC2842711.

Brenner S. The Impact of Society on Science. *Science*. 1998; 282:1411-1412. doi: 10.1126/science.282.5393.1411.

Brenner S. History of science. The revolution in the life sciences. *Science*. 2012 Dec 14; 338(6113): 1427-8. doi: 10.1126/science.1232919. PMID: 23239722.

C. elegans Sequencing Consortium. Genome sequence of the nematode C. elegans: a platform for investigating biology. *Science*. 1998 Dec 11; 282(5396):2012-8. doi: 10.1126/science.282.5396.2012. Erratum in: *Science* 1999 Jan 1; 283(5398):35. Erratum in: *Science* 1999 Mar 26; 283(5410):2103. Erratum in: *Science* 1999 Sep 3; 285(5433):1493. PMID: 9851916.

Cobb M. 60 years ago, Francis Crick changed the logic of biology. *PLoS Biology*. 2017 Sep 18; 15(9):e2003243. doi: 10.1371/journal.pbio.2003243 . PMID: 28922352; PMCID: PMC5602739.

Crick FH. On protein synthesis. *Symposia of the Society for Experimental Biology*. 1958; 12:138-63. PMID: 13580867.

Descartes R. *Discourses on method and meditations on first philosophy*. Fourth Edition. 1637. Translated by Donald. A. Cress. Hackett Publishing Company (1998).

Descartes R. *Meditations on First Philosophy* (Fourth Edition). Translated by Michael Moriarty. Oxford University Press (2008).

Hume, D. *Dialogues concerning natural religion.* William Strahan, London. 1779.

Jacob F, Monod J. Genetic regulatory mechanisms in the synthesis of proteins. *Journal of Molecular Biology*. 1961 Jun; 3:318–56. doi: 10.1016/s0022-2836(61)80072-7. PMID: 13718526.

Kenyon C. Sydney Brenner (1927-2019). *Science*. 2019 May 17; 364(6441): 638. doi: 10.1126/science.aax8563. PMID: 31097656.

Lofting H. *The Story of Doctor Dolittle*. New York: Frederick A. Stokes Company (1920).

Offray de La Mettrie, J. *La Mettrie's L'Homme Machine. A Study in the Origins of an Idea*. Critical Edition with an Introductory Monograph and Notes by Aram Vartanian. Princeton University Press, 1960.

Shapin S. One peculiar nut. *London Review of Books*. 23 January 2003; 25:2.

Steinbeck, J. *East of Eden*. Viking, New York. 1952.

Theodoris CV. Learning the language of DNA. *Science*. 2024 Nov 15; 386(6723):729–730. doi: 10.1126/science.adt3007. Epub 2024 Nov 14. PMID: 39541478.

Turing AM. On computable numbers with an application to the entscheidungsproblem. *Proceedings of the London Mathematical Society*. 1936; S2-42:230–265.

von Neumann, J. *Theory of Self-Reproducing Automata*. edited and completed by Arthus W. Burks. University of Illinois Press. Urbana and London. 1966.

Zhang B et al. A near-complete chromosome-level genome assembly of looseleaf lettuce (Lactuca sativa var. crispa). *Scientific Data*. 2024 Sep 4; 11(1):961. doi: 10.1038/s41597-024-03830-y. PMID: 39231996; PMCID: PMC11375085.

CHAPTER 1

Attar N. Raymond Gosling: the man who crystallized genes. *Genome Biology*. 2013 Apr 25; 14(4):402. doi: 10.1186/gb-2013-14-1-402. PMID: 23651528; PMCID: PMC3663117.

Baker TA. Obituary: Arthur Kornberg (1918-2007). *Nature*. 2007 Dec 6; 450(7171):809. doi: 10.1038/450809a. Erratum in: *Nature*. 2007 Dec 20; 450(7173):1176. PMID: 18063999.

Berg P, Lehman IR. Retrospective: Arthur Kornberg (1918-2007). *Science*. 2007 Dec 7; 318(5856):1564. doi: 10.1126/science.1152989. PMID: 18063778.

Berman HA. *Restitution: Rosalind Franklin and DNA*. Anne Sayre. Norton, New York, 1975. Review in *Science*. 1975; 190:665. doi: 10.1126/science.190.4215.665.a.

Bernal J. Dr. Rosalind E. Franklin. *Nature.* 1958; 182:154. https://doi.org/10.1038/182154a0.

Bessman MJ, Kornberg A, Lehman IR, Simms ES. Enzymic synthesis of deoxyribonucleic acid. *Biochimica et Biophysica Acta.* 1956 Jul; 21(1):197–8. doi: 10.1016/0006-3002(56)90127-5. PMID: 13363894.

Biffi G, Tannahill D, McCafferty J, Balasubramanian S. Quantitative visualization of DNA G-quadruplex structures in human cells. *Nature Chemistry.* 2013 Mar; 5(3):182–6. doi: 10.1038/nchem.1548. Epub 2013 Jan 20. PMID: 23422559; PMCID: PMC3622242.

Chargaff E. Chemical specificity of nucleic acids and mechanism of their enzymatic degradation. *Experientia.* 1950 Jun 15; 6(6):201–9. doi: 10.1007/BF02173653. PMID: 15421335.

Chargaff E. Preface to a grammar of biology. A hundred years of nucleic acid research. *Science.* 1971 May 14; 172(3984):637–42. doi: 10.1126/science.172.3984.637. PMID: 4929532.

Chargaff E. A Quick Climb Up Mount Olympus: *The Double Helix. A Personal Account of the Discovery of the Structure of DNA.* James D. Watson. Atheneum, New York, 1968. Review in *Science.* 1968; 159:1448–1449. doi: 10.1126/science.159.3822.1448.

Chargaff E. What really is DNA? Remarks on the changing aspects of a scientific concept. *Progress in Nucleic Acid Research and Molecular Biology.* 1968; 8:297–333. doi: 10.1016/s0079-6603(08)60549-8. PMID: 4874234.

Clarke T. DNA's family tree. *Nature.* 2003. https://doi.org/10.1038/news030421-5.

Cobb M, Comfort N. What Watson and Crick really took from Franklin. *Nature.* 2023 Apr; 616(7958):657–660. doi: 10.1038/d41586-023-01313-5. PMID: 37100935.

Cobb M. Crick: *A Mind in Motion from DNA to the Brain.* Profile Books, London EC1A 7JQ. 2025. ISBN 978 1 80081 105 8. eISBN 978 80081 106 5.

Crick FH. On the genetic code. *Science.* 1963 Feb 8; 139(3554):461–4. doi: 10.1126/science.139.3554.461. PMID: 14023853.

Dahm R. Discovering DNA: Friedrich Miescher and the early years of nucleic acid research. *Human Genetics.* 2008 Jan; 122(6):565–81. doi: 10.1007/s00439-007-0433-0. Epub 2007 Sep 28. PMID: 17901982.

Dahm R. Friedrich Miescher and the discovery of DNA. *Developmental Biology.* 2005 Feb 15; 278(2):274–88. doi: 10.1016/j.ydbio.2004.11.028. PMID: 15680349.

Dahm R. A slip in the date of DNA's discovery. *Nature.* 2010 Dec 16; 468(7326):897. doi: 10.1038/468897d. PMID: 21164468.

Dahm R, Banerjee M. How We Forgot Who Discovered DNA: Why It Matters How You Communicate Your Results. *BioEssays.* 2019 Apr; 41(4):e1900029. doi: 10.1002/bies.201900029. PMID: 30919468.

Dickens C. *A Christmas Carol*. London: Chapman & Hall (1843).

Editorial. How Rosalind Franklin was let down by DNA's dysfunctional team. *Nature*. 2023 Apr; 616(7958):630. doi: 10.1038/d41586-023-01390-6. PMID: 37100947.

Editorial. Rosalind Franklin: celebrating an inspirational legacy. *Nature*. 2020 Jul 23; 583(7817):492. doi: 10.1038/d41586-020-02144-4.

Ferry G. The structure of DNA. *Nature*. 2019 Nov; 575(7781):35-36. doi: 10.1038/d41586-019-02554-z. PMID: 31686042.

Friedberg EC. The eureka enzyme: the discovery of DNA polymerase. *Nature Reviews Molecular Cell Biology*. 2006 Feb; 7(2):143-7. doi: 10.1038/nrm1787. PMID: 16493419.

Fuller RS. A tribute to Arthur Kornberg 1918-2007. *Nature Structural and Molecular Biology*. 2008 Jan; 15(1):2-17. doi: 10.1038/nsmb0108-2. PMID: 18176549.

Goldstein B. On Francis Crick, the genetic code, and a clever kid. *Current Biology*. 2018 Apr 2; 28(7):R305. doi: 10.1016/j.cub.2018.02.058. PMID: 29614285.

Goulian M, Kornberg A, Sinsheimer RL. Enzymatic synthesis of DNA, XXIV. Synthesis of infectious phage phi-X174 DNA. *Proceedings of the National Academy of Sciences USA*. 1967 Dec; 58(6):2321-8. doi: 10.1073/pnas.58.6.2321. PMID: 4873588; PMCID: PMC223838.

Griffith F. The Significance of Pneumococcal Types. *Journal of Hygiene (London)*. 1928 Jan; 27(2):113-59. doi: 10.1017/s0022172400031879. PMID: 20474956; PMCID: PMC2167760.

Haeckel E. *Generelle Morphologie der Organismen*. Berlin: G. Reimer (1866).

Hall K. Florence Bell: an unsung heroine of DNA. An interview with Kersten Hall, University of Leeds. *The Naked Scientists*. 17 June 2022. https://www.thenakedscientists.com/articles/interviews/florence-bell-unsung-heroine-dna.

Hall K. William Astbury and the biological significance of nucleic acids, 1938-1951. *Studies in History and Philosophy of Biological and Biomedical Sciences*. 2011 Jun; 42(2):119-28. doi: 10.1016/j.shpsc.2010.11.018. Epub 2011 Feb 5. PMID: 21486649.

Hall K. Florence Bell – The 'housewife' with X-Ray vision. *Notes and Records: Royal Society Journal of the History of Science*. 2022; 76:619-631. https://doi.org/10.1098/rsnr.2020.0064.

Hershey AD, Chase M. Independent functions of viral protein and nucleic acid in growth of bacteriophage. *Journal of General Physiology*. 1952 May; 36(1):39-56. doi: 10.1085/jgp.36.1.39. PMID: 12981234; PMCID: PMC2147348.

His W et al. (eds). *Die Histochemischen und Physiologischen Arbeiten von Friedrich Miescher – Aus dem wissenschaftlichen Briefwechsel von F. Miescher*. Vol. 1. Leipzig, Germany: FCW Vogel (1869). Letter I, February 26, 1869, pp. 33-38; Letter I, August 21, 1869, p. 39; Letter XXX, September 20,

1873, p. 73; Letter LXVI, July 03, 1890, p. 107; Letter LXXV, December 17, 1892, pp. 116–17.

Kennedy E. A life in biochemistry: for the love of enzymes. *Science*. 1989 May 19; 244(4906):852–3. doi: 10.1126/science.244.4906.852. PMID: 17802261.

Klug A. Rosalind Franklin and the discovery of the structure of DNA. *Nature*. 1968 Aug 24; 219(5156):808–10 passim. doi: 10.1038/219808a0. PMID: 4876935.

Kornberg A. Biologic synthesis of deoxyribonucleic acid. *Science*. 1960 May 20; 131(3412):1503–8. doi: 10.1126/science.131.3412.1503. PMID: 14411056.

Kornberg A. The Universal Language. *Nature Biotechnology*. 1987; 5:520. https://doi.org/10.1038/nbt0587-520.

Kornberg A, Lehman IR, Bessman MJ, Simms ES. Enzymic synthesis of deoxyribonucleic acid. 1956. *Biochimica et Biophysica Acta*. 1989; 1000: 57–8. PMID: 2673407.

Krause R. Maclyn McCarty (1911–2005). *Nature*. 2005; 433:372. https://doi.org/10.1038/433372a.

Lamm E, Harman O, Veigl SJ. Before Watson and Crick in 1953 came Friedrich Miescher in 1869. *Genetics*. 2020 Jun; 215(2):291-296. doi: 10.1534/genetics.120.303195. PMID: 32487691; PMCID: PMC7268995.

Lofting, J. *The story of Dr Doolittle*. Frederick A. Stokes Company, New York, 1920.

Maddox, B. *Rosalind Franklin: The Dark Lady of DNA* (HarperCollins, 2002).

McCarty M, Avery OT. Studies on the chemical nature of the substance inducing transformation of pneumococcal types; effect of desoxyribonuclease on the biological activity of the transforming substance. *Journal of Experimental Medicine*. 1946 Feb; 83:89-96. PMID: 21010046.

McCarty M. Discovering genes are made of DNA. *Nature*. 2003 Jan 23; 421(6921):406. doi: 10.1038/nature01398. PMID: 12540908.

Miescher JF. Ueber die chemische Zusammensetzung der Eiterzellen. *Medicinisch-Chemische Untersuchungen*. 1871; 441-460.

Miescher JF. Ueber das Leben des Rheinlachses im Süsswasser. *Archives of Anatomy and Physiology*. Anat. Abt. 1881; 193-218.

Miescher JF. Statistische und biologische Beiträge zur Kenntniss vom Leben des Rheinlachses im Süsswasser. In W His et al. (eds). *Die Histochemischen und Physiologischen Arbeiten von Friedrich Miescher*. Vol. 2, 116-191. Leipzig: FCW Vogel (1897).

Mirsky AE. The discovery of DNA. *Scientific American*. 1968 Jun; 218(6): 78-88. doi: 10.1038/scientificamerican0668-78. PMID: 5648255.

Morrison KL, Weiss GA. The origins of chemical biology. *Nature Chemical Biology*. 2006 Jan; 2(1):3-6. doi: 10.1038/nchembio0106-3. PMID: 16408079.

Neidle S. On the origins of the conflict between Rosalind Franklin and Maurice Wilkins. *Nature Reviews Chemistry*. 2023 Nov; 7(11):747-748. doi: 10.1038/s41570-023-00551-5. PMID: 37828114.

Watson JD, Crick FH. Molecular structure of nucleic acids; a structure for deoxyribose nucleic acid. *Nature*. 1953 Apr 25;171(4356):737-8. doi: 10.1038/171737a0. PMID: 13054692.

Olby R, Posner E. An early reference to genetic coding. *Nature*. 1967 Jul 29; 215(5100):556. doi: 10.1038/215556a0. PMID: 4862227.

CHAPTER 2

Abraham-Juárez MJ et al. The arches and spandrels of maize domestication, adaptation, and improvement. *Current Opinion in Plant Biology*. 2021 Dec; 64:102124. doi: 10.1016/j.pbi.2021.102124. Epub 2021 Oct 26. PMID: 34715472.

Arnold F. The Library of Maynard-Smith: My Search for Meaning in the Protein Universe. *Microbe*. July 2011; 6(7):316-318. Doi :10.1128 /microbe.6.316.1.

Arnold, F. Innovation by Evolution: Bringing New Chemistry to Life. Nobel Lecture. NobelPrize.org Nobel Prize Outreach 2025. Fri 7 Nov 2025. https://www.nobelprize.org/prizes/chemistry/2018/arnold /lecture/

Auden WH. *The Dyer's Hand*. New York: Random House (1962).

Bacon F. *New Atlantis*. London: William Rawley (1627).

Baedeker K. *Eastern Alps including the Bavarian Highlands, The Tyrol, Salzkammergut, Styria, and Carinthia. Handbook for Travelers*. Leipsic (1879).

Ball P. The patent threat to designer biology. *Nature*. 2007. https://doi .org/10.1038/news070618-17.

Baltimore D. Paul Berg (1926-2023). *Science*. 2023 Mar 17; 379(6637):1095. doi: 10.1126/science.adh2943. Epub 2023 Mar 16. PMID: 36927020.

Berg P. Dissections and reconstructions of genes and chromosomes. *Science*. 1981 Jul 17; 213(4505):296-303. doi: 10.1126/science.6264595. PMID: 6264595.

Berg P. Fred Sanger: a memorial tribute. *Proceedings of the National Academy of Sciences USA*. 2014 Jan 21; 111(3):883-4. doi: 10.1073/pnas.1323264111. Epub 2014 Jan 21. PMID: 24449828; PMCID: PMC3903207.

Berg P et al. Potential biohazards of recombinant DNA molecules. *Science*. 1974 Jul 26; 185(4148):303. PMID: 11661080.

Bibikova M, Beumer K, Trautman JK, Carroll D. Enhancing gene targeting with designed zinc finger nucleases. *Science*. 2003 May 2; 300(5620):764. doi: 10.1126/science.1079512. PMID: 12730594.

Boddy J. Catching ancient maize domestication in the act. *Science*. 2016 Nov 25; 354(6315):953-954. doi: 10.1126/science.354.6315.953. PMID: 27884985.

Borges JL. La Biblioteca de Babel (The Library of Babel) in *El Jardín de senderos que se bifurcan* (The Garden of Forking Paths). Buenos Aires, Argentina: Editorial Sur (1941).

Bottke WF, Vokrouhlický D, Nesvorný D. An asteroid breakup 160 Myr ago as the probable source of the K/T impactor. *Nature*. 2007 Sep 6; 449(7158):48–53. doi: 10.1038/nature06070. PMID: 17805288.

Bragg WL. The Diffraction of X-rays by Crystals. Nobel Lecture, 6 September 1922. *Zeitschrift für Physikalische Chemie*. 1923; 104: 337-348. doi: 10.1515/zpch-2014-9026.

Bragg W. The Reflection of X-Rays by Crystals. *Nature*. 1913; 91(477). https://doi.org/10.1038/091477b0.

Bragg W. The Specular Reflection of X-rays. *Nature*. 1912; 90(410). https://doi.org/10.1038/090410b0.

Brenner S. *Fred Sanger, Double Nobel Laureate: A Biography*, by George G. Brownlee. Cambridge University Press (2014). Review in *RNA*. 2016 Mar; 22(3):317. doi: 10.1261/rna.055590.115. PMCID: PMC4748809.

Brown DM, Kornberg H. Alexander Robertus Todd, O.M., Baron Todd of Trumpington. 2 October 1907 - 10 January 1997. *Biographical Memoirs of Fellows of the Royal Society of London*. 1 November 2000; 46:515–532. https/doi.org/10.1098/rsbm.1999.0099.

Brownlee GG. Frederick Sanger (1918–2013). *Current Biology*. 2013 Dec 16; 23(24):R1074–6. doi: 10.1016/j.cub.2013.11.037. PMID: 24501770.

Brownlee GG. The Legacy of Fred Sanger – 100 Years on from 1918. *Journal of Molecular Biology*. 2018 Aug 17; 430(17):2661–2669. doi: 10.1016/j.jmb.2018.05.034. Epub 2018 May 29. PMID: 29857002.

Brownlee GG. The (chain) terminators. *Nature Reviews Molecular Cell Biology*. 2019 Jan; 20(1):2. doi: 10.1038/s41580-018-0063-5. PMID: 30228347.

Capecchi MR. Altering the genome by homologous recombination. *Science*. 1989 Jun 16; 244(4910):1288–92. doi: 10.1126/science.2660260. PMID: 2660260.

Carlson A, Hermann J. Muller: *Genes, Radiation and Society. The Life and Work of H. J. Muller*. Ithaca: Cornell University Press (1981).

Carrol, L. *Alice's Adventures In Wonderland*. MacMillan and Co, London, 1879.

Chargaff E. Engineering a molecular nightmare. *Nature*. 1987 May 21–27; 327(6119):199–200. doi: 10.1038/327199a0. PMID: 3472081.

Chargaff E, Simring FR. On the dangers of genetic meddling. *Science*. 1976 Jun 4; 192(4243):938+. PMID: 11643312.

Cleaver JE. Cambridge Laboratory of Molecular Biology. *Science*. 2003 Jun 20; 300(5627):1875. doi: 10.1126/science.300.5627.1875c. PMID: 12817124.

Cohen J. The Birth of CRISPR Inc. *Science*. 2017 Feb 17; 355(6326): 680–684. doi: 10.1126/ science.355.6326.680. PMID: 28209854.

Cohen SN. DNA cloning: a personal view after 40 years. *Proceedings of the National Academy of Sciences USA*. 2013 Sep 24; 110(39):15521-9. doi: 10.1073/pnas.1313397110. Epub 2013 Sep 16. PMID: 24043817; PMCID: PMC3785787.

Cohen SN, Chang AC, Boyer HW, Helling RB. Construction of biologically functional bacterial plasmids in vitro. *Proceedings of the National Academy of Sciences USA*. 1973 Nov; 70(11):3240-4. doi: 10.1073/pnas.70.11.3240. PMID: 4594039; PMCID: PMC427208.

Crea R, Kraszewski A, Hirose T, Itakura K. Chemical synthesis of genes for human insulin. *Proceedings of the National Academy of Sciences USA*. 1978 Dec; 75(12):5765-9. doi: 10.1073/pnas.75.12.5765. PMID: 282602; PMCID: PMC393054.

Crow JF. Timeline: Hermann Joseph Muller, evolutionist. *Nature Reviews Genetics*. 2005 Dec; 6(12):941-5. doi: 10.1038/nrg1728. PMID: 16341074.

Dennett DC. *Darwin's Dangerous Idea: Evolution and the meanings of Life*. Allen Lane (1995).

Chargaff E. Engineering a molecular nightmare. Nature. 1987 May 21-27;327(6119):199-200. doi: 10.1038/327199a0. PMID: 3472081. Chargaff E. Engineering a molecular nightmare. Nature. 1987 May 21-27;327(6119):199-200. doi: 10.1038/327199a0. PMID: 3472081.Top of Form

Doebley J. Plant science. Unfallen grains: how ancient farmers turned weeds into crops. *Science*. 2006 Jun 2; 312(5778):1318-9. doi: 10.1126/science.1128836. PMID: 16741100.

Editorial. Our Enzymology Correspondent. Enzymes: Enzymes to Order. *Nature*. 1969; 221(316). https://doi.org/10.1038/221316a0.

Everts, S. Understanding the workings of life. *Chemical and Engineering News*. 9 September 2013; 1991:36. https://cen.acs.org/articles/91/i36/Understanding-Workings-Life.html.

Greenberg DS. The synthesis of DNA: how they spread the good news. *Science*. 1967 Dec 22; 158(3808):1548-50. doi: 10.1126/science.158.3808.1548. PMID: 6060357.

Henderson R. Max Perutz (1914-2002). Great scientist and modest leader. *Structure*. 2002 Apr; 10(4):455-8. doi: 10.1016/s0969-2126(02)00753-0. PMID: 11937050.

Heywood P. The quagga and science: what does the future hold for this extinct zebra? *Perspectives in Biology and Medicine*. 2013 Winter; 56(1): 53-64. doi: 10.1353/pbm.2013.0008. PMID: 23748526.

Higuchi R, Bowman B, Freiberger M, Ryder OA, Wilson AC. DNA sequences from the quagga, an extinct member of the horse family. *Nature*. 1984 Nov 15-21; 312(5991):282-4. doi: 10.1038/312282a0. PMID: 6504142.

Holley RW et al. Structure of a Ribonucleic Acid. *Science*. 1965 Mar 19; 147(3664):1462-5. doi: 10.1126/science.147.3664.1462. PMID: 14263761.

Hoose A, Vellacott R, Storch M, Freemont PS, Ryadnov MG. DNA synthesis technologies to close the gene writing gap. *Nature Reviews Chemistry*. 2023; 7(3):144–161. doi: 10.1038/s41570-022-00456-9. Epub 2023 Jan 23. Erratum in: *Nature Reviews Chemistry*. 2023 Aug; 7(8):590. doi: 10.1038/s41570-023-00521-x. PMID: 36714378; PMCID: PMC9869848.

Ingram V. The birth of molecular biology. *Nature*. 2002; 419(669–670). https://doi.org/10.1038/419669a.

Itakura K et al. Expression in Escherichia coli of a chemically synthesized gene for the hormone somatostatin. *Science*. 1977 Dec 9; 198(4321): 1056-63. doi: 10.1126/science.412251. PMID: 412251.

Jackson DA, Symons RH, Berg P. Biochemical method for inserting new genetic information into DNA of Simian Virus 40: circular SV40 DNA molecules containing lambda phage genes and the galactose operon of Escherichia coli. *Proceedings of the National Academy of Sciences USA*. 1972 Oct; 69(10):2904-9. doi: 10.1073/pnas.69.10.2904. PMID: 4342968; PMCID: PMC389671.

Kevles DJ. *Hermann J. Muller: Genes, Radiation and Society. The Life and Work of H. J. Muller*. Elof Axel Carlson. Ithaca, NY: Cornell University Press (1981), xiv. Review in *Science*. 1981; 214:1232-1233. doi: 10.1126/science.214.4526.1232.

Kim YG, Cha J, Chandrasegaran S. Hybrid restriction enzymes: zinc finger fusions to Fok I cleavage domain. *Proceedings of the National Academy of Sciences USA*. 1996 Feb 6; 93(3):1156-60. doi: 10.1073/pnas.93.3.1156. PMID: 8577732; PMCID: PMC40048.

Kistler L et al. Multiproxy evidence highlights a complex evolutionary legacy of maize in South America. *Science*. 2018 Dec 14; 362(6420):1309-1313. doi: 10.1126/science.aav0207. PMID: 30545889.

Klug A. The discovery of zinc fingers and their applications in gene regulation and genome manipulation. *Annual Review of Biochemistry*. 2010; 79:213-31. doi: 10.1146/annurev-biochem-010909-095056. PMID: 20192761.

Knuuttila T, Loettgers A. (Un)Easily Possible Synthetic Biology. *Philosophy of Science*. 2022; 89(5):908-917. doi: 10.1017/psa.2022.60.

Köhler G, Milstein C. Continuous cultures of fused cells secreting antibody of predefined specificity. *Nature*. 1975 Aug 7; 256(5517):495-7. doi: 10.1038/256495a0. PMID: 1172191.

Macaulay, TB. *History of England from the Accession of James II*. Longman, Brown, Green, and Longmans (1848-1861).

Maddox J. Alex Todd (1907-97). *Nature*. 1997 Feb 6; 385(6616):492. doi: 10.1038/385492a0. PMID: 9020354.

Mangan RJ et al. Adaptive sequence divergence forged new neurodevelopmental enhancers in humans. *Cell*. 2022 Nov 23; 185(24): 4587-4603.e23. doi: 10.1016/j.cell.2022.10.016. PMID: 36423581; PMCID: PMC10013929.

Martienssen R. The origin of maize branches out. *Nature*. 1997 Apr 3; 386(6624):443, 445. doi: 10.1038/386443a0. PMID: 9087398.

McElheny VK. Laboratory of Molecular Biology, Cambridge. *Science*. 1964 Apr 24; 144(3617):398-400. doi: 10.1126/science.144.3617.398. PMID: 14169327.

McElheny VK. Laboratory of Molecular Biology, Cambridge, II. *Science*. 1964 Jul 3; 145(3627):36-8. doi: 10.1126/science.145.3627.36. PMID: 14162689.

Mehr SA et al. Universality and diversity in human song. *Science*. 2019 Nov 22; 366(6468):eaax0868. doi: 10.1126/science.aax0868. PMID: 31753969; PMCID: PMC7001657.

Michelson AM, Todd, AR. Nucleotides part XXXII. Synthesis of a dithymidine dinucleotide containing a 3': 5'-internucleotidic linkage. *Journal of the Chemical Society* (resumed). 1955; 2632-2638. https://doi.org/10.1039/JR9550002632.

Milstein C. From the structure of antibodies to the diversification of the immune response. Nobel Lecture. *Bioscience Reports*. 2004 Aug-Oct; 24(4-5):280-301. doi: 10.1007/s10540-005-2735-6. PMID: 16134016.

Muller HJ. Artificial Transmutation of the Gene. *Science*. 1927 Jul 22; 66(1699):84-7. doi: 10.1126/science.66.1699.84. PMID: 17802387.

Muller HJ. The production of mutations. 12 December 1946. Nobel Prize Lecture. https://www.nobelprize.org/prizes/medicine/1946/muller/lecture/.

Muller HJ. The Problem of Genetic Modification. *Zeit ind. Abst. und Vereb.* 1927; Supp. 1:234-260.

Nirenberg MW. Will society be prepared? *Science*. 1967 Aug 11; 157(3789):633. doi: 10.1126/science.157.3789.633. PMID: 17792839.

Nogales E, Mahamid J. Bridging structural and cell biology with cryo-electron microscopy. *Nature*. 2024 Apr; 628(8006):47-56. doi: 10.1038/s41586-024-07198-2. Epub 2024 Apr 3. PMID: 38570716; PMCID: PMC11211576.

Olby R. Quiet debut for the double helix. *Nature*. 2003 Jan 23; 421(6921):402-5. doi: 10.1038/nature01397. PMID: 12540907.

Olmert MD. Genes unleashed: How the Victorians engineered our dogs. *Nature*. 16 October 2018; 562, 336-337. doi: https://doi.org/10.1038/d41586-018-07039-z.

Pääbo S. Neolithic genetic engineering. *Nature*. 1999 Mar 18; 398(6724): 194-5. doi: 10.1038/18315. PMID: 10094040.

Pederson T. The double helix: 'Photo 51' revisited. *FASEB Journal*. 2020 Feb; 34(2):1923-1927. doi: 10.1096/fj.202000119. PMID: 32046470.

Pennisi E. A hothouse of molecular biology. *Science*. 2003 Apr 11; 300(5617):278-82. doi: 10.1126/science.300.5617.278. PMID: 12690185.

Perutz MF. X-ray analysis of hemoglobin. *Science*. 1963 May 24; 140(3569):863-9. doi: 10.1126/science.140.3569.863. PMID: 13942632.

Pigafetta A. *Relazione del Primo Viaggio Intorno al Mondo* (The First Voyage Around the World).

Potter P. The fragrance of the Heifer's breath. *Emerging Infectious Diseases*. 2011 Apr; 17(4):763-4. doi: 10.3201/eid1704.ac1704. PMID: 21470487; PMCID: PMC3377429.

Press G. On thinking machines, machine learning, and how AI took over statistics. *Forbes*. 15 June 2021. https://www.forbes.com/sites/gilpress/2021/05/28/on-thinking-machines-machine-learning-and-how-ai-took-over-statistics/.

Prüfer K et al. The complete genome sequence of a Neanderthal from the Altai Mountains. *Nature*. 2014 Jan 2; 505(7481):43-9. doi: 10.1038/nature12886. Epub 2013 Dec 18. PMID: 24352235; PMCID: PMC4031459.

Retallack G, Leahy GD. Cretaceous-tertiary dinosaur extinction. *Science*. 1986 Dec 5; 234(4781):1170-1. doi: 10.1126/science.234.4781.1170. PMID: 17777986.

Sanger F. The arrangement of amino acids in proteins. *Advances in Protein Chemistry*. 1952; 7:1-67. doi: 10.1016/s0065-3233(08)60017-0. PMID: 14933251.

Sanger F. Chemistry of insulin. *British Medical Bulletin*. 1960 Sep; 16:183-8. doi: 10.1093/oxfordjournals.bmb.a069832. PMID: 13746240.

Sanger F et al. Nucleotide sequence of bacteriophage phi X174 DNA. *Nature*. 1977 Feb 24; 265(5596):687-95. doi: 10.1038/265687a0. PMID: 870828.

Sanger F. Determination of nucleotide sequences in DNA. *Science*. 1981 Dec 11; 214(4526):1205-10. doi: 10.1126/science.7302589. PMID: 7302589.

Sanger F, Coulson AR. A rapid method for determining sequences in DNA by primed synthesis with DNA polymerase. *Journal of Molecular Biology*. 1975 May 25; 94(3):441-8. doi: 10.1016/0022-2836(75)90213-2. PMID: 1100841.

Sanger F, Nicklen S, Coulson AR. DNA sequencing with chain-terminating inhibitors. *Proceedings of the National Academy of Sciences USA*. 1977 Dec; 74(12):5463-7. doi: 10.1073/pnas.74.12.5463. PMID: 271968; PMCID: PMC431765.

Sanger F, Coulson AR, Hong GF, Hill DF, Petersen GB. Nucleotide sequence of bacteriophage lambda DNA. *Journal of Molecular Biology*. 1982 Dec 25; 162(4):729-73. doi: 10.1016/0022-2836(82)90546-0. PMID: 6221115.

Sanger F. The early days of DNA sequences. *Nature Medicine*. 2001 Mar; 7(3):267-8. doi: 10.1038/85389. PMID: 11231611.

Shampo MA, Kyle RA, Steensma DP. Alexander Todd – British Nobel laureate. *Mayo Clinic Proceedings*. 2012 Mar; 87(3):e19. doi: 10.1016/j.mayocp.2011.12.009. PMID: 22386189; PMCID: PMC3498063.

Singer MF. In vitro synthesis of DNA: a perspective on research. *Science*. 1967 Dec 22; 158(3808):1550-1. doi: 10.1126/science.158.3808.1550. PMID: 6060358.

Smith JM. Natural selection and the concept of a protein space. *Nature*. 1970 Feb 7; 225(5232):563-4. doi: 10.1038/225563a0. PMID: 5411867.

Smith TJ et al. Thiamine deficiency disorders: a clinical perspective. *Annals of the New York Academy of Sciences*. 2021 Aug; 1498(1):9-28. doi: 10.1111/nyas.14536. Epub 2020 Dec 10. PMID: 33305487; PMCID: PMC8451766.

Todd A. Synthesis in the study of nucleotides; basic work on phosphorylation opens the way to an attack on nucleic acids and nucleotide coenzymes. *Science*. 1958 Apr 11; 127(3302):787-92. doi: 10.1126/science.127.3302.787. PMID: 13543331.

Tolstoy L. *War and Peace*. 1886. Bell C (translated from French). New York: Gottsberger.

Tolstoy L. *War and Peace*. 1922. Maude L & Maude A (translated from Russian). Oxford University Press.

Thomas JM. Centenary: The birth of X-ray crystallography. *Nature*. 2012 Nov 8; 491(7423):186-7. doi: 10.1038/491186a. PMID: 23135450.

Urnov FD et al. Highly efficient endogenous human gene correction using designed zinc-finger nucleases. *Nature*. 2005 Jun 2; 435(7042): 646-51. doi: 10.1038/nature03556. Epub 2005 Apr 3. PMID: 15806097.

Van der Valk T et al. Million-year-old DNA sheds light on the genomic history of mammoths. *Nature*. 2021 Mar; 591(7849):265-269. doi: 10.1038/s41586-021-03224-9. Epub 2021 Feb 17. PMID: 33597750; PMCID: PMC7116897.

Walker J. Frederick Sanger (1918-2013). *Nature*. 2014 Jan 2; 505(7481):27. doi: 10.1038/505027a. PMID: 24380948.

Wang H et al. The origin of the naked grains of maize. *Nature*. 2005 Aug 4; 436(7051):714-9. doi: 10.1038/nature03863. PMID: 16079849; PMCID: PMC1464477.

Worboys M, Strange J-M, Pemberton N. *The Invention of the Modern Dog: Breed and Blood in Victorian Britain*. Johns Hopkins University Press (2018).

CHAPTER 3

Agarwal KL et al. Total synthesis of the gene for an alanine transfer ribonucleic acid from yeast. *Nature*. 1970 Jul 4; 227(5253):27-34. doi: 10.1038/227027a0. PMID: 5422620.

van Aken J. Risks of resurrecting 1918 flu virus outweigh benefits. *Nature*. 2006 Jan 19; 439(7074):266. doi: 10.1038/439266a. PMID: 16421546.

van Aken J. Ethics of reconstructing Spanish flu: is it wise to resurrect a deadly virus? *Heredity (Edinburgh)*. 2007 Jan; 98(1):1-2. doi: 10.1038/sj.hdy.6800911. Epub 2006 Oct 11. PMID: 17035950.

Ansari AZ, Rosner MR, Adler J. Har Gobind Khorana 1922-2011. *Cell*. 2011 Dec 23; 147(7):1433-5. doi: 10.1016/j.cell.2011.12.008. PMID: 22355806.

Aradhyam GK, Jagannathan NR. Biophysical Reviews contribution call: an issue focus on the life and works of Prof. Har Gobind Khorana on the occasion of the 100th anniversary of the year of his birth. *Biophysical Reviews*. 2022 May 23; 14(3):611-612. doi: 10.1007/s12551-022-00958-2. PMID: 35791388; PMCID: PMC9250578.

Aradhyam GK, Jagannathan NR. Gobind: an inspiring enigma. *Biophysical Reviews*. 2023 Feb 13; 15(1):71-73. doi: 10.1007/s12551-023-01045-w. PMID: 36909957; PMCID: PMC9995614.

Arita I, Nakane M, Fenner F. Public health. Is polio eradication realistic? *Science*. 2006 May 12; 312(5775):852-4. doi: 10.1126/science.1124959. PMID: 16690846.

Ball P. Genome stitched together by hand. *Nature*. 2008. https://doi.org/10.1038/news.2008.522.

Ball P. Synthetic biology: designs for life. *Nature*. 2007 Jul 5; 448(7149):32-3. doi: 10.1038/448032a. PMID: 17611530.

Ball P. Smallest genome clocks in at 182 genes. *Nature*. 2006. https://doi.org/10.1038/news061009-10.

Ball P. Genome transplant makes species switch. *Nature*. 2007. https://doi.org/10.1038/news070625-9.

Behbehani AM. The smallpox story: life and death of an old disease. *Microbiology Reviews*. 1983 Dec; 47(4):455-509. doi: 10.1128/mr.47.4.455-509.1983. PMID: 6319980; PMCID: PMC281588.

Bell E. Who was Edward Jenner? *Nature Reviews Immunology*. 2003; 3:90. https://doi.org/10.1038/nri1016.

Block SM. A not-so-cheap stunt. *Science*. 2002 Aug 2; 297(5582):769-70. Author reply 769-70. doi: 10.1126/science.297.5582.769b. PMID: 12162317.

von Bubnoff A. Deadly flu virus can be sent through the mail. *Nature*. 2005 Nov 10; 438(7065):134-5. doi: 10.1038/438134a. PMID: 16280992.

Callaway E. What it would take to bring back the dodo. *Nature*. 2023 Feb; 614(7948):402. doi: 10.1038/d41586-023-00379-5. PMID: 36765251.

Caruthers M, Wells R. Retrospective. Har Gobind Khorana (1922–2011). *Science*. 2011 Dec 16; 334(6062):1511. doi: 10.1126/science.1217138. PMID: 22174242.

Caruthers MH. Gene synthesis with H G Khorana. *Resonance*. 2012; 17, 1143–1156. https://doi.org/10.1007/s12045-012-0131-7.

Caruthers MH. The chemical synthesis of DNA/RNA: our gift to science. *Journal of Biological Chemistry*. 2013 Jan 11; 288(2):1420-7. doi: 10.1074/jbc.X112.442855. Epub 2012 Dec 6. PMID: 23223445; PMCID: PMC3543024.

Cello J, Paul AV, Wimmer E. Chemical synthesis of poliovirus cDNA: generation of infectious virus in the absence of natural template. *Science*. 2002 Aug 9; 297(5583):1016-8. doi: 10.1126/science.1072266. Epub 2002 Jul 11. PMID: 12114528.

Check Hayden, E. Scientists devise new way to modify organisms. *Nature*. 2009. https://doi.org/10.1038/news.2009.847.

Cole F. Edward Jenner, Naturalist. *Nature*. 1952; 169, 4. https://doi.org/10.1038/169004a0.

Couzin J. Virology. Active poliovirus baked from scratch. *Science*. 2002 Jul 12; 297(5579):174-5. doi: 10.1126/science.297.5579.174b. PMID: 12114601.

Crea R, Kraszewski A, Hirose T, Itakura K. Chemical synthesis of genes for human insulin. *Proceedings of the National Academy of Sciences USA*. 1978 Dec; 75(12):5765-9. doi: 10.1073/pnas.75.12.5765. PMID: 282602; PMCID: PMC393054.

De Cock K. The Eradication of Smallpox: Edward Jenner and The First and Only Eradication of a Human Infectious Disease. *Nature Medicine*. 2001; 7, 15–16. https://doi.org/10.1038/83286.

Dorin A, Stepney S. What Is Artificial Life Today, and Where Should It Go? *Artificial Life*. 2024 Feb 1; 30(1):1-15. doi: 10.1162/artl_e_00435. PMID: 38537175.

Editorial. Edward Jenner. *Nature*. 1923; 111, 69–70. https://doi.org/10.1038/111069a0.

Editorial. The three original publications on vaccination against smallpox by Edward Jenner. Jennerhttps://biotech.law.lsu.edu/cphl/history/articles/jenner.htm.

Editorial. The 1918 flu virus is resurrected. *Nature*. 2005 Oct 6; 437(7060): 794-5. doi: 10.1038/437794a. Erratum in: *Nature*. 2005 Oct 13; 437(7061): 940. PMID: 16208326; PMCID: PMC7095040.

Editorial. This giant extinct salmon had tusks like a warthog. *Nature*. 24 April 2024; 629, 11.

Editorial. The persistence of polio. *Nature Medicine*. 2012 Mar 6; 18(3):323. doi: 10.1038/nm.2708. PMID: 22395675.

Editorial. Smallpox should be saved. *Nature*. 2011 Jan 20; 469(7330):265. doi: 10.1038/469265a. PMID: 21248793.

Editorial. End of the road for poliomyelitis? *Nature*. 1995 Apr 20; 374(6524): 663. doi: 10.1038/374663a0. PMID: 7715710.

Egli M. Robert Letsinger and the Evolution of Oligonucleotide Synthesis. *ACS Omega*. 2023 Sep 1; 8(36):32222-32230. doi: 10.1021/acsomega.3c05177. PMID: 37720801; PMCID: PMC10500693.

Eisenstein M. Why is it so hard to rewrite a genome? *Nature*. 2025 Feb; 638(8051):848-850. doi: 10.1038/d41586-025-00462-z. PMID: 39966637.

Ellis T. What is synthetic genomics anyway? *Biochemistry (London)* 1 June 2019; 41 (3): 6–9. doi: https://doi.org/10.1042/BIO04103006.

Empson J. Country doctor and speckled monster. *Nature*. 1996 May 2; 381(6577):26. doi: 10.1038/381026b0. PMID: 8609981.
Endy D. Genomics. Reconstruction of the genomes. *Science*. 2008 Feb 29; 319(5867):1196-7. doi: 10.1126/science.1155749. PMID: 18309068.
Enserink M. Bioterrorism. How devastating would a smallpox attack really be? *Science*. 2002 May 31; 296(5573):1592-5. doi: 10.1126/science.296.5573.1592. Erratum in: *Science* 2002 Jul 26; 297(5581):522. PMID: 12040157.
Ferguson D. How Mary Wortley Montagu's bold experiment led to smallpox vaccine – 75 years before Jenner. *Guardian*. 28 March 2021. https://www.theguardian.com/society/2021/mar/28/how-mary-wortley-montagus-bold-experiment-led-to-smallpox-vaccine-75-years-before-jenner.
Gellene, DH. Gobind Khorana, 89, Nobel-Winning Scientist Dies. *New York Times*. 14 November 2011. https://www.nytimes.com/2011/11/14/us/h-gobind-khorana-1968-nobel-winner-for-rna-research-dies.html.
Gibson DG et al. Complete chemical synthesis, assembly, and cloning of a Mycoplasma genitalium genome. *Science*. 2008 Feb 29; 319(5867):1215-20. doi: 10.1126/science.1151721. Epub 2008 Jan 24. PMID: 18218864.
Gibson DG et al. Creation of a bacterial cell controlled by a chemically synthesized genome. *Science*. 2010 Jul 2; 329(5987):52-6. doi: 10.1126/science.1190719. Epub 2010 May 20. PMID: 20488990.
Glass JI. Synthetic genomics and the construction of a synthetic bacterial cell. *Perspectives in Biology and Medicine*. 2012; 55(4):473-89. doi: 10.1353/pbm.2012.0040. PMID: 23502559.
Harbinger J. Are manmade viruses the next big terrorist threat? *Newsweek*. 24 October 2019. https://www.newsweek.com/2019/11/01/synthetic-biology-manmade-virus-terrorism-1467 569.html.
Holmes, EC. 1918 and All That. *Science*. 2004; 303:1787-1788. doi: 10.1126/science.1096550.
Holt RA. Synthetic genomes brought closer to life. *Nature Biotechnology*. 2008 Mar; 26(3):296-7. doi: 10.1038/nbt0308-296. PMID: 18327239.
Hughes RA, Ellington AD. Synthetic DNA Synthesis and Assembly: Putting the Synthetic in Synthetic Biology. *Cold Spring Harbor Perspectives in Biology*. 2017 Jan 3; 9(1):a023812. doi: 10.1101/cshperspect.a023812. PMID: 28049645; PMCID: PMC5204324.
Hull HF. The future of polio eradication. *Lancet Infectious Diseases*. 2001 Dec; 1(5):299-303. doi: 10.1016/S1473-3099(01)00143-8. PMID: 11871802.
Hume JP, Martill DM, Dewdney C. Palaeobiology: Dutch diaries and the demise of the dodo. *Nature*. 2004 Jun 10; 429(6992):1 p following 621. doi: 10.1038/nature02688. PMID: 15190921.
Itakura K et al. Expression in Escherichia coli of a chemically synthesized gene for the hormone somatostatin. *Science*. 1977 Dec 9; 198(4321):1056-63. doi: 10.1126/science.412251. PMID: 412251.

Jenner E. *On the origin of the vaccine inoculation*. London: DN Shury (1801).

Jenner E. *An Inquiry into the causes and effects of the variolae vaccinae, a disease discovered in some of the western counties of England, particularly Gloucestershire, and known by the name of Cow Pox*. London: Sampson Low (1798).

Jenner E. Observations on the Natural History of the Cuckoo. By Mr. Edward Jenner. In a Letter to John Hunter, Esq. F. R. S. *Philosophical Transactions of the Royal Society of London*. Vol. 78. 1788: 219–37. JSTOR, http://www.jstor.org/stable/106657. Accessed 14 July 2024.

Jenner E. Some observations on the migration of birds. By the late Edward Jenner, M.D. F. R. S.; with an introductory letter to Sir Humphry Davy, Bart. Pres. R. S. By the Rev. G. C. JennerPhil. *Transactions of the Royal Society*. 11411–44. http://doi.org/10.1098/rstl.1824.0005.

Kaiser J. Virology. Resurrected influenza virus yields secrets of deadly 1918 pandemic. *Science*. 2005 Oct 7; 310(5745):28-9. doi: 10.1126/science.310.5745.28. PMID: 16210501.

Katsnelson A. Synthetic genome resets biotech goals. *Nature*. 2010 May 27; 465(7297):406. doi: 10.1038/465406a. PMID: 20505702.

Khorana HG. Total synthesis of a gene. *Science*. 1979 Feb 16; 203(4381): 614-25. doi: 10.1126/science.366749. PMID: 366749.

Khorana HG. Synthesis in the study of nucleic acids. The Fourth Jubilee Lecture. *Biochemical Journal*. 1968 Oct; 109(5):709-25. doi: 10.1042/bj1090709c. PMID: 4880351; PMCID: PMC1187021.

Khorana HG. Nucleic acid synthesis in the study of the genetic code. Nobel lecture. 12 December 1968. https://www.nobelprize.org/uploads/2018/06/khorana-lecture.pdf.

Khorana HG et al. Total synthesis of the structural gene for the precursor of a tyrosine suppressor transfer RNA from Escherichia coli. 1. General introduction. *Journal of Biological Chemistry*. 1976 Feb 10; 251(3):565-70. PMID: 765327.

Khorana HG. *Chemical Biology. Selected Papers of H. Gobind Khorana (with introductions)*. 25 May 2000. World Scientific Publishing Company, Pte Ltd.

Kosuri S, Church GM. Large-scale de novo DNA synthesis: technologies and applications.

Nature Methods. 2014 May; 11(5):499-507. doi: 10.1038/nmeth.2918. PMID: 24781323; PMCID: PMC7098426.

Kupferschmidt K. Labmade smallpox is possible, study shows. *Science*. 2017 Jul 14; 357(6347):115-116. doi: 10.1126/science.357.6347.115. PMID: 28706017.

Kushner D. Synthetic biology could bring a pox on us all. *Wired*. 25 March 2019. https://www.wired.com/story/synthetic-biology-vaccines-viruses-horsepox/.

Kwok R. Genomics: DNA's master craftsmen. *Nature*. 2010 Nov 4; 468(7320):22-5. doi: 10.1038/468022a. PMID: 21048740.

Lartigue C et al. Genome transplantation in bacteria: changing one species to another. *Science*. 2007 Aug 3; 317(5838):632–8. doi: 10.1126/science.1144622. Epub 2007 Jun 28. PMID: 17600181.

Makri A. After smallpox, can other diseases be eradicated? *Nature Medicine*. 2022 Sep; 28(9):1726–1729. doi: 10.1038/s41591-022-01914-z. PMID: 35986219.

Maugh TH. The artificial gene: it's synthesized and it works in cells. *Science*. 1976 Oct 1; 194(4260):44. doi: 10.1126/science.11643334. PMID: 11643334.

Mueller S, Coleman JR, Wimmer E. Putting synthesis into biology: a viral view of genetic engineering through de novo gene and genome synthesis. *Chemical Biology*. 2009 Mar 27; 16(3):337–47. doi: 10.1016/j.chembiol.2009.03.002. PMID: 19318214; PMCID: PMC2728443.

Nelson MI, Ghedin E. 100-year-old pandemic flu viruses yield new genomes. *Nature*. 2022 Jul; 607(7918):244–245. doi: 10.1038/d41586-022-01741-9. PMID: 35794385.

Noyce RS, Lederman S, Evans DH. Construction of an infectious horsepox virus vaccine from chemically synthesized DNA fragments. *PLoS One*. 2018 Jan 19; 13(1):e0188453. doi: 10.1371/journal.pone.0188453. PMID: 29351298; PMCID: PMC5774680.

Ostrov N et al. Technological challenges and milestones for writing genomes. *Science*. 2019 Oct 18; 366(6463):310–312. doi: 10.1126/science.aay0339. PMID: 31624201.

Pennisi E. Genetics. Replacement genome gives microbe new identity. *Science*. 2007 Jun 29;316(5833):1827. doi: 10.1126/science.316.5833.1827a. PMID: 17600190.

Pennisi E. Genomics. Synthetic genome brings new life to bacterium. *Science*. 2010 May 21; 328(5981):958–9. doi: 10.1126/science.328.5981.958. PMID: 20488994.

Pollack A. Traces of terror: The Science; Scientists create a live polio virus. *New York Times*. 12 July 2002. https://www.nytimes.com/2002/07/12/us/traces-of-terror-the-science-scientists-create-a-live-polio-virus.html.

Porter R. A little man made good. *Nature*. 1991; 352, 203. https://doi.org/10.1038/352203a0.

RajBhandary UL. Har Gobind Khorana (1922–2011). *Nature*. 2011 Dec 14; 480(7377):322. doi: 10.1038/480322a. PMID: 22170673.

Reardon S. How machine learning could keep dangerous DNA out of terrorists' hands. *Nature*. 2019 Feb; 566(7742):19. doi: 10.1038/d41586-019-00277-9. PMID: 30723344.

Riedel S. Edward Jenner and the history of smallpox and vaccination. *Baylor University Medical Center Proceedings*. 2005 Jan; 18(1):21–5. doi: 10.1080/08998280.2005.11928028. PMID: 16200144; PMCID: PMC1200696.

Rourke MF, Phelan A, Lawson C. Access and benefit-sharing following the synthesis of horsepox virus. *Nature Biotechnology*. 2020 May; 38(5):537-539. doi: 10.1038/s41587-020-0518-z. PMID: 32322089.

Sakmar TP. Har Gobind Khorana (1922-2011): chemical biology pioneer. *ACS Chemical Biology*. 2012 Feb 17; 7(2):250-1. doi: 10.1021/cb3000293. PMID: 22339934.

Schmeck Jr, HM. Substance usually made in brain grown in bacteria. *New York Times*. 3 November 1977. https://www.nytimes.com/1977/11/03/archives/substance-usually-made-in-brain-grown-in-bacteria.html.

Sekiya T et al. Total synthesis of a tyrosine suppressor tRNA gene. XV. Synthesis of the promoter region. *Journal of Biological Chemistry*. 1979 Jul 10; 254(13):5781-6. PMID: 376519.

Shampo MA, Kyle RA. Har Gobind Khorana – Nobel Prize for physiology or medicine. *Mayo Clinic Proceedings*. 2006 Mar; 81(3):284. doi: 10.4065/81.3.284. PMID: 16529128.

Singer E. Synthesizing a genome from scratch. *MIT Technology Review*. https://www. technologyreview. com/2008/ 01/25/ 128261/synthesizing-a-genome-from-scratch/.

Smith GL, McFadden G. Smallpox: anything to declare? *Nature Reviews Immunology*. 2002 Jul; 2(7):521-7. doi: 10.1038/nri845. PMID: 12094226.

Smith HO, Hutchison CA 3rd, Pfannkoch C, Venter JC. Generating a synthetic genome by whole genome assembly: phiX174 bacteriophage from synthetic oligonucleotides. *Proceedings of the National Academy of Sciences USA*. 2003 Dec 23; 100(26):15440-5. doi: 10.1073/pnas.2237126100. Epub 2003 Dec 2. PMID: 14657399; PMCID: PMC307586.

Sullivan W. Complete synthesis of gene reported. *New York Times*. 3 June 1970. https://www.nytimes.com/1970/06/03/archives/complete-synthesis-of-gene-reported-total-synthesis-of-gene.html.

Taubenberger JK, Kash JC, Morens DM. The 1918 influenza pandemic: 100 years of questions answered and unanswered. *Science Translational Medicine*. 2019 Jul 24; 11(502):eaau5485. doi: 10.1126/scitranslmed.aau5485. PMID: 31341062; PMCID: PMC11000447.

Taubenberger JK, Morens DM. The 1918 Influenza Pandemic and Its Legacy *Cold Spring Harbor Perspectives in Medicine*. 2020 Oct 1; 10(10):a038695. doi: 10.1101/cshperspect.a038695. PMID: 31871232; PMCID: PMC7528857.

Thèves C, Biagini P, Crubézy E. The rediscovery of smallpox. *Clinical Microbiology and Infection*. 2014 Mar; 20(3):210-8. doi: 10.1111/1469-0691.12536. PMID: 24438205.

Thiel V. Synthetic viruses – Anything new? *PLoS Pathogens*. 2018 Oct 4; 14(10):e1007019. doi: 10.1371/journal.ppat.1007019. PMID: 30286176; PMCID: PMC6171941.

Tumpey TM, Belser JA. Resurrected pandemic influenza viruses. *Annual Review of Microbiology*. 2009; 63:79-98. doi: 10.1146/annurev.micro.091208.073359. PMID: 19385726.

Tumpey TM et al. Characterization of the reconstructed 1918 Spanish influenza pandemic virus. *Science*. 2005 Oct 7; 310(5745):77-80. doi: 10.1126/science.1119392. PMID: 16210530.

Wang Y, Shen Y, Gu Y, Zhu S, Yin Y. Genome Writing: Current Progress and Related Applications. *Genomics Proteomics Bioinformatics*. 2018 Feb; 16(1):10-16. doi: 10.1016/j.gpb.2018.02.001. Epub 2018 Feb 21. PMID: 29474887; PMCID: PMC6000237.

WHO Advisory Committee on Variola Virus Research. Report of the Eighteenth Meeting. Geneva, Switzerland. 2-3 November 2016. https://www.who.int/publications/i/item/who-advisory-committee-on-variola-virus-research-2016.

Willis NJ. Edward Jenner and the eradication of smallpox. *Scottish Medical Journal*. 1997 Aug; 42(4):118-21. doi: 10.1177/003693309704200407. PMID: 9507590.

Wimmer E. The test-tube synthesis of a chemical called poliovirus. The simple synthesis of a virus has far-reaching societal implications. *EMBO Reports*. 2006 Jul; 7 Spec No:S3-9. doi: 10.1038/sj.embor.7400728. PMID: 16819446; PMCID: PMC1490301.

Wimmer E, Mueller S, Tumpey TM, Taubenberger JK. Synthetic viruses: a new opportunity to understand and prevent viral disease. *Nature Biotechnology*. 2009 Dec; 27(12):1163-72. doi: 10.1038/nbt.1593. PMID: 20010599; PMCID: PMC2819212.

York A. Resurrection of a poxvirus causes alarm. *Nature Reviews Microbiology*. 2018 Apr; 16(4):184. doi: 10.1038/nrmicro.2018.31. Epub 2018 Mar 5. PMID: 29503458.

Zhang H, Xiong Y, Xiao W, Wu Y. Investigation of Genome Biology by Synthetic Genome Engineering. *Bioengineering (Basel)*. 2023 Feb 20; 10(2):271. doi: 10.3390/bioengineering10020271. PMID: 36829765; PMCID: PMC9952402.

CHAPTER 4

Airhart M. New era at UT Austin begins for famous Long-Term evolution experiment. *The University of Texas, Austin. Department of Molecular Biosciences*. https://molecularbiosci.utexas.edu/news/research/new-era-ut-austin-begins-famous-long-term-evolution-experiment.

Albertin CB, Ragsdale CW. More than one way to a central nervous system. *Nature*. 2018 Jan 4; 553(7686):34-36. doi: 10.1038/d41586-017-08195-4. PMID: 29300031.

Ball P. What distinguishes the elephant from *E. coli*: Causal spreading and the biological principles of metazoan complexity. *Journal of Biosciences*. 2023; 48:14. PMID: 37194562.

Ball P. What is life? *Nature*. 2018 Aug 30; 560:548-550. doi: https://doi.org/10.1038/d41586-018-06034-8.

Beccaloni GW, Smith VS. Celebrations for Darwin downplay Wallace's role. *Nature*. 2008 Feb 28; 451(7182):1050. doi: 10.1038/4511050d. PMID: 18305520.

Beccaloni G. The other evolutionist. *Nature*. 2004 7 October; 431, 630. doi: 10.1038/431630a.

Beccaloni G. The 1858 Darwin-Wallace paper. *The Alfred Russel Wallace Website*. https://wallacefund.myspecies.info/content/1858-darwin-wallace-paper.

Berry A. Shipwrecked science. *Nature*. 2023 5 January; 613, 22-24. doi: https://doi.org/10.1038/d41586-022-04507-5.

Berry A, Browne J. The other beetle-hunter. *Nature*. 2008 Jun 26; 453(7199):1188-90. doi: 10.1038/4531188a. PMID: 18580934.

Berry A. Natural selection: The evolutionary struggle. *Nature*. 2012; 485:171–172. https://doi.org/10.1038/485171a.

Berry A. Alfred Russel Wallace: evolution's red-hot radical. *Nature*. 2013 Apr 11; 496(7444):162-4. doi: 10.1038/496162a. PMID: 23579663.

Blount ZD, Barrick JE, Davidson CJ, Lenski RE. Genomic analysis of a key innovation in an experimental Escherichia coli population. *Nature*. 2012 Sep 27; 489(7417):513-8. doi: 10.1038/nature11514. Epub 2012 Sep 19. PMID: 22992527; PMCID: PMC3461117.

Blount ZD, Lenski RE, Losos JB. Contingency and determinism in evolution: Replaying life's tape. *Science*. 2018 Nov 9; 362(6415):eaam5979. doi: 10.1126/science.aam5979. PMID: 30409860.

Booth D, King N. Gene regulation in transition. *Nature*. 2016; 534:482–483.

Bottjer DJ, Davidson EH, Peterson KJ, Cameron RA. Paleogenomics of echinoderms. *Science*. 2006 Nov 10; 314(5801):956-60. doi: 10.1126/science.1132310. PMID: 17095693.

Brenner S. History of science. The revolution in the life sciences. *Science*. 2012 Dec 14; 338(6113):1427-8. doi: 10.1126/science.1232919. PMID: 23239722.

Bridgham JT. Predicting the basis of convergent evolution. *Science*. 2016 Oct 21; 354(6310):289. doi: 10.1126/science.aai7394. PMID: 27846519.

Buckling A, Craig Maclean R, Brockhurst MA, Colegrave N. The Beagle in a bottle. *Nature*. 2009 Feb 12; 457(7231):824-9. doi: 10.1038/nature07892. PMID: 19212400.

Burkhardt FH, Smith S. *The Correspondence of Charles Darwin*. Vol. 7 (1858-59). Cambridge: Cambridge University Press (1992).

Callaway E. How humans lost their tails – and why the discovery took 2.5 years to publish. *Nature*. 2024 Mar; 627(8002):15-16. doi: 10.1038/d41586-024-00610-x. PMID: 38418734.

Callaway E. Jelly genome mystery. *Nature*. 2014 May 22; 509(7501):411. doi: 10.1038/509411a. PMID: 24848042.

Camerini J. The other man to discover evolution. *Nature*. 2001; 413: 357–358. https://doi.org/10.1038/35096612.

Cappelluti MA et al. Durable and efficient gene silencing in vivo by hit-and-run epigenome editing. *Nature*. 2024 Mar; 627(8003):416-423. doi: 10.1038/s41586-024-07087-8. Epub 2024 Feb 28. PMID: 38418872; PMCID: PMC10937395.

Conway Morris S. *The crucible of creation: The Burgess Shale and the rise of animals*. Oxford University Press (1998).

Couce A et al. Changing fitness effects of mutations through long-term bacterial evolution. *Science*. 2024 Jan 26; 383(6681):eadd1417. doi: 10.1126/science.add1417. Epub 2024 Jan 26. PMID: 38271521.

Darwin C. *On the Origin of Species, by Means of Natural Selection, or the preservation of favoured races in the struggle for life*. London: John Murray (24 November 1859).

Darwin C. *The Life and Letters of Charles Darwin*. London: John Murray (1887).

Darwin F. *Life and Letters of Charles Darwin*. 2 vols. New York: Appleton (1898). I, 473 (Darwin to Lyell, June 18, 1858).

Davidson EH. The sea urchin genome: where will it lead us? *Science*. 2006 Nov 10; 314(5801):939-40. doi: 10.1126/science.1136252. PMID: 17095689.

Davies R. How Charles Darwin received Wallace's Ternate paper 15 days earlier than he claimed: a comment on van Wyhe and Roomaaker. *Biological Journal of the Linnean Society*. 2012; 105:472–477.

Davies R. 1 July 1858: what Wallace knew; what Lyell thought he knew; what both he and Hooker took on trust; and what Charles Darwin never told them. *Biological Journal of the Linnean Society*. 2013; 109(3):725-736. https://doi.org/10.1111/bij.12081.

Dawkins R. *River Out of Eden*. Basic Books (1995).

Dawkins R. *The blind watchmaker*. W.W. Norton & Company (1986).

Del Valle RP, McLaughlin RN Jr. Stealing genes and facing consequences. *Science*. 2022 Oct 28; 378(6618):356-357. doi: 10.1126/science.ade4942. Epub 2022 Oct 27. PMID: 36302006.

Dunn CW, Ryan JF. The evolution of animal genomes. *Current Opinion in Genetics and Development*. 2015 Dec; 35:25-32. doi: 10.1016/j.gde.2015.08.006. Epub 2015 Sep 9. PMID: 26363125.

E.B.P. Alfred Russel Wallace. *Nature*. 1913; 2299(92):347-349. https://doi.org/10.1038/092347c0.

Editorial. A little fern packs a lot of DNA. *Science*. 7 June 2024; 384(6700): 1051. doi: 10.1126/science.z31ri16.

Editorial. Hunting spiders lose web skills. *Nature*. 22 June 2017; 546:455. doi: https://doi.org/10.1038/d41586-017-00812-6.

Editorial. Light sensing protein found in the brain. *Science*. 12 January 1998. doi: 10.1126/article.39742.

Editorial. The master genes that sculpt tentacles and legs alike. *Nature*. 18 June 2019; 570(278). doi: https://doi.org/10.1038/d41586-019-01829-9.

Editorial. King Alexander of Greece dead; Succumbs to wounds inflicted by his pet monkey early this month. *New York Times*. 26 October 1920. https://www.nytimes.com/1920/10/26/archives/king-alexander-of-greece-dead-succumbs-to-wounds-inflicted-by-his.html.

Emlen DJ, Marangelo J, Ball B, Cunningham CW. Diversity in the weapons of sexual selection: horn evolution in the beetle genus Onthophagus (Coleoptera: Scarabaeidae). *Evolution*. 2005 May; 59(5):1060-84. PMID: 16136805.

Favate JS et al. Linking genotypic and phenotypic changes in the *E. coli* long-term evolution experiment using metabolomics. *Elife*. 2023 Nov 22; 12:RP87039. doi: 10.7554/eLife.87039. PMID: 37991493; PMCID: PMC10665018.

Feynman R. Found as a handwritten chalk note on the blackboard in his laboratory at the time of his death. *Caltech Images Collection*. https://digital.archives.caltech.edu/collections/Images/1.10-29/.

Fitch WT, Popescu T. The world in a song. *Science*. 2019 Nov 22; 366(6468):944-945. doi: 10.1126/science.aay2214. PMID: 31753980.

Fortey R. Shock lobsters. Review of *The Crucible of Creation: The Burgess Shale and the rise of animals*, by Simon Conway Morris. *London Review of Books*. 1 October 1998; 20(19).

Frank JA et al. Evolution and antiviral activity of a human protein of retroviral origin. *Science*. 2022 Oct 28; 378(6618):422-428. doi: 10.1126/science.abq7871. Epub 2022 Oct 27. PMID: 36302021; PMCID: PMC10542854.

Friedmann HC. From 'butyribacterium' to 'E. coli': an essay on unity in biochemistry. *Perspectives in Biology and Medicine*. 2004 Winter; 47(1): 47-66. doi: 10.1353/pbm.2004.0007. PMID: 15061168.

Galilei G. *Sidereus Nuncius*. Apud Thomas Baglioni, Republic of Venice (13 March 1610).

Gall JG. Human genome sequence. *Science*. 1986 Sep 26; 233(4771):1367-8. doi: 10.1126/science.233.4771.1367-e. PMID: 3749880.

Good BH, McDonald MJ, Barrick JE, Lenski RE, Desai MM. The dynamics of molecular evolution over 60,000 generations. *Nature*. 2017 Nov 2; 551(7678):45-50. doi: 10.1038/nature24287. Epub 2017 Oct 18. PMID: 29045390; PMCID: PMC5788700.

Gould SJ. *Wonderful life: The Burgess Shale and the history of nature*. New York: Norton (1989).

Hall BG. Predicting evolution by in vitro evolution requires determining evolutionary pathways. *Antimicrobial Agents and Chemotherapy*. 2002 Sep; 46(9):3035-8. doi: 10.1128/AAC.46.9.3035-3038.2002. PMID: 12183265; PMCID: PMC127434.

Hawlitschek O et al. New estimates of genome size in Orthoptera and their evolutionary implications. *PLoS One*. 2023 Mar 15; 18(3):e0275551. doi: 10.1371/journal.pone.0275551. PMID: 36920952; PMCID: PMC10016648.

Hejnol A. Evolutionary biology: Excitation over jelly nerves. *Nature*. 2014 Jun 5; 510(7503):38-9. doi: 10.1038/nature13340. Epub 2014 May 21. PMID: 24847878.

Hendrickson H, Rainey PB. Evolution: How the unicorn got its horn. *Nature*. 2012 Sep 27; 489(7417):504-5. doi: 10.1038/nature11487. Epub 2012 Sep 19. PMID: 22992522.

Hoy RR. Evolution. Convergent evolution of hearing. *Science*. 2012 Nov 16; 338(6109):894-5. doi: 10.1126/science.1231169. PMID: 23161985.

Huang W, Traulsen A, Werner B, Hiltunen T, Becks L. Dynamical trade-offs arise from antagonistic coevolution and decrease intraspecific diversity. *Nature Communications*. 2017 Dec 12; 8(1):2059. doi: 10.1038/s41467-017-01957-8. PMID: 29233970; PMCID: PMC5727225.

Hull D. Rich man, poor man. *Nature*. 1984; 308: 798–799. https://doi.org/10.1038/308798a0.

Jacob F. Evolution and tinkering. *Science*. 1977 Jun 10; 196(4295):1161-6. doi: 10.1126/science.860134. PMID: 860134.

Jarvis J. *The Gutenberg Parenthesis: The age of print and its lessons for the age of the internet*. Bloomsbury (2023).

Kachroo AH et al. Evolution. Systematic humanization of yeast genes reveals conserved functions and genetic modularity. *Science*. 2015 May 22; 348(6237):921-5. doi: 10.1126/science.aaa0769. PMID: 25999509; PMCID: PMC4718922.

Katana R et al. Chromophore-Independent Roles of Opsin Apoproteins in Drosophila Mechanoreceptors. *Current Biology* 2019 Sep 9; 29(17):2961-2969.e4. doi: 10.1016/j.cub.2019.07.036. Epub 2019 Aug 22. PMID: 31447373.

Kauffman SA. *The origins of order: self-organization and selection in evolution*. Oxford University Press (1993).

Kryazhimskiy S, Draghi JA, Plotkin JB. Evolution. In evolution, the sum is less than its parts. *Science*. 2011 Jun 3; 332(6034):1160-1. doi: 10.1126/science.1208072. PMID: 21636764.

Lenski RE. Experimental evolution and the dynamics of adaptation and genome evolution in microbial populations. *ISME Journal*. 2017 Oct; 11(10):2181-2194. doi: 10.1038/ismej.2017.69. Epub 2017 May 16. PMID: 28509909; PMCID: PMC5607360.

Lenski RE. Twice as natural. *Nature*. 2001 Nov 15; 414(6861):255. doi: 10.1038/35104715. PMID: 11713507.

Lucas A, Plesters J. 'Titian's "Bacchus and Ariadne"'. *National Gallery Technical Bulletin* Vol. 2. 1978;25–47. http://www.nationalgallery.org.uk/technical-bulletin/lucas_plesters1978.

Martín-Durán JM et al. Convergent evolution of bilaterian nerve cords. *Nature*. 2018 Jan 4; 553(7686):45-50. doi: 10.1038/nature25030. Epub 2017 Dec 13. PMID: 29236686; PMCID: PMC5756474.

Maruyama M, Parker J. Deep-Time Convergence in Rove Beetle Symbionts of Army Ants. *Current Biology*. 2017 Mar 20; 27(6):920-926. doi: 10.1016/j.cub.2017.02.030. Epub 2017 Mar 9. PMID: 28285995.

Maynard Smith J. Tinkering. *London Review of Books*. 17 September 1981; 3:17. https://www.lrb.co.uk/the-paper/v03/n17/john-maynard-smith/tinkering.

Mehr SA et al. Universality and diversity in human song. *Science*. 2019 Nov 22; 366(6468):eaax0868. doi: 10.1126/science.aax0868. PMID: 31753969; PMCID: PMC7001657.

Meyer A. How was Wallace led to the Discovery of Natural Selection? *Nature*. 1895; 52, 415. https://doi.org/10.1038/052415a0.

Moczek AP. Evolutionary biology: the origins of novelty. *Nature*. 2011 May 5; 473(7345):34-5. doi: 10.1038/473034a. PMID: 21544136.

Mohrig JR et al. Importance of historical contingency in the stereochemistry of hydratase-dehydratase enzymes. *Science*. 1995 Jul 28; 269(5223):527-9. doi: 10.1126/science.7624773. PMID: 7624773.

Monod J. *Chance and Necessity: An essay on the natural philosophy of modern biology*. Collins (12 September 1972).

Montealegre-Z F, Jonsson T, Robson-Brown KA, Postles M, Robert D. Convergent evolution between insect and mammalian audition. *Science*. 2012 Nov 16; 338(6109):968-71. doi: 10.1126/science.1225271. PMID: 23162003.

Moroz LL et al. The ctenophore genome and the evolutionary origins of neural systems. *Nature*. 2014 Jun 5; 510(7503):109-14. doi: 10.1038/nature13400. Epub 2014 May 21. PMID: 24847885; PMCID: PMC4337882.

Moyers BT. Is Genetic Evolution Predictable? *Plant Cell*. 2018 Jun; 30(6):1171-1172. doi: 10.1105/tpc.18.00438. Epub 2018 Jun 12. PMID: 29895569; PMCID: PMC6048784.

New AM, Lehner B. Systems biology: Network evolution hinges on history. *Nature*. 2015 Jul 16; 523(7560):297-8. doi: 10.1038/nature14537. Epub 2015 Jul 8. PMID: 26153857.

Nutman AP, Bennett VC, Friend CR, Van Kranendonk MJ, Chivas AR. Rapid emergence of life shown by discovery of 3,700-million-year-old microbial structures. *Nature*. 2016 Sep 22; 537(7621):535-538. doi: 10.1038/nature19355. Epub 2016 Aug 31. PMID: 27580034.

Okasha S. Does diversity always grow? *Nature*. 2010; 466, 318. https://doi.org/10.1038/466318a.

Padian K. Evolution: Parallel lives. *Nature*. 2017; 548, 156–157. https://doi.org/10.1038/548156a.

Paley W. *Natural theology: or, Evidences of the Existence and Attributes of the Deity, Collected from the Appearances of Nature*. London (1802).

Partridge D. The famous Linnean Society meeting: from old errors to new insights. *Biological Journal of the Linnean Society*. November 2022; 137, 3, 556–567. https://doi.org/10.1093/biolinnean/blac108.

Pascal B. *Pensées*. Paris: Guillaume Desprez (1670).

Pennisi E. Lizard man. *Science*. 2020 Jul 31; 369(6503):496-499. doi: 10.1126/science.369.6503.496. PMID: 32732406.

Pennisi E. A modular backbone aided the rise of mammals. *Science*. 2018 Sep 21; 361(6408):1176. doi: 10.1126/science.361.6408.1176. PMID: 30237334.

Pennisi E. Genomics. 'Simple' animal's genome proves unexpectedly complex. *Science*. 2008 Aug 22; 321(5892):1028-9. doi: 10.1126/science.321.5892.1028b. PMID: 18719257.

Pennisi E. Genomics. Sea anemone provides a new view of animal evolution. *Science*. 2007 Jul 6; 317(5834):27. doi: 10.1126/science.317.5834.27. PMID: 17615309.

Pennisi E. Genetics. Sea urchin genome confirms kinship to humans and other vertebrates. *Science*. 2006 Nov 10; 314(5801):908-9. doi: 10.1126/science.314.5801.908. PMID: 17095665.

Pennisi E. A new evolutionary classic. *Science*. 2016 Nov 18; 354(6314):813. doi: 10.1126/science.354.6314.813. PMID: 27856856.

Pennisi E. Physiology. Opsins: not just for eyes. *Science*. 2013 Feb 15; 339(6121):754-5. doi: 10.1126/science.339.6121.754. PMID: 23413333.

Penny D. Defining moments. *Nature*. 2006; 442, 745–746. https://doi.org/10.1038/442745a.

Phillips ML. Surprises in sea anemone genome. *The Scientist*. 4 July 2007. https://www.the-scientist.com/surprises-in-sea-anemone-genome-46315.

Piatigorsky J, Wistow G. The recruitment of crystallins: new functions precede gene duplication. *Science*. 1991 May 24; 252(5009):1078-9. doi: 10.1126/science.252.5009.1078. PMID: 2031181.

Plotkin JB. Molecular evolution: No escape from the tangled bank. *Nature*. 2017 Nov 2; 551(7678):42-43. doi: 10.1038/nature24152. Epub 2017 Oct 18. PMID: 29045393.

Poskett J. Letters of Alfred Russel Wallace go online. *Nature*. 2013. https://doi.org/10.1038/nature.2013.12300.

Putnam NH et al. Sea anemone genome reveals ancestral eumetazoan gene repertoire and genomic organization. *Science*. 2007 Jul 6; 317(5834): 86-94. doi: 10.1126/science.1139158. PMID: 17615350.

Reznick D, Travis J. Is evolution predictable? *Science*. 2018 Feb 16; 359(6377):738-739. doi: 10.1126/science.aas9043. PMID: 29449475.

Roush W. Sizing Up Dung Beetle Evolution. *Science*. 1997; 277, 184-184. doi: 10.1126/science.277.5323.184.

Scharping, N. How a 30-year experiment has fundamentally changed our view of how evolution works. *Discover*. 11 November 2019. https://www.discovermagazine.com/planet-earth/how-a-30-year-experiment-has-fundamentally-changed-our-view-of-how.

Shubin N, Tabin C, Carroll S. Deep homology and the origins of evolutionary novelty. *Nature*. 2009 Feb 12; 457(7231):818-23. doi: 10.1038/nature07891. PMID: 19212399.

Sleigh C. Putting evolution in context. *Nature*. 2002; 416, 790–791. https://doi.org/10.1038/416790a.

Sokol J. Cracking the Cambrian. *Science*. 2018 Nov 23; 362(6417):880-884. doi: 10.1126/science.362.6417.880. PMID: 30467152.

Sorrells TR, Booth LN, Tuch BB, Johnson AD. Intersecting transcription networks constrain gene regulatory evolution. *Nature*. 2015 Jul 16; 523(7560):361-5. doi: 10.1038/nature14613. Epub 2015 Jul 8. PMID: 26153861; PMCID: PMC4531262.

Srivastava M et al. The Trichoplax genome and the nature of placozoans. *Nature*. 2008 Aug 21; 454(7207):955-60. doi: 10.1038/nature07191. PMID: 18719581.

Starr TN, Picton LK, Thornton JW. Alternative evolutionary histories in the sequence space of an ancient protein. *Nature*. 2017 Sep 21; 549(7672):409-413. doi: 10.1038/nature23902. Epub 2017 Sep 13. PMID: 28902834; PMCID: PMC6214350.

Tarazona OA, Lopez DH, Slota LA, Cohn MJ. Evolution of limb development in cephalopod mollusks. *Elife*. 2019 Jun 18; 8:e43828. doi: 10.7554/eLife.43828. PMID: 31210127; PMCID: PMC6581508.

Tenaillon O et al. The molecular diversity of adaptive convergence. *Science*. 2012 Jan 27; 335(6067):457-61. doi: 10.1126/science.1212986. PMID: 22282810.

Tenaillon O et al. Tempo and mode of genome evolution in a 50,000-generation experiment. *Nature*. 2016 Aug 11; 536(7615):165-70. doi: 10.1038/nature18959. Epub 2016 Aug 1. PMID: 27479321; PMCID: PMC4988878.

The Alfred Russel Wallace Correspondence Project. https://wallaceletters.myspecies.info.

Travisano M, Mongold JA, Bennett AF, Lenski RE. Experimental tests of the roles of adaptation, chance, and history in evolution. *Science*. 1995 Jan 6; 267(5194):87-90. doi: 10.1126/science.7809610. PMID: 7809610.

Van Oss CJ. Computer prehistory. *Nature*. 1975; 258, 192. https://doi.org/10.1038/258192a0.

Venkatesh B, Gilligan P, Brenner S. Fugu: a compact vertebrate reference genome. *FEBS Letters*. 2000 Jun 30; 476(1-2):3-7. doi: 10.1016/s0014-5793(00)01659-8. PMID: 10878239.

Wallace AR. *A Narrative of Travels on the Amazon and Rio Negro* (1853). Reprint, New York: Haskell House (1969).

Wallace AR. On the monkeys of the Amazon. *Proceedings of the Zoological Society of London* 1852; part 20: 107-110.

Wallace AR. Man and natural selection. *Nature*. 3 Nov 1870; 8-9.

Wallace AR. *Contributions to the Theory of Natural Selection*. Macmillan (1970).
Wallace AR. On the tendency of varieties to depart indefinitely from the original type: instability of varieties supposed to prove the permanent distinctiveness of species. *Journal of the proceedings of the Linnean Society of London (Zoology)*. 1898; 3:53–62.
Wei T et al. Whole-genome resequencing of 445 Lactuca accessions reveals the domestication history of cultivated lettuce. *Nature Genetics*. 2021 May; 53(5):752–760. doi: 10.1038/s41588-021-00831-0. Epub 2021 Apr 12. PMID: 33846635.
Wills M. Evolution's highest branches. *Nature* 2006; 443, 633. https://doi.org/10.1038/443633a.
Woolfson A. Evolution on other worlds. *Science*. 2021; 371, 895. doi: 10.1126/science.abe8625.
Woolfson A. How to make a Mermaid: A theology of evolution. Review of *Life's solution: Inevitable humans in a lonely universe*, by Simon Conway Morris (2003). *London Review of Books*. 5 February 2004.
Xia B et al. On the genetic basis of tail-loss evolution in humans and apes. *Nature*. 2024 Feb; 626(8001):1042–1048. doi: 10.1038/s41586-024-07095-8. Epub 2024 Feb 28. PMID: 38418917; PMCID: PMC10901737.
Yamamoto S, Maruyama M, Parker J. Evidence for social parasitism of early insect societies by Cretaceous rove beetles. *Nature Communications*. 2016 Dec 8; 7:13658. doi: 10.1038/ncomms13658. PMID: 27929066; PMCID: PMC5155144.

CHAPTER 5

Alberts B. The cell as a collection of protein machines: preparing the next generation of molecular biologists. *Cell*. 1998 Feb 6; 92(3):291–4. doi: 10.1016/s0092-8674(00)80922-8. PMID: 9476889.
Avise JC. Colloquium paper: footprints of nonsentient design inside the human genome. *Proceedings of the National Academy of Sciences USA*. 2010 May 11; 107 Suppl 2(Suppl 2):8969–76. doi: 10.1073/pnas.0914609107. Epub 2010 May 5. PMID: 20445101; PMCID: PMC3024021.
Baden T, Nilsson DE. Is our retina really upside down? *Current Biology*. 2022 Apr 11; 32(7):R300–R303. doi: 10.1016/j.cub.2022.02.065. PMID: 35413251.
Ball P. What a shoddy piece of work is man. *Nature*. 2010. https://doi.org/10.1038/news.2010.215.
Bar-Even A et al. The moderately efficient enzyme: evolutionary and physicochemical trends shaping enzyme parameters. *Biochemistry*. 2011 May 31; 50(21):4402–10. doi: 10.1021/bi2002289. Epub 2011 May 4. PMID: 21506553.
Bartholomew M. James Lind's Treatise of the Scurvy (1753). *Postgraduate Medical Journal*. 2002 Nov; 78(925):695–6. doi: 10.1136/pmj.78.925.695. PMID: 12496338; PMCID: PMC1742547.

Bassett DS, Bullmore ET. Small-World Brain Networks Revisited. *Neuroscientist*. 2017 Oct; 23(5):499-516. doi: 10.1177/1073858416667720. Epub 2016 Sep 21. PMID: 27655008; PMCID: PMC5603984.

Bennett MR, Hasty J. Systems biology: genome rewired. *Nature*. 2008 Apr 17; 452(7189):824-5. doi: 10.1038/452824a. PMID: 18421342.

Berenbrink M. Extinct proteins resurrected to reconstruct the evolution of vertebrate haemoglobin. *Nature*. 2020 May; 581(7809):388-389. doi: 10.1038/d41586-020-01287-8. PMID: 32433630.

Boomsma JJ. Fifty years of illumination about the natural levels of adaptation. *Current Biology*. 2016 Dec 19; 26(24):R1250-R1255. doi: 10.1016/j.cub.2016.11.034. PMID: 27997830.

Chapman RW. The genome is the perfect imperfect machine. *Proceedings of the National Academy of Sciences USA*. 2010 Jul 20; 107(29):E119; author reply E120. doi: 10.1073/pnas.1006896107. Epub 2010 Jun 30. PMID: 20616064; PMCID: PMC2919970.

Chapple CE et al. Extreme multifunctional proteins identified from a human protein interaction network. *Nature Communications*. 2015 Jun 9; 6:7412. doi: 10.1038/ncomms8412. PMID: 26054620; PMCID: PMC4468855.

Cheng L, Leung KS. Identification and characterization of moonlighting long non-coding RNAs based on RNA and protein interactome. *Bioinformatics*. 2018 Oct 15; 34(20):3519-3528. doi: 10.1093/bioinformatics/bty399. PMID: 29771280.

Chong L, Ray LB. Whole-istic biology. *Science*. 2002 March 1; 295(5660): 1161. doi: 10.1126/science.295.5560.1661.

Comfort N. Genetics: The genetic watchmaker. *Nature*. 2013; 502:436–437. https://doi.org/10.1038/502436a.

Copley SD. Moonlighting is mainstream: paradigm adjustment required. *Bioessays*. 2012 Jul; 34(7):578-88. doi: 10.1002/bies.201100191. PMID: 22696112.

Copley SD. An evolutionary perspective on protein moonlighting. *Biochemical Society Transactions*. 2014 Dec; 42(6):1684-91. doi: 10.1042/BST20140245. PMID: 25399590; PMCID: PMC4405106.

Csete ME, Doyle JC. Reverse engineering of biological complexity. *Science*. 2002 Mar 1; 295(5560):1664-9. doi: 10.1126/science.1069981. PMID: 11872830.

Danchin A, Sekowska A. The logic of metabolism and its fuzzy consequences. *Environmental Microbiology*. 2014 Jan; 16(1):19-28. doi: 10.1111/1462-2920.12270. Epub 2013 Oct 6. PMID: 24387040.

D'Ari R, Casadesús J. Underground metabolism. *Bioessays*. 1998 Feb; 20(2):181-6. doi: 10.1002/(SICI)1521-1878(199802)20:2<181::AID-BIES10>3.0.CO;2-0. PMID: 9631663.

Darwin, C. *On the origins of species, by means of natural selection, or the preservation of favoured races in the struggle for life*. 1859. 24 November, John Murray, Albemarle Street, London.

Dishman AF et al. Evolution of fold switching in a metamorphic protein. *Science*. 2021 Jan 1; 371(6524):86-90. doi: 10.1126/science.abd8700. PMID: 33384377; PMCID: PMC8017559.

Drouin G, Godin JR, Pagé B. The genetics of vitamin C loss in vertebrates. *Current Genomics*. 2011 Aug; 12(5):371-8. doi: 10.2174/138920211796429736. PMID: 22294879; PMCID: PMC3145266.

Espinosa-Cantú A, Cruz-Bonilla E, Noda-Garcia L, DeLuna A. Multiple Forms of Multifunctional Proteins in Health and Disease. *Frontiers in Cell and Developmental Biology*. 2020 Jun 10; 8:451. doi: 10.3389/fcell.2020.00451. PMID: 32587857; PMCID: PMC7297953.

Finnigan GC, Hanson-Smith V, Stevens TH, Thornton JW. Evolution of increased complexity in a molecular machine. *Nature*. 2012 Jan 9; 481(7381):360-4. doi: 10.1038/nature10724. PMID: 22230956; PMCID: PMC3979732.

Franco-Serrano L et al. Multifunctional Proteins: Involvement in Human Diseases and Targets of Current Drugs. *Protein Journal*. 2018 Oct; 37(5):444-453. doi: 10.1007/s10930-018-9790-x. PMID: 30123928; PMCID: PMC6132618.

Gibbons A. Human evolution: Gain came with pain. *Science*. 16 February 2013. doi: 10.1126/article.26387.

Gold DA et al. The genome of the jellyfish Aurelia and the evolution of animal complexity. *Nature Ecology and Evolution*. 2019 Jan; 3(1):96-104. doi: 10.1038/s41559-018-0719-8. Epub 2018 Dec 3. PMID: 30510179.

Gould SJ, Lewontin RC. The spandrels of San Marco and the Panglossian paradigm: a critique of the adaptationist programme. *Proceedings of the Royal Society B: Biological Sciences*. 1979 Sep 21; 205(1161):581-98. doi: 10.1098/rspb.1979.0086. PMID: 42062.

Gould SJ. The exaptive excellence of spandrels as a term and prototype. *Proceedings of the National Academy of Sciences USA*. 1997 Sep 30; 94(20):10750-5. doi: 10.1073/pnas.94.20.10750. PMID: 11038582; PMCID: PMC23474.

Greenspan RJ. The flexible genome. *Nature Reviews Genetics*. 2001 May; 2(5):383-7. doi: 10.1038/35072018. PMID: 11331904.

Grinnell F. Discovery in the lab: Plato's paradox and Delbrück's principle of limited sloppiness. *FASEB Journal*. 2009 Jan; 23(1):7-9. doi: 10.1096/fj.09-0102ufm. PMID: 19118078.

Haldane JBS. *On being the right size and other essays* (ed. John Maynard Smith). Oxford University Press (1991).

Henderson B, Martin AC. Protein moonlighting: a new factor in biology and medicine. *Biochemical Society Transactions*. 2014 Dec; 42(6):1671-8. doi: 10.1042/BST20140273. PMID: 25399588.

Heyde A, Guo L, Jost C, Theraulaz G, Mahadevan L. Self-organized biotectonics of termite nests. *Proceedings of the National Academy of Sciences USA*.

2021 Feb 2; 118(5):e2006985118. doi: 10.1073/pnas.2006985118. PMID: 33468628; PMCID: PMC7865135.

Hume D. *A Treatise of Human Nature: Being an attempt to introduce the experimental method of reasoning into moral subjects.* Vol I: Of the understanding. London: John Noon (1739).

Hume D. *Dialogues Concerning Natural Religion.* London: William Strahan (1779).

Huttlin EL et al. Architecture of the human interactome defines protein communities and disease networks. *Nature.* 2017 May 25; 545(7655): 505-509. doi: 10.1038/nature22366. Epub 2017 May 17. PMID: 28514442; PMCID: PMC5531611.

Kaszubowska J. Masterpiece story: Sagrada Familia by Antoni Gaudí. *Daily Art Magazine.* 25 June 2024. https://www.dailyartmagazine.com/sagrada-familia/.

Kestenbaum D. The mystery of the Tappan Zee: Why build a bridge where the river is wide? All things considered. *NPR.* 14 May 2014. https://www.npr.org/2014/05/14/312523746/the-mystery-of-tappan-zee-why-build-a-bridge-where-the-rivers-wide.

Khersonsky O, Tawfik DS. Enzyme promiscuity: a mechanistic and evolutionary perspective. *Annual Review of Biochemistry.* 2010; 79:471-505. doi: 10.1146/annurev-biochem-030409-143718. PMID: 20235827.

Kimura M. Evolutionary rate at the molecular level. *Nature.* 1968 Feb 17; 217(5129):624-6. doi: 10.1038/217624a0. PMID: 5637732.

Kipling J. *Just So Stories for Little Children.* MacMillan & Co, London. 11 August 1902.

Kirschner M, Gerhart J, Mitchison T. Molecular 'vitalism'. *Cell.* 2000 Jan 7; 100(1):79-88. doi: 10.1016/s0092-8674(00)81685-2. PMID: 10647933.

Kozubek J. How gene editing could ruin human evolution. *Time.* 09 January 2017. https://time.com/4626571/crispr-gene-modification-evolution/.

Kuzmin E et al. Systematic analysis of complex genetic interactions. *Science.* 2018 Apr 20; 360(6386):eaao1729. doi: 10.1126/science.aao1729. PMID: 29674565; PMCID: PMC6215713.

de La Mettrie JO. *L'Homme Machine.* Leiden: Luzac (1748) (Published anonymously). *Man a Machine.* London: W. Owen (1750).

Landry CR, Lemos B, Rifkin SA, Dickinson WJ, Hartl DL. Genetic properties influencing the evolvability of gene expression. *Science.* 2007 Jul 6; 317(5834):118-21. doi: 10.1126/science.1140247. Epub 2007 May 24. PMID: 17525304.

Ledford H. Human genes are multitaskers. *Nature.* 2008 Nov 6; 456(7218):9. doi: 10.1038/news.2008.1199. PMID: 19004070.

Leibniz GW. *Essais de theodicée sur la bonté de Dieu, la liberté de l'homme, et l'origine du mal.* Amsterdam: Troyel (1710).

Leslie M. Red blood cells may be immune sentinels. *Science*. 2021 Oct 22; 374(6566):383. doi: 10.1126/science.acx9389. Epub 2021 Oct 21. PMID: 34672723.

Lin MT, Salihovic H, Clark FK, Hanson MR. Improving the efficiency of Rubisco by resurrecting its ancestors in the family Solanaceae. *Science Advances*. 2022 Apr 15; 8(15):eabm6871. doi: 10.1126/sciadv.abm6871. Epub 2022 Apr 15. PMID: 35427154; PMCID: PMC9012466.

Lind JA. *Treatise of the Scurvy. In three parts*. London: A Millarin (1753).

Lipsh-Sokolik R et al. Combinatorial assembly and design of enzymes. *Science*. 2023 Jan 13; 379(6628):195-201. doi: 10.1126/science.ade9434. Epub 2023 Jan 12. PMID: 36634164.

Liu H, Jeffery CJ. Moonlighting Proteins in the Fuzzy Logic of Cellular Metabolism. *Molecules*. 2020 Jul 29; 25(15):3440. doi: 10.3390/molecules 25153440. PMID: 32751110; PMCID: PMC7435893.

Loison L. Why did Jacques Monod make the choice of mechanistic determinism? *Comptes Rendus Biologies*. 2015 Jun; 338(6):391-7. doi: 10.1016/j .crvi.2015.03.008. Epub 2015 Apr 22. PMID: 25912960.

Macosko EZ, McCarroll SA. Genetics. Our fallen genomes. *Science*. 2013 Nov 1; 342(6158):564-5. doi: 10.1126/science.1246942. PMID: 24179207.

Manhart M, Morozov AV. Protein folding and binding can emerge as evolutionary spandrels through structural coupling. *Proceedings of the National Academy of Sciences USA*. 2015 Feb 10; 112(6):1797-802. doi: 10.1073/pnas.1415895112. Epub 2015 Jan 26. PMID: 25624494; PMCID: PMC4330747.

Malphigi, M. Quoted in Marco Piccolino, 'Biological machines: from mills to molecules,' *Nature Reviews Molecular Cell Biology*, Vol. 1:149–153 (Nov. 2000).

Mani M et al. MoonProt: a database for proteins that are known to moonlight. *Nucleic Acids Research*. 2015 Jan; 43(Database issue):D277-82. doi: 10.1093/nar/gku954. Epub 2014 Oct 16. PMID: 25324305; PMCID: PMC4384022.

Mazur J. On the right path. *Nature*. 2007; 445:25. https://doi.org/10.1038 /445025a.

Michaelis AC et al. The social and structural architecture of the yeast protein interactome. *Nature*. 2023 Dec; 624(7990):192-200. doi: 10.1038 /s41586-023-06739-5. Epub 2023 Nov 15. PMID: 37968396; PMCID: PMC10700138.

Milgram S. The small world problem. *Psychology Today*. 1 May 1967; 60-67.

Miller A. *Death of a Salesman*. New York: Viking Press (1949).

Millican P. *Dialogues Concerning Natural Religion* (1779). Hume texts online. https://davidhume.org/texts/d/notes.

Morange M. Monod and the spirit of molecular biology. *Comptes Rendus Biologies*. 2015 Jun; 338(6):380-4. doi: 10.1016/j.crvi.2015.03.005. Epub 2015 Apr 15. PMID: 25890787.

Nesse R, Williams GC. *Evolution and Healing: The new science of Darwinian medicine*. JM Dent & Sons (1996).

Nicholson DJ. On being the right size, revisited. In *Philosophical perspectives on the engineering approach in biology* ed. Holm S and Serban M. Routledge (2020).

Nicholson DJ. Is the cell really a machine? *Journal of Theoretical Biology*. 2019 Sep 21; 477:108–126. doi: 10.1016/j.jtbi.2019.06.002. Epub 2019 Jun 4. PMID: 31173758.

Offray de La Mettrie, J. *La Mettrie's L'Homme Machine. A Study in the Origins of an Idea*. Critical Edition with an Introductory Monograph and Notes by Aram Vartanian. Princeton University Press, 1960.

Oliveri P, Davidson EH. Development. Built to run, not fail. *Science*. 2007 Mar 16; 315(5818):1510–1. doi: 10.1126/science.1140979. PMID: 17363653.

Pagel M. Rise of the digital machine. *Nature*. 2008 Apr 10; 452(7188):699. doi: 10.1038/452699a. PMID: 18401394.

Palazzo AF, Kejiou NS. Non-Darwinian Molecular Biology. *Frontiers in Genetics*. 2022 Feb 16; 13:831068. doi: 10.3389/fgene.2022.831068. PMID: 35251134; PMCID: PMC8888898.

Peng GS, Tan SY, Wu J, Holme P. Trade-offs between robustness and small-world effect in complex networks. *Scientific Reports*. 2016 Nov 17; 6:37317. doi: 10.1038/srep37317. PMID: 27853301; PMCID: PMC5112524.

Piccolino M. Biological machines: from mills to molecules. *Nature Reviews Molecular Cell Biology*. 2000 Nov; 1(2):149–53. doi: 10.1038/35040097. PMID: 11253368.

Pigliucci M, Kaplan J. The fall and rise of Dr Pangloss: adaptationism and the Spandrels paper 20 years later. *Trends in Ecology and Evolution*. 2000 Feb; 15(2):66–70. doi: 10.1016/s0169-5347(99)01762-0. PMID: 10652558.

Pillai AS et al. Origin of complexity in haemoglobin evolution. *Nature*. 2020 May; 581(7809):480–485. doi: 10.1038/s41586-020-2292-y. Epub 2020 May 20. Erratum in: *Nature*. 2020 Jul; 583(7816):E26. doi: 10.1038/s41586-020-2472-9. PMID: 32461643; PMCID: PMC8259614.

Porcar M, Danchin A, de Lorenzo V. Confidence, tolerance, and allowance in biological engineering: the nuts and bolts of living things. *Bioessays*. 2015 Jan; 37(1):95–102. doi: 10.1002/bies.201400091. Epub 2014 Oct 27. PMID: 25345679.

Porcar M, Peretó J. Nature versus design: synthetic biology or how to build a biological non-machine. *Integrative Biology (Cambridge)*. 2016 Apr 18; 8(4):451–5. doi: 10.1039/c5ib00239g. Epub 2015 Nov 30. PMID: 26616724.

Ridley M. Dreadful beasts. Review of *Wonderful Life* by Stephen Jay Gould. *London Review of Books*. 28 June 1990; 12(12):11–12.

Rudolph J, Luger K. The secret life of histones. *Science*. 2020 Jul 3; 369(6499):33. doi: 10.1126/science.abc8242. PMID: 32631882.

Simons DJ, Chabris CF. Gorillas in our midst: sustained inattentional blindness for dynamic events. *Perception*. 1999; 28(9):1059–74. doi: 10.1068/p281059. PMID: 10694957.

Smith CU. Julien Offray de la Mettrie (1709–1751). *Journal of the History of the Neurosciences*. 2002 Jun; 11(2):110–24. doi: 10.1076/jhin.11.2.110.15188. PMID: 12122804.

Striedter G. Brain botch. *Nature*. 2007; 447:640. https://doi.org/10.1038/447640a.

Sutherland W. The best solution. *Nature*. 2005; 435:569. https://doi.org/10.1038/435569a.

Swift G. *Travels into several remote nations of the world, by Lemuel Gulliver, first a surgeon, and then a captain of several ships [Gulliver's Travels]*. London (1726).

Taipale J. Informational limits of biological organisms. *EMBO Journal*. 2018 May 15; 37(10):e96114. doi: 10.15252/embj.201696114. Epub 2018 Apr 18. PMID: 29669861; PMCID: PMC5978287.

Tattersall I. Evolution and inevitability. *Annals of the New York Academy of Sciences*. 2018 Nov; 1432(1):72–75. doi: 10.1111/nyas.13881. Epub 2018 Jun 20. PMID: 29923611.

Tawfik DS. Messy biology and the origins of evolutionary innovations. *Nature Chemical Biology*. 2010 Oct; 6(10):692–6. doi: 10.1038/nchembio.441. PMID: 20852602.

Tesei G et al. Conformational ensembles of the human intrinsically disordered proteome. *Nature*. 2024 Feb; 626(8000):897–904. doi: 10.1038/s41586-023-07004-5. Epub 2024 Jan 31. PMID: 38297118.

Tompa P, Szász C, Buday L. Structural disorder throws new light on moonlighting. *Trends in Biochemical Sciences*. 2005 Sep; 30(9):484–9. doi: 10.1016/j.tibs.2005.07.008. PMID: 16054818.

Voltaire. *Candide, ou L'optimisme*. Geneva (1759).

Wallisch P. Unleashing the beast within. *Science*. 2016; 351:232. doi: 10.1126/science.aad7138.

Wang C. The 'unintelligent design' of the recurrent laryngeal nerve. *McGill University*. 25 May 2022. https://www.mcgill.ca/oss/article/student-contributors-did-you-know-general-science/unintelligent-design-recurrent-laryngeal-nerve.

Wedel MJ. A monument of inefficiency: The presumed course of the recurrent laryngeal nerve in the Sauropod Dinosaur. *Acta Palaeontologica Polonica*. 2011; 57(2):251–256. https://doi.org/10.4202/app.2011.0019.

Weiss MA et al. Protein structure and the spandrels of San Marco: insulin's receptor-binding surface is buttressed by an invariant leucine essential for its stability. *Biochemistry*. 2002 Jan 22; 41(3):809–19. doi: 10.1021/bi011839+. PMID: 11790102.

Williams GC, Nesse RM. The dawn of Darwinian medicine. *Quarterly Review of Biology*. 1991 Mar; 66(1):1–22. doi: 10.1086/417048. PMID: 2052670.

Williams GC. *Adaptation and natural selection*. Princeton University Press (1966).

Woolfson A. Inevitable or improbable? *Science*. 2017; 357:362. doi: 10.1126/science.aan8380.

Yan KK, Fang G, Bhardwaj N, Alexander RP, Gerstein M. Comparing genomes to computer operating systems in terms of the topology and evolution of their regulatory control networks. *Proceedings of the National Academy of Sciences USA*. 2010 May 18; 107(20):9186–91. doi: 10.1073/pnas.0914771107. Epub 2010 May 3. PMID: 20439753; PMCID: PMC2889091.

CHAPTER 6

Abid C, Alizadeh V, Kessentini M, Ferreira TDN, Dig D. 30 years of software refactoring research: A systematic literature review. *IEEE Transactions of Software Engineering*. 1 June 2020; 1:1–23. https://doi.org/10.48550/arXiv.2007.02194.

Andrianantoandro E, Basu S, Karig DK, Weiss R. Synthetic biology: new engineering rules for an emerging discipline. *Molecular Systems Biology*. 2006; 2:2006.0028. doi: 10.1038/msb4100073. Epub 2006 May 16. PMID: 16738572; PMCID: PMC1681505.

Baker M. Synthetic genomes: The next step for the synthetic genome. *Nature*. 2011 May 19; 473(7347):403, 405–8. doi: 10.1038/473403a. PMID: 21593873.

Ball P. *How Life Works: A user's guide to the new biology*. Chicago: University of Chicago Press (2022).

Brenner S. Turing centenary: Life's code script. *Nature*. 2012 Feb 22; 482(7386):461. doi: 10.1038/482461a. PMID: 22358811.

Brenner S. The architecture of biological complexity. Lecture given at the Salk Institute for Biological Sciences, San Diego, CA. 18 October 2012. https://www.icts.res. in/lectures /architecturebiocompl2012.

Byron, Lord. (Initially published anonymoulsy). *Don Juan*. London: Thomas Davison [i.e., John Murray], 1819–21 (Cantos I–V, Volumes I–II); John Hunt (Cantos VI–XVI, Volumes III–VI), 1823–24.

Callaway E. 'Minimal' cell raises stakes in race to harness synthetic life. *Nature*. 2016; 531:557–558. https://doi.org/10.1038/531557a.

Callaway E. Entire yeast genome squeezed into one chromosome. *Nature*. 1 August 2018. doi: https://doi.org/10.1038/d41586-018-05857-9.

Campos L. Our synthetic moment. *Science*. 2017 Mar 17; 355(6330):1136. doi: 10.1126/science.aam6893. PMID: 28302815.

Chan LY, Kosuri S, Endy D. Refactoring bacteriophage T7. *Molecular Systems Biology*. 2005; 1:2005.0018. doi: 10.1038/msb4100025. Epub 2005 Sep 13. PMID: 16729053; PMCID: PMC1681472.

Cherfas J. Seeking the soul of an old machine. *Science*. 1991 Jun 7; 252(5011):1370–1. doi: 10.1126/science.252.5011.1370. PMID: 17772901.

Coradini ALV, Hull CB, Ehrenreich IM. Building genomes to understand biology. *Nature Communications*. 2020 Dec 2; 11(1):6177. doi: 10.1038/s41467-020-19753-2. PMID: 33268788; PMCID: PMC7710724.

Danchin A. Bacteria as computers making computers. *FEMS Microbiology Reviews*. 2009 Jan; 33(1):3-26. doi: 10.1111/j.1574-6976.2008.00137.x. Epub 2008 Nov 7. PMID: 19016882; PMCID: PMC2704931.

Dickens C. *Dombey and Son*. Bradbury & Evans (1848).

Dostoyevsky F. *Crime and Punishment*. Russian Messenger (1866).

Editorial. Charles Babbage. *Nature*. 1871; 5:28–29. https://doi.org/10.1038/005028a0.

Eisenstein M. Pursuing the simple life. *Nature Methods*. 2017 Jan 31; 14(2):117-121. doi: 10.1038/nmeth.4158. PMID: 28139672.

Ellman R. *James Joyce*. Oxford University Press (1959).

Endy D. Foundations for engineering biology. *Nature*. 2005 Nov 24; 438(7067):449-53. doi: 10.1038/nature04342. PMID: 16306983.

Estrada S, Weiss B. Wall Street 'Cobol Cowboys' are spread thin fixing legacy tech – but AI may soon ride to the rescue. *Fortune*. 13 October 2023. https://fortune.com/2023/10/13/wall-street-cobol-cowboys-legacy-tech-ai/.

Fleischmann RD et al. Whole-genome random sequencing and assembly of Haemophilus influenzae Rd. *Science*. 1995 Jul 28; 269(5223):496-512. doi: 10.1126/science.7542800. PMID: 7542800.

Fraser CM et al. The minimal gene complement of Mycoplasma genitalium. *Science*. 1995 Oct 20; 270(5235):397-403. doi: 10.1126/science.270.5235.397. PMID: 7569993.

Fuegi J, Francis J. Lovelace & Babbage and the creation of the 1843 'Notes'. *IEEE Annals of the History of Computing*. 2003; 25(4):16-26. doi: 10.1109/MAHC.2003.1253887.

Ghosh D, Kohli AG, Moser F, Endy D, Belcher AM. Refactored M13 bacteriophage as a platform for tumor cell imaging and drug delivery. *ACS Synthetic Biology*. 2012 Dec 21; 1(12):576-582. doi: 10.1021/sb300052u. Epub 2012 Sep 24. PMID: 23656279; PMCID: PMC3905571.

Gong Li. The hidden influencers who code our world. *Nature*. 2 April 2019; 568: 30-31. doi: https://doi.org/10.1038/d41586-019-01037-5.

Good M. Living systems engineered. *Nature*. 8 November 2018; 563: 188-189.

Gregory R. Ada and the engines. *Nature*. 1986; 320: 224. https://doi.org/10.1038/320224a0.

Griswold WG. Program restructuring: as an aid to software maintenance. PhD thesis. University of Washington. 1992.

Holmes R. Computer science: Enchantress of abstraction. *Nature*. 2015; 525:30–31. https://doi.org/10.1038/525030a.

Hutchison CA 3rd et al. Design and synthesis of a minimal bacterial genome. *Science*. 2016 Mar 25; 351(6280):aad6253. doi: 10.1126/science

.aad6253. Erratum in: *ACS Chemical Biology*. 2016 May 20; 11(5):1463. doi: 10.1021/acschembio.6b00374. PMID: 27013737.

Istrail S, De-Leon SB, Davidson EH. The regulatory genome and the computer. *Developmental Biology*. 2007 Oct 15; 310(2):187–95. doi: 10.1016/j.ydbio.2007.08.009. Epub 2007 Aug 10. PMID: 17822690.

Itaya M, Tsuge K, Koizumi M, Fujita K. Combining two genomes in one cell: stable cloning of the Synechocystis PCC6803 genome in the Bacillus subtilis 168 genome. *Proceedings of the National Academy of Sciences USA*. 2005 Nov 1; 102(44):15971-6. doi: 10.1073/pnas.0503868102. Epub 2005 Oct 18. PMID: 16236728; PMCID: PMC1276048.

Jaschke PR, Lieberman EK, Rodriguez J, Sierra A, Endy D. A fully decompressed synthetic bacteriophage øX174 genome assembled and archived in yeast. *Virology*. 2012 Dec 20; 434(2):278–84. doi: 10.1016/j.virol.2012.09.020. Epub 2012 Oct 15. PMID: 23079106.

Jaschke PR, Dotson GA, Hung KS, Liu D, Endy D. Definitive demonstration by synthesis of genome annotation completeness. *Proceedings of the National Academy of Sciences USA*. 2019 Nov 26; 116(48):24206–24213. doi: 10.1073/pnas.1905990116. Epub 2019 Nov 12. PMID: 31719208; PMCID: PMC6883844.

Joyce J. *Dubliners*. London: Grant Richards Ltd (1914).

Kolisnychenko V et al. Engineering a reduced Escherichia coli genome. *Genome Research*. 2002 Apr; 12(4):640–7. doi: 10.1101/gr.217202. PMID: 11932248; PMCID: PMC187512.

Kwok R. Five hard truths for synthetic biology. *Nature*. 2010 Jan 21; 463(7279):288–90. doi: 10.1038/463288a. PMID: 20090726.

Lander ES, Weinberg RA. Genomics: journey to the center of biology. *Science*. 2000 Mar 10; 287(5459):1777–82. doi: 10.1126/science.287.5459.1777. PMID: 10755930.

Lane A. Lord Byron was more than just Byronic. *New Yorker*. 26 February 2024.

Leroi AM. *Mutants: On genetic variety and the human body*.Viking Press (2004).

Leslie M. Researchers coax stripped-down cells to grow normally. *Science*. 2021 Apr 2; 372(6537):18. doi: 10.1126/science.372.6537.18. PMID: 33795440.

Linshiz G, Goldberg A, Konry T, Hillson NJ. The fusion of biology, computer science, and engineering: towards efficient and successful synthetic biology. *Perspectives in Biology and Medicine*. 2012; 55(4):503–20. doi: 10.1353/pbm.2012.0044. PMID: 23502561.

Liti G. Chromosomes get together. *Nature*. 16 August 2018; 560:317–318.

Lovelace AAB. Sketch of the analytical machine invented by Charles Babbage. (1843).

Lowe D. Engineering biology, for real? *Science*. 12 Nov 2018. https://www.science.org/content/blog-post/engineering-biology-real.

Lu T, Purcell O. Cells that compute come closer to reality. *Scientific American*. April 2016; 315(5):42–29.

Luo J, Sun X, Cormack BP, Boeke JD. Karyotype engineering by chromosome fusion leads to reproductive isolation in yeast. *Nature*. 2018 Aug; 560(7718):392–396. doi: 10.1038/s41586-018-0374-x. Epub 2018 Aug 1. PMID: 30069047; PMCID: PMC8223741.

Menabrea LF. *Notions sur la machine analytique de M. Charles Babbage*. Paris: Anselin (1842).

Menabrea LF. *Sketch of the Analytical Engine invented by Charles Babbage Esq.* (1843).

Morrison P, Morrison E. *Charles Babbage and his calculating engines: Selected writings by Charles Babbage and others*. Dover Publications (1961).

Muka S. Making and remaking life. *Science*. 2021; 373,499–499.doi: 10.1126/science.abj2437.

Mushegian AR, Koonin EV. A minimal gene set for cellular life derived by comparison of complete bacterial genomes. *Proceedings of the National Academy of Sciences USA*. 1996 Sep 17; 93(19):10268–73. doi: 10.1073/pnas.93.19.10268. PMID: 8816789; PMCID: PMC38373.

Naughton J. Why it's dangerous to liken DNA to computer code. *Guardian*. 22 September 2019.

Opdyke WF. Refactoring: An aid in designing application frameworks and evolving object-oriented systems. *Proceedings of SOOPPA'90* (Symposium on Object-Oriented Programming Emphasizing Practical Applications). 1990.

Opdyke WF. Refactoring object-oriented frameworks. Thesis. University of Illinois at Urbana-Champaign. 1992.

Pennisi E. Synthetic biology. Synthetic biology remakes small genomes. *Science*. 2005 Nov 4; 310(5749):769–70. doi: 10.1126/science.310.5749.769. PMID: 16272096.

Pósfai G et al. Emergent properties of reduced-genome Escherichia coli. *Science*. 2006 May 19; 312(5776):1044–6. doi: 10.1126/science.1126439. Epub 2006 Apr 27. PMID: 16645050.

Schneier B, Rudenko L. What digital nerds and bio geeks have to worry about. *CNN Opinion*. 13 September 2019. https://www.cnn.com/2019/09/12/opinions/digital-nerds-bio-geeks-worry-about-opinion-schneier-rudenko/index.html.

Science museum. *Charles Babbage's Difference Engines and the Science Museum*. 18 July 2023. https://www.sciencemuseum.org.uk/objects-and-stories/charles-babbages-difference-engines-and-science-museum

Service RF. Synthetic Biology. Synthetic microbe has fewest genes, but many mysteries. *Science*. 2016 Mar 25; 351(6280):1380–1. doi: 10.1126/science.351.6280.1380. PMID: 27013708.

Singer E. Better drug-producing bacteria. *MIT Technology Review*. 02 May 2006. https://www.technologyreview.com/2006/05/02/229185/better-drug-producing-bacteria/.

Stauffer A. *Byron: A Life in Ten Letters*. Cambridge: Cambridge University Press (2024).

Stein D. Ada: A life and legacy. *Nature*. 20 March 1986; 320:224.

Tang L. Does the number of chromosomes matter? *Nature Methods*. 2018 Oct; 15(10):761. doi: 10.1038/s41592-018-0164-2. PMID: 30275595.

Thompson C. *Coders: The making of a new tribe and the remaking of the world*. Penguin (2019).

Weiss B. Can AI fix Wall Street's 'spaghetti code' crisis? Microsoft and IBM are betting that it can. *Fortune*. 9 October 2023. https://fortune.com/2023/10/09/generative-ai-cobol-code-wall-street-ibm-microsoft/.

Wieschaus E, Nüsslein-Volhard C. The Heidelberg Screen for Pattern Mutants of Drosophila: A Personal Account. *Annual Review of Cell and Developmental Biology*. 2016 Oct 6; 32:1-46. doi: 10.1146/annurev-cell bio-113015-023138. Epub 2016 Aug 3. PMID: 27501451.

Witzany G, Baluška F. Life's code script does not code itself. The machine metaphor for living organisms is outdated. *EMBO Reports*. 2012 Dec; 13(12):1054-6. doi: 10.1038/embor.2012.166. Epub 2012 Nov 13. PMID: 23146891; PMCID: PMC3512409.

Wolfram S. Untangling the tale of Ada Lovelace. 10 December 2015. *Stephen Wolfram writings*. https://writings.stephenwolfram.com/2015/12/untangling-the-tale-of-ada-lovelace/#:~: text=Then%2C%20on%20Sept.,'%20not%20'Numbers'.

Xu X et al. Trimming the genomic fat: minimising and re-functionalising genomes using synthetic biology. *Nature Communications*. 2023 Apr 8; 14(1):1984. doi: 10.1038/s41467-023-37748-7. PMID: 37031253; PMCID: PMC10082837.

Yan KK, Fang G, Bhardwaj N, Alexander RP, Gerstein M. Comparing genomes to computer operating systems in terms of the topology and evolution of their regulatory control networks. *Proceedings of the National Academy of Sciences USA*. 2010 May 18; 107(20):9186-91. doi: 10.1073/pnas.0914771107. Epub 2010 May 3. PMID: 20439753; PMCID: PMC2889091.

CHAPTER 7

AlQuraishi M. Protein-structure prediction revolutionized. *Nature*. 2021 Aug; 596(7873):487-488. doi: 10.1038/d41586-021-02265-4. PMID: 34426694.

Amis M. *Invasion of the Space Invaders: An addict's guide to battle tactics, big scores, and the best machines*. London: Hutchinson & Co (1982).

Appenzeller T. The AI revolution in science. *Science*. 7 July 2017. doi: 10.1126/science.aan7064.

Barrio-Hernandez I et al. Clustering predicted structures at the scale of the known protein universe. *Nature*. 2023 Oct; 622(7983):637-645. doi:

10.1038/s41586-023-06510-w. Epub 2023 Sep 13. PMID: 37704730; PMCID: PMC10584675.

Benzer S. On the topography of the genetic fine structure. *Proceedings of the National Academy of Sciences USA*. 1961 Mar; 47(3):403-15. doi: 10.1073/pnas.47.3.403. PMID: 16590840; PMCID: PMC221592.

Benzer S. Fine structure of a genetic region in bacteriophage. *Proceedings of the National Academy of Sciences USA*. 1955 Jun 15; 41(6):344-54. doi: 10.1073/pnas.41.6.344. PMID: 16589677; PMCID: PMC528093.

Berg J. Banking on protein structural data. *Science*. 2021 Aug 20; 373(6557):835. doi: 10.1126/science.abl8151. PMID: 34413213.

Brown S. *Machine learning explained*. MIT Management Sloan School. 21 April 2021.

Callaway E. Major AlphaFold upgrade offers boost for drug discovery. *Nature*. 2024 May; 629(8012):509-510. doi: 10.1038/d41586-024-01383-z. PMID: 38719965.

Callaway E. 'The entire protein universe': AI predicts shape of nearly every known protein. *Nature*. 2022 Aug; 608(7921):15-16. doi: 10.1038/d41586-022-02083-2. PMID: 35902752.

Callaway E. AlphaFold's new rival? Meta AI predicts shape of 600 million proteins. *Nature*. 2022 Nov; 611(7935):211-212. doi: 10.1038/d41586-022-03539-1. PMID: 36319775.

Callaway E. DeepMind's AI predicts structures for a vast treasure trove of proteins. *Nature*. 29 July 2021; 595:635.

Callaway E. 'It will change everything': DeepMind's AI makes gigantic leap in solving protein structures. *Nature*. 2020 Dec; 588(7837):203-204. doi: 10.1038/d41586-020-03348-4. PMID: 33257889.

Chan D. In the Age of Google DeepMind, do young Go prodigies of Asia have a future? *New Yorker*. 11 March 2016.

Chandrasekaran R. Kasparov proves no match for computer. *Washington Post*. 12 May 1997. A01.

Chargaff E. Preface to a grammar of biology. A hundred years of nucleic acid research. *Science*. 1971 May 14; 172(3984):637-42. doi: 10.1126/science.172.3984.637. PMID: 4929532.

Chomsky N. The false promise of ChatGPT. *New York Times*. Opinion. Guest Essay. 8 March 2023.

Chomsky N. A review of Verbal Behavior. *Language*. 1959; 35(1):26-58.

Chomsky N. *Aspects of the theory of syntax*. Massachusetts Institute of Technology, Massachusetts: MIT Press (May 1965).

Comfort N. Genetics: Dawkins, redux. *Nature*. 2015; 525:184–185. https://doi.org/10.1038/525184a.

Dill KA, MacCallum JL. The protein-folding problem, 50 years on. *Science*. 2012 Nov 23; 338(6110):1042-6. doi: 10.1126/science.1219021. PMID: 23180855.

Durairaj J et al. Uncovering new families and folds in the natural protein universe. *Nature*. 2023 Oct; 622(7983):646-653. doi: 10.1038/s41586-023-06622-3. Epub 2023 Sep 13. PMID: 37704037; PMCID: PMC10584680.

Editorial. Chess grandmaster Garry Kasparov on what happens when machines 'reach the level that is impossible for humans to compete'. *Business Insider India*. 29 December 2017. https://www.kasparov.com/kasparov-on-what-happens-when-machines-reach-the-level-that-is-impossible-for-humans-to-compete-business-insider-dec-29th2017/.

Editorial. Artificial intelligence: Google's AlphaGo beats Go master Lee Se-dol. *BBC News*. 12 March 2016. https://www.bbc.com/news/technology-35785875.

Editorial. How the game of Go explains China's aggression towards India. *Economist*. 11 November 2021. https://www.economist.com/asia/2021/11/11/how-the-game-of-go-explains-chinas-aggression-towards-india.

Editorial. ChatGPT is a black box: how AI can crack it open. *Nature*. 25 July 2023; 619:671-672. doi: https://doi.org/10.1038/d41586-023-02366-2.

Editorial. Artificial intelligence in structural biology is here to stay. *Nature*. 29 July 2021; 595:625-626. doi: https://doi.org/10.1038/d41586-021-02037-0.

Editorial. Jensen Huang says Moore's Law is dead. Not quite yet. *Economist*. 13 December 2023. https://www.economist.com/science-and-technology/2023/12/13/jensen-huang-says-moores-law-is-dead-not-quite-yet.

Fernandez AA, Burchardt LS, Nagy M, Knörnschild M. Babbling in a vocal learning bat resembles human infant babbling. *Science*. 2021 Aug 20; 373(6557):923-926. doi: 10.1126/science.abf9279. PMID: 34413237.

Freeman W. Space Invaders at 40: 'I tried soldiers but shooting people was frowned upon'. *Guardian*. 4 June 2018. https://www.theguardian.com/games/2018/jun/04/space-invaders-at-40-tomohiro-nishikado-interview.

Gibney E. Google AI algorithm masters ancient game of Go. *Nature*. 2016 Jan 28; 529(7587):445-6. doi: 10.1038/529445a. PMID: 26819021.

Goldsmith J. Linguistics: Sound sculptor. *Nature*. 2012; 483:538–539. https://doi.org/10.1038/483538a.

Grannell C. The Space Invaders creator reveals the game's origin story. *Wired*. 12 April 2023. https://www.wired.com/story/space-invaders-45-years-tomohiro-nishikado/.

Harris WA. Seymour Benzer 1921-2007. The man who took us from genes to behavior. *PLOS Biology*. 12 February 2008. https://doi.org/10.1371/journal.pbio.0060041.

Hassabis D. AlphaFold reveals the structure of the protein universe. *Google DeepMind*. 28 July 2022. https://deepmind.google/discover/blog/alphafold-reveals-the-structure-of-the-protein-universe/.

Hassabis, D. Introducing Isomorphic Labs. https://www.isomorphiclabs.com/articles/introducing-isomorphic-labs.

Hassabis D. Artificial Intelligence: Chess match of the century. *Nature*. 2017; 544:413–414. https://doi.org/10.1038/544413a.

Hayes B. The manifest destiny of artificial intelligence. *American Scientist*. 2012; 100:282-287. https://www.americanscientist.org/article/the-manifest-destiny-of-artificial-intelligence.

Hern A. Google AI in landmark victory over Go grandmaster. *Guardian*. 27 January 2016. https://www.theguardian.com/technology/2016/jan/27/google-hits-ai-milestone-as-computer-beats-go-grandmaster.

House P. AlphaGo, Lee Sedol, and the reassuring future of humans and machines. *New Yorker*. 15 March 2016. https://www.newyorker.com/tech/annals-of-technology/alphago-lee-sedol-and-the-reassuring-future-of-humans-and-machines.

Jóźwik KM. What AI can learn from the biological brain. *Science*. 2021; 372:798. doi: 10.1126/science.abi4889.

Jumper J et al. Highly accurate protein structure prediction with AlphaFold. *Nature*. 2021 Aug; 596(7873):583-589. doi: 10.1038/s41586-021-03819-2. Epub 2021 Jul 15. PMID: 34265844; PMCID: PMC8371605.

Kasparov G. Chess, a Drosophila of reasoning. *Science*. 2018 Dec 7; 362(6419):1087. doi: 10.1126/science.aaw2221. PMID: 30523085.

Kissinger H. *On China*. London: Penguin (2011).

Lai D. Learning from the stones: A Go approach to mastering China's strategic concept. *SHI*. May 2004. https://man.fas.org/eprint/lai.pdf.

Latson J. Did Deep Blue beat Kasparov because of a system glitch? *Time*. 17 February 2015. https://time.com/3705316/deep-blue-kasparov/.

Levinthal C. How to fold graciously. Mössbauer Spectroscopy in Biological Systems Proceedings. *University of Illinois Bulletin*. 1969; 67(41):22-24. https://faculty.cc.gatech.edu/~turk/bio_sim/articles/proteins_levinthal_1969.pdf.

Levy S. Big Blue's hand of God. *Newsweek*. 18 May 1997. https://www.newsweek.com/big-blues-hand-god-173076.

Levy S. The brain's last stand. *Newsweek*. Front page. 4 May 1997. https://www.newsweek.com/man-vs-machine-173038.

Lofting H. *The Story of Doctor Dolittle*. New York: Frederick A. Stokes Company (1920).

McCarthy J. AI as Sport. *Science*. 1997; 276:1518-1519. doi: 10.1126/science.276.5318.1518.

Metro-Goldwyn-Mayer (MGM). *The Wizard of Oz*. 1939. Adaptation of Baum, LF. *The Wonderful Wizard of Oz*. George M. Hill Company (1900).

Metz C. A.I. Predicts the Shape of Nearly Every Protein Known to Science. *New York Times*. 28 July 2022.

Mitchell M. Debates on the nature of artificial general intelligence. *Science*. 2024 Mar 22; 383(6689):eado7069. doi: 10.1126/science.ado7069. Epub 2024 Mar 21. PMID: 38513022.

Moor J. The Dartmouth College Artificial Intelligence Conference: The Next Fifty Years. *AI Magazine*. January 2006; 27:87-91.

Moore GE. Cramming more components onto integrated circuits. *Electronics*. 19 April 1965; 38:8.

Newport C. What kind of mind does CHatGPT have? *New Yorker*. 13 April 2023. https://www.newyorker.com/science/annals-of-artificial-intelligence/what-kind-of-mind-does-chatgpt-have.

O'Grady C. Baby bats babble, much like human infants. *Science*. 19 August 2021. doi: 10.1126/science.abm0126.

Parkin S. The Space Invader. *New Yorker*. 13 October 2017. https://www.newyorker.com/tech/annals-of-technology/the-space-invader.

Pennisi E. Protein structure prediction now easier, faster. *Science*. 2021 Jul 16; 373(6552):262-263. doi: 10.1126/science.373.6552.262. PMID: 34437129.

Raina R, Madhavan A, Ng AY. Large-scale deep unsupervised learning using graphics processors. *ICML '09: Proceedings of the 26th Annual International Conference on Machine Learning*. June 2009; 873–880. https://doi.org/10.1145/1553374.1553486.

Samuel AL. Some studies in machine learning using the game of checkers. *IBM Journal*. 1959; 3:3.

Sanderson K. GPT-4 is here: what scientists think. *Nature*. 2023 Mar; 615(7954):773. doi: 10.1038/d41586-023-00816-5. PMID: 36928404.

Sang-Hun C, Markoff J. Master of Go board game is walloped by Google computer program. *New York Times*. 9 March 2016. https://www.nytimes.com/2016/03/10/world/asia/google-alphago-lee-se-dol.html.

de Saussure F. *Mémoire sur le système primitif des voyelles dans les langues indo-européenes*. Leipzig (1878).

de Saussure F. *Cours de linguistique générale*. Lausanne & Paris: Payot (1916).

Schaeffer J et al. Checkers is solved. *Science*. 2007 Sep 14; 317(5844):1518-22. doi: 10.1126/science.1144079. Epub 2007 Jul 19. PMID: 17641166.

Service RF. Huge protein structure database could transform biology. *Science*. 2021 Jul 30; 373(6554):478. doi: 10.1126/science.373.6554.478. PMID: 34326216.

Service RF. AI reveals structures of protein complexes. *Science*. 2021 Nov 12; 374(6569):804. doi: 10.1126/science.acx9610. Epub 2021 Nov 11. PMID: 34762473.

Service RF. 'The game has changed.' AI triumphs at protein folding. *Science*. 2020 Dec 4; 370(6521):1144-1145. doi: 10.1126/science.370.6521.1144. PMID: 33273077.

Shannon CE. XXII. Programming a computer for playing chess. *Philosophical Magazine*. March 1950; 7(41):314. https://vision.unipv.it/IA1/ProgrammingaComputerforPlayingChess.pdf.

Shuja U. Is deep learning the Big Bang moment for AI? *Forbes*. 20 March 2018. https://www.forbes.com/sites/forbestechcouncil/2018/03/20/is-deep-learning-the-big-bang-moment-for-ai/.

Silver D et al. Mastering the game of Go without human knowledge. *Nature*. 2017 Oct 18; 550(7676):354-359. doi: 10.1038/nature24270. PMID: 29052630.

Silver D et al. Mastering the game of Go with deep neural networks and tree search. *Nature*. 2016 Jan 28;529(7587):484-9. doi: 10.1038/nature16961. PMID: 26819042.

Somers J. How the Artificial-Intelligence program AlphaZero mastered its games. *New Yorker*. 28 December 2018. https://www.newyorker.com/science/elements/how-the-artificial-intelligence-program-alphazero-mastered-its-games.

Thompson C. What the future of AI tells us about its future. *MIT Technology Review*. 18 February 2022. https://www.technologyreview.com/2022/02/18/1044709/ibm-deep-blue-ai-history/.

Tunyasuvunakool K et al. Highly accurate protein structure prediction for the human proteome. *Nature*. 2021 Aug; 596(7873):590-596. doi: 10.1038/s41586-021-03828-1. Epub 2021 Jul 22. PMID: 34293799; PMCID: PMC8387240.

Tsuboyama K et al. Mega-scale experimental analysis of protein folding stability in biology and design. *Nature* 2023; 620:434–444. https://doi.org/10.1038/s41586-023-06328-6.

Von Bardeleben C. Zur Eröffnung Caro-Kann. *Deutsche Schachzeitung*. July 1890; 193-195.

Warburton CW. *My chess adventures*. Chicago (1980).

Weber B. Swift and slashing, computer topples Kasparov. *New York Times*. 12 May 1997. https://www.nytimes.com/1997/05/12/nyregion/swift-and-slashing-computer-topples-kasparov.html.

Wells HG. *The War of the Worlds*. London: William Heinemann (1898).

Wiederhold G, McCarthy J. Arthur Samuel: Pioneer in machine learning. *IBM Journal of Research and Development*. May 1992; 36(3):329-331. doi: 10.1147/rd.363.0329.

Winter E. The Caro-Kann defence. *Chess History*. https://www.chesshistory.com/winter /extra/carokann.html.

Witt S. How Jensen Huang's Nvidia is powering the A.I. revolution. *New Yorker*. 27 November 2023. https://www.newyorker.com/magazine/2023/12/04/how-jensen-huangs-nvidia-is-powering-the-ai-revolution.

Yao D. 25 Years Ago Today: How Deep Blue vs. Kasparov changed AI forever. *AI Business*. 11 May 2022. https://aibusiness.com/ml/25-years-ago-today-how-deep-blue-vs-kasparov-changed-ai-forever.

Zimmer C. Seymour Benzer, geneticist, is dead at 86. *New York Times*. 8 December 2007. https://www.nytimes.com/2007/12/08/science/08benzer.html.

CHAPTER 8

Akbarian S, Won H. Chromosomal contacts change with age. *Science*. 2023 Sep 8; 381(6662):1049-1050. doi: 10.1126/science.adk0961. Epub 2023 Sep 7. PMID: 37676934.

Allou L et al. Non-coding deletions identify Maenli lncRNA as a limb-specific En1 regulator. *Nature*. 2021 Apr; 592(7852):93-98. doi: 10.1038/s41586-021-03208-9. Epub 2021 Feb 10. PMID: 33568816.

Amaral P et al. The status of the human gene catalogue. *Nature*. 2023 Oct; 622(7981):41-47. doi: 10.1038/s41586-023-06490-x. Epub 2023 Oct 4. PMID: 37794265; PMCID: PMC10575709.

Annaluru N, Ramalingam S, Chandrasegaran S. Rewriting the blueprint of life by synthetic genomics and genome engineering. *Genome Biology*. 2015 Jun 16; 16(1):125. doi: 10.1186/s13059-015-0689-y. Erratum in: *Genome Biology*. 2015 Aug 07; 16:159. doi: 10.1186/s13059-015-0703-4. PMID: 26076868; PMCID: PMC4469412.

Annaluru N et al. Total synthesis of a functional designer eukaryotic chromosome. *Science*. 2014 Apr 4; 344(6179):55-8. doi: 10.1126/science.1249252. Epub 2014 Mar 27. Erratum in: *Science*. 2014 May 23; 344(6186):816. PMID: 24674868; PMCID: PMC4033833.

Arc Institute. AI can now model and design the genetic code of all domains of life with Evo 2. https://search.app/k9xb5557peimsWTW7. 19 February 2025.

Bacon F. *Novum Organum*. 1620. In Bacon F. *Instauratio magna: Novum organum sive Indicia vera de interpretatione naturae*. London: Joannem Billium (1620).

Bai N, Smith D. The mysterious 98%: Scientists look to shine light on our dark genome. *UCSF*. https://www.ucsf.edu/news/2017/02/405686/mysterious-98-scientists-look-shine-light-our-dark-genome.

Baker M. Synthetic genomes: The next step for the synthetic genome. *Nature*. 2011 May 19; 473(7347):403, 405-8. doi: 10.1038/473403a. PMID: 21593873.

Ball P. Revolutionary genetics research shows RNA may rule our genome. *Scientific American*. 14 May 2024.

Bartley BA, Beal J, Karr JR, Strychalski EA. Organizing genome engineering for the gigabase scale. *Nature Communications*. 2020 Feb 4; 11(1):689. doi: 10.1038/s41467-020-14314-z. PMID: 32019919; PMCID: PMC7000699.

Basu A et al. Measuring DNA mechanics on the genome scale. *Nature*. 2021 Jan; 589(7842):462-467. doi: 10.1038/s41586-020-03052-3. Epub 2020 Dec 16. PMID: 33328628; PMCID: PMC7855230.

Benegas G, Batra SS, Song YS. DNA language models are powerful predictors of genome-wide variant effects. *Proceedings of the National Academy of Sciences USA*. 26 October 2023. 20 (44) e2311219120. https://doi.org/10.1073/pnas.2311219120.

Blaxter M. Genetics. Revealing the dark matter of the genome. *Science*. 2010 Dec 24; 330(6012):1758-9. doi: 10.1126/science.1200700. Epub 2010 Dec 22. PMID: 21177977.

de Boer CG, Taipale J. Hold out the genome: a roadmap to solving the cis-regulatory code. *Nature*. 2024 Jan; 625(7993):41-50. doi: 10.1038/s41586-023-06661-w. Epub 2023 Dec 13. PMID: 38093018.

Bourzac K. Engineered yeast breaks new record: a genome with over 50% synthetic DNA. *Nature*. 16 November 2023; 623:469. doi: 10.1038/d41586-023-03495-4. Epub ahead of print. PMID: 37940692.

Bower G et al. Conserved *Cis*-Acting Range Extender Element Mediates Extreme Long-Range Enhancer Activity in Mammals. *bioRxiv* [Preprint]. 2024.05.26.595809. doi: 10.1101/2024.05.26.595809. PMID: 38826394; PMCID: PMC11142232.

Brandes N, Goldman G, Wang CH, Ye CJ, Ntranos V. Genome-wide prediction of disease variant effects with a deep protein language model. *Nature Genetics*. 2023 Sep; 55(9):1512-1522. doi: 10.1038/s41588-023-01465-0. Epub 2023 Aug 10. PMID: 37563329; PMCID: PMC10484790.

Brixi G et al. Genome modelling and design across all domains of life with Evo 2. *bioRxiv*. 2025.02.18.638918; doi: https://doi.org/10.1101/2025.02.18. 638918.

Brooks AN et al. Transcriptional neighborhoods regulate transcript isoform lengths and expression levels. *Science*. 2022 Mar 4; 375(6584):1000-1005. doi: 10.1126/science.abg0162. Epub 2022 Mar 3. PMID: 35239377; PMCID: PMC7613581.

Bustin M, Misteli T. Nongenetic functions of the genome. *Science*. 2016 May 6; 352(6286):aad6933. doi: 10.1126/science.aad6933. PMID: 27151873; PMCID: PMC6312727.

Cable J et al. Noncoding RNAs: biology and applications – a Keystone Symposia report. *Annals of the New York Academy of Sciences*. 2021 Dec; 1506(1):118-141. doi: 10.1111/nyas.14713. Epub 2021 Nov 17. Erratum in: *Annals of the New York Academy of Sciences*. 2022 Jan; 1507(1):171. doi: 10.1111/nyas.14751. PMID: 34791665; PMCID: PMC9808899.

Callaway E. First synthetic yeast chromosome revealed. *Nature*. 27 March 2014. https://doi.org/ 10.1038/nature.2014.14941.

Carr PA, Church GM. Genome engineering. *Nature Biotechnology*. 2009 Dec; 27(12):1151-62. doi: 10.1038/nbt.1590. PMID: 20010598.

Chi K. The dark side of the human genome. *Nature*. 2016; 538:275–277. https://doi.org/ 10.1038/538275a.

Crick F. General Model for the Chromosomes of Higher Organisms. *Nature*. 1971; 234:25–27. https://doi.org/10.1038/234025a0.

Csete ME, Doyle JC. Reverse engineering of biological complexity. *Science*. 2002 Mar 1; 295(5560):1664-9. doi: 10.1126/science.1069981. PMID: 11872830.

Dekker C, Haering CH, Peters JM, Rowland BD. How do molecular motors fold the genome? *Science*. 2023 Nov 10; 382(6671):646-648. doi: 10.1126/science.adi8308. Epub 2023 Nov 9. PMID: 37943927.

Dymond JS et al. Synthetic chromosome arms function in yeast and generate phenotypic diversity by design. *Nature*. 2011 Sep 14; 477(7365):471-6. doi: 10.1038/nature10403. PMID: 21918511; PMCID: PMC3774833.

Eisenstein M. How to build a genome. *Nature*. 2020 Feb; 578(7796): 633-635. doi: 10.1038/d41586-020-00511-9. PMID: 32094921.

Enyeart P, Ellington A. A yeast for all reasons. *Nature*. 2011; 477:413–414. https://doi.org/ 10.1038/477413a.

Finch JT, Klug A. Solenoidal model for superstructure in chromatin. *Proceedings of the National Academy of Sciences USA*. 1976 Jun; 73(6):1897-901. doi: 10.1073/pnas.73.6.1897. PMID: 1064861; PMCID: PMC430414.

Flemming W. *Zellsubstanz, Kern und Zelltheilung*. FCW Vogel (1882).

Foo JL, Chang MW. Yeast shuffles towards a diverse future. *Nature*. 31 May 2018; 557(7707): 647-648. doi: 10.1038/d41586-018-05164-3.

Gibson DG, Venter JC. Synthetic biology: Construction of a yeast chromosome. *Nature*. 2014 May 8; 509(7499):168-9. doi: 10.1038/509168a. PMID: 24805340.

Janes KA et al. An engineering design approach to systems biology. *Integrative Biology (Cambridge)*. 2017 Jul 17; 9(7):574-583. doi: 10.1039 /c7ib00014f. PMID: 28590470; PMCID: PMC6534349.

Jones S. SCRaMbLE does the yeast genome shuffle. *Nature Biotechnology* 2018; 36:503. https:// doi.org/10.1038/nbt.4164.

Kannan K, Gibson DG. Yeast genome, by design. *Science*. 2017 Mar 10; 355(6329):1024-1025. doi: 10.1126/science.aam9739. PMID: 28280169.

Kavanagh, K. World's first AI-designed viruses a step towards AI-generated life. *Nature* 646, 16 (2025). doi: https://doi.org/10.1038/d41586-025 -03055-y.

Kesner JS et al. Noncoding translation mitigation. *Nature*. 2023 May; 617(7960):395-402. doi: 10.1038/s41586-023-05946-4. Epub 2023 Apr 12. PMID: 37046090; PMCID: PMC10560126.

King, SH, Driscoll, CL, Li DB, Guo D, Merchant AT, Brixi G, Wilkinson ME, Hie BL. Generative design of novel bacteriophages with genome language models bioRxiv 2025.09.12.675911; doi: https://doi .org/10.1101/2025.09.12.675911

King, S. Hie B. How We Built The First AI-Generated Genomes. 17 September, 2025. Arc Institute. https://arcinstitute.org/news/hie-king -first-synthetic-phage

Kornberg RD. Chromatin structure: a repeating unit of histones and DNA. *Science*. 1974 May 24; 184(4139):868–71. doi: 10.1126/science.184.4139.868. PMID: 4825889.

Kornberg RD. The molecular basis of eukaryotic transcription. *Proceedings of the National Academy of Sciences USA*. 2007 Aug 7; 104(32):12955–61. doi: 10.1073/pnas.0704138104. Epub 2007 Aug 1. PMID: 17670940; PMCID: PMC1941834.

Kuhlman TE. Repetitive DNA regulates gene expression. *Science*. 2023 Sep 22; 381(6664):1289–1290. doi: 10.1126/science.adk2055. Epub 2023 Sep 21. PMID: 37733865.

Kutyna DR et al. Construction of a synthetic Saccharomyces cerevisiae pan-genome neo-chromosome. *Nature Communications*. 2022 Jun 24; 13(1):3628. doi: 10.1038/s41467-022-31305-4. PMID: 35750675; PMCID: PMC9232646.

Kwok R. Five hard truths for synthetic biology. *Nature*. 2010 Jan 21; 463(7279):288–90. doi: 10.1038/463288a. PMID: 20090726.

Kwok R. Yeast thrives with partially synthetic genome. *Nature*. 2011. https://doi.org/ 10.1038/news.2011.537.

Lang RJ. *Origami Design Secrets: Mathematical methods for an ancient art*. Natick, MA/A.K. Peters (2003).

Lawrence J, Hall L. Exceptionally long-lived nuclear RNAs. *Science*. 2024 Apr 5; 384(6691):31–32. doi: 10.1126/science.ado5751. Epub 2024 Apr 4. PMID: 38574156.

Leroi AM. *Mutants: On Genetic Variety and the Human Body*. Viking (2003).

Leslie M. Researchers close in on fully artificial yeast genome. *Science*. 2023 Nov 10; 382(6671):631. doi: 10.1126/science.adm8232. Epub 2023 Nov 9. PMID: 37943934.

Li S, Hannenhalli S, Ovcharenko I. De novo human brain enhancers created by single-nucleotide mutations. *Science Advances*. 2023 Feb 15; 9(7):eadd2911. doi: 10.1126/sciadv.add2911. Epub 2023 Feb 15. PMID: 36791193; PMCID: PMC9931207.

Li S, Vogt DM, Rus D, Wood RJ. Fluid-driven origami-inspired artificial muscles. *Proceedings of the National Academy of Sciences USA*. 27 November 2017; 114(50):13132–13137. https://doi.org/10.1073/pnas.1713450114.

Liang J, Luo Y, Zhao H. Synthetic biology: putting synthesis into biology. *Wiley Interdisciplinary Reviews: Systems Biology and Medicine*. 2011 Jan-Feb; 3(1):7–20. doi: 10.1002/wsbm.104. PMID: 21064036; PMCID: PMC3057768.

Lin, F. AI Designs Viable Bacteriophage Genomes, Combats Antibiotic Resistance. *Genetic Engineering & Biotechnology News*. 17 September, 2025. https://www.genengnews.com/topics/artificial-intelligence/ai-designs-viable-bacteriophage-genomes-combats-antibiotic-resistance/

Loyfer N et al. A DNA methylation atlas of normal human cell types. *Nature*. 2023 Jan; 613(7943):355-364. doi: 10.1038/s41586-022-05580-6. Epub 2023 Jan 4. PMID: 36599988; PMCID: PMC9811898.

Marshall N. Scientists create baker's yeast with more than 50% synthetic DNA. *Independent*. 8 November 2023. https://www.the-independent.com/news/science/dna-scientists-cell-university-of-nottingham-university-of-manchester-b2443899.html.

Mattick JS et al. Long non-coding RNAs: definitions, functions, challenges and recommendations. *Nature Reviews Molecular Cell Biology*. 2023 Jun; 24(6):430-447. doi: 10.1038/s41580-022-00566-8. Epub 2023 Jan 3. PMID: 36596869; PMCID: PMC10213152.

Mattick JS. The genomic foundation is shifting. *Science*. 2011; 331:874. doi: 10.1126/science.1203703.

McCarty, N. *AI-Designed Phages*. Asimov Press (2025). https://doi.org/10.62211/21er-45fg.

Mercy G et al. 3D organization of synthetic and scrambled chromosomes. *Science*. 2017 Mar 10; 355(6329):eaaf4597. doi: 10.1126/science.aaf4597. PMID: 28280150; PMCID: PMC5679085.

NASA. James Webb Space Telescope. https://webb.nasa.gov.

Nguyen E et al. Sequence modeling and design from molecular to genome scale with Evo. *Science*. 2024 Nov 15; 386(6723):eado9336. doi: 10.1126/science.ado9336. Epub 2024 Nov 15. PMID: 39541441.

Oudelaar AM et al. Between form and function: the complexity of genome folding. *Human Molecular Genetics*. 2017 Oct 1; 26(R2):R208-R215. doi: 10.1093/hmg/ddx306. PMID: 28977451; PMCID: PMC5886466.

Park DS et al. High-throughput Oligopaint screen identifies druggable 3D genome regulators. *Nature*. 2023 Aug; 620(7972):209-217. doi: 10.1038/s41586-023-06340-w. Epub 2023 Jul 12. PMID: 37438531.

Pennisi E. Human genome is much more than just genes. *Science*. 5 September 2012. doi: 10.1126 /article.26976.

Pennisi E. Surprise RNA paints colorful patterns on butterfly wings. *Science*. 2024 Mar 8; 383(6687):1039-1040. doi: 10.1126/science.adp0471. Epub 2024 Mar 7. PMID: 38452086.

Pennisi E. Massive RNA sequencing effort proposed. *Science*. 2024 Mar 29; 383(6690):1398. doi: 10.1126/science.adp4706. Epub 2024 Mar 28. PMID: 38547270.

Pennisi E. Building the ultimate yeast genome. *Science*. 2014 Mar 28; 343(6178):1426-9. doi: 10.1126/science.343.6178.1426. PMID: 24675935.

Perelman PL et al. Three-dimensional genome architecture persists in a 52,000-year-old woolly mammoth skin sample. *Cell*. 2024 Jul 11; 187(14):3541-3562.e51. doi: 10.1016/j.cell.2024.06.002. PMID: 38996487.

Perkel JM. Genome engineering: writing a better genome. *Biotechniques*. 2012 Oct; 53(4):213, 215, 217. doi: 10.2144/000113934. PMID: 23046504.

Rao SS et al. A 3D map of the human genome at kilobase resolution reveals principles of chromatin looping. *Cell*. 2014 Dec 18; 159(7):1665–80. doi: 10.1016/j.cell.2014.11.021. Epub 2014 Dec 11. Erratum in: *Cell*. 2015 Jul 30; 162(3):687–8. PMID: 25497547; PMCID: PMC5635824.

Richard G. et al. ChatNT: A Multimodal Conversational Agent for DNA, RNA and Protein Tasks. *bioRxiv*. 2024.04.30.591835; doi: https://doi.org/10.1101/2024.04.30.591835.

Richardson SM et al. Design of a synthetic yeast genome. *Science*. 2017 Mar 10; 355(6329):1040–1044. doi: 10.1126/science.aaf4557. PMID: 28280199.

Row TS. *Geometric Exercises in Paper Folding*. Madras: Addison & Co. (1893).

Turner D, Kropinski AM, Adriaenssens EM. A Roadmap for Genome-Based Phage Taxonomy. Viruses. 2021 Mar 18;13(3):506. doi: 10.3390/v13030506. PMID: 33803862; PMCID: PMC8003253.

Sandoval-Velasco M et al. Histone modifications regulate pioneer transcription factor cooperativity. *Nature*. 2023 Jul; 619(7969):378–384. doi: 10.1038/s41586-023-06112-6. Epub 2023 May 24. PMID: 37225990; PMCID: PMC10338341.

van Schoonhoven A, Stadhouders R. A base-pair view of interactions between genes and their enhancers. *Nature*. 2021 Jul; 595(7865):36–37. doi: 10.1038/d41586-021-01494-x. PMID: 341 08721.

Statello L, Guo CJ, Chen LL, Huarte M. Gene regulation by long non-coding RNAs and its biological functions. *Nature Reviews Molecular Cell Biology*. 2021 Feb; 22(2):96–118. doi: 10.1038/s41580-020-00315-9. Epub 2020 Dec 22. Erratum in: Nat Rev Mol Cell Biol. 2021 Feb;22(2):159. doi: 10.1038/s41580-021-00330-4. PMID: 33353982; PMCID: PMC7754182.

Szymanski E, Calvert J. Designing with living systems in the synthetic yeast project. *Nature Communications*. 2018 Jul 27; 9(1):2950. doi: 10.1038/s41467-018-05332-z. PMID: 30054478; PMCID: PMC6063962.

Taghon GJ, Strychalski EA. Rise of synthetic yeast: Charting courses to new applications. *Cell Genomics*. 2023 Nov 9; 3(11):100438. doi: 10.1016/j.xgen.2023.100438. PMID: 38020966; PMCID: PMC10667549.

Tan L et al. Lifelong restructuring of 3D genome architecture in cerebellar granule cells. *Science*. 2023 Sep 8; 381(6662):1112–1119. doi: 10.1126/science.adh3253. Epub 2023 Sep 7. PMID: 37676945; PMCID: PMC11059189.

Theodoris CV. Learning the language of DNA. *Science*. 2024 Nov; 386(6723):729–780. doi: 10.1126/science.adt3007.

Tian W et al. Single-cell DNA methylation and 3D genome architecture in the human brain. *Science*. 2023 Oct 13; 382(6667):eadf5357. doi: 10.1126/science.adf5357. Epub 2023 Oct 13. PMID: 37824674; PMCID: PMC10572106.

Uckelmann M, Davidovich C. An added layer of repression for human genes. *Nature*. 2022 Apr; 604(7904):41–42. doi: 10.1038/d41586-022-00519-3. PMID: 35354969.

Vaishnav ED et al. The evolution, evolvability and engineering of gene regulatory DNA. *Nature*. 2022; 603:455–463. https://doi.org/10.1038/s41586-022-04506-6.

Wagner A. An oracle for gene regulation. *Nature*. 17 March 2022; 603:399-340. doi: https://doi.org/10.1038/d41586-022-00384-0.

Walsh B. In the future we won't edit genomes – we'll just print out new ones. *MIT Technology Review*. 6 February 2018.

Wang S, Qiao J, Feng S. Prediction of lncRNA and disease associations based on residual graph convolutional networks with attention mechanism. *Scientific Reports*. 2024 Mar 2; 14(1):5185. doi: 10.1038/s41598-024-55957-y. PMID: 38431702.

Winkle M, El-Daly SM, Fabbri M, Calin GA. Noncoding RNA therapeutics – challenges and potential solutions. *Nature Reviews Drug Discovery*. 2021 Aug; 20(8):629-651. doi: 10.1038/s41573-021-00219-z. Epub 2021 Jun 18. PMID: 34145432; PMCID: PMC8212082.

de Wit E, Nora EP. New insights into genome folding by loop extrusion from inducible degron technologies. *Nature Reviews Genetics*. 2023 Feb; 24(2):73-85. doi: 10.1038/s41576-022-00530-4. Epub 2022 Sep 30. PMID: 36180596.

Wu Y et al. Bug mapping and fitness testing of chemically synthesized chromosome X. *Science*. 2017 Mar 10; 355(6329):eaaf4706. doi: 10.1126/science.aaf4706. PMID: 28280152; PMCID: PMC5679077.

Yan J et al. Systematic analysis of binding of transcription factors to noncoding variants. *Nature*. 2021 Mar; 591(7848):147-151. doi: 10.1038/s41586-021-03211-0. Epub 2021 Jan 27. PMID: 33505025; PMCID: PMC9367673.

Yang JH, Hansen AS. Enhancer selectivity in space and time: from enhancer-promoter interactions to promoter activation. *Nature Reviews Molecular Cell Biology*. 2024 Jul; 25(7):574-591. doi: 10.1038/s41580-024-00710-6. Epub 2024 Feb 27. PMID: 38413840.

Yu M, Ren B. The Three-Dimensional Organization of Mammalian Genomes *Annual Review of Cell and Developmental Biology*. 2017 Oct 6; 33:265-289. doi: 10.1146/annurev-cellbio-100616-060531. Epub 2017 Aug 7. PMID: 28783961; PMCID: PMC5837811.

Zande VP, Hill MS, Wittkopp PJ. Pleiotropic effects of trans-regulatory mutations on fitness and gene expression. *Science*. 2022 Jul; 377(6601):105-109. doi: 10.1126/science.abj7185. Epub 2022 Jun 30. PMID: 35771906; PMCID: PMC9569154.

Zhou T et al. GAGE-seq concurrently profiles multiscale 3D genome organization and gene expression in single cells. *Nature Genetics*. 2024 May 14. doi: 10.1038/s41588-024-01745-3. Epub ahead of print. PMID: 38744973.

Zuin J et al. Nonlinear control of transcription through enhancer-promoter interactions. *Nature*. 2022 Apr; 604(7906):571-577. doi:

10.1038/s41586-022-04570-y. Epub 2022 Apr 13. PMID: 35418676; PMCID: PMC 9021019.

Zuin J, Roth G, Zhan Y, Cramard J, Redolfi J, Piskadlo E, Mach P, Kryzhanovska M, Tihanyi G, Kohler H, Eder M, Leemans C, van Steensel B, Meister P, Smallwood S, Giorgetti L. Nonlinear control of transcription through enhancer-promoter interactions. Nature. 2022 Apr;604(7906):571-577. doi: 10.1038/s41586-022-04570-y. Epub 2022 Apr 13. PMID: 35418676; PMCID: PMC9021019.

CHAPTER 9

All of Us Research Program Genomics Investigators. Genomic data in the All of Us Research Program. *Nature*. 2024 Mar; 627(8003):340-346. doi: 10.1038/s41586-023-06957-x. Epub 2024 Feb 19. PMID: 38374255; PMCID: PMC10937371.

Allentoft ME. Population genomics of post-glacial western Eurasia. *Nature*. 2024 Jan; 625(7994):301-311. doi: 10.1038/s41586-023-06865-0. Epub 2024 Jan 10. Erratum in: *Nature*. 2024 Feb; 626(7997):E3. doi: 10.1038/s41586-024-07044-5. PMID: 38200295; PMCID: PMC10781627.

Allison AC. Protection afforded by sickle-cell trait against subtertian malarial infection. *British Medical Journal*. 1954 Feb 6; 1(4857):290-4. doi: 10.1136/bmj.1.4857.290. PMID: 13115700; PMCID: PMC 2093356.

Allison AC. The distribution of the sickle-cell trait in East Africa and elsewhere, and its apparent relationship to the incidence of subtertian malaria. *Transactions of the Royal Society of Tropical Medicine and Hygiene*. 1954 Jul; 48(4):312-8. doi: 10.1016/0035-9203(54)90101-7. PMID: 13187561.

Apgar TL, Sanders CR. Compendium of causative genes and their encoded proteins for common monogenic disorders. *Protein Science*. 2022 Jan; 31(1):75-91. doi: 10.1002/pro.4183. Epub 2021 Sep 21. PMID: 34515378; PMCID: PMC8740837.

Arnold GL. Inborn errors of metabolism in the 21st century: past to present. *Annals of Translational Medicine*. 2018 Dec; 6(24):467. doi: 10.21037/atm.2018.11.36. PMID: 30740398; PMCID: PMC6331363.

Backman JD et al. Exome sequencing and analysis of 454,787 UK Biobank participants. *Nature*. 2021 Nov; 599(7886):628-634. doi: 10.1038/s41586-021-04103-z. Epub 2021 Oct 18. PMID: 34662886; PMCID: PMC8596853.

Bateson W. *Mendel's Principles of Heredity: A Defence*. Cambridge University Press (1902).

Beethoven L van. *Heiligenstadt Testament*. University of Hamburg Library (Catalogue record 877868875).

Begg TJA et al. Genomic analyses of hair from Ludwig van Beethoven. *Current Biology*. 2023 Apr 24; 33(8):1431-1447.e22. doi: 10.1016/j.cub.2023.02.041. Epub 2023 Mar 22. PMID: 36958333.

Benton ML et al. The influence of evolutionary history on human health and disease. *Nature Reviews Genetics.* 2021; 22:269–283. https://doi.org/10.1038/s41576-020-00305-9.

Bhalla US, Iyengar R. Emergent properties of networks of biological signaling pathways. *Science.* 1999 Jan 15; 283(5400):381-7. doi: 10.1126/science.283.5400.381. PMID: 9888852.

Blain T. A guide to Beethoven's Heiligenstadt Testament and what drove him to write it. *Classical Music.* 29 December 2022. https://www.classical-music.com/features/composers/heiligenstadt-testament.

Bourzac K. Gene therapy: Erasing sickle-cell disease. *Nature.* 2017; 549:S28–S30. https://doi.org/10.1038/549S28a.

Boyle EA, Li YI, Pritchard JK. An Expanded View of Complex Traits: From Polygenic to Omnigenic. *Cell.* 2017 Jun 15; 169(7):1177-1186. doi: 10.1016/j.cell.2017.05.038. PMID: 28622505; PMCID: PMC5536862.

Bray D. Protein molecules as computational elements in living cells. *Nature.* 1995 Jul 27; 376(6538):307-12. doi: 10.1038/376307a0. Erratum in: *Nature.* 1995 Nov 23; 378(6555):419. PMID: 7630396.

Bubna AK, Veeraraghavan M, Anandan S, Rangarajan S. Congenital Generalized Hypertrichosis, Gingival Hyperplasia, a Coarse Facies with Constriction Bands: A Rare Association. *International Journal of Trichology.* 2015 Apr-Jun; 7(2):67-71. doi: 10.4103/0974-7753.160113. PMID: 26180451; PMCID: PMC4502477.

Callaway E. New concerns raised over value of genome-wide disease studies. *Nature.* 2017; 546:463. https://doi.org/10.1038/nature.2017.22152.

Callaway E. Neanderthals had outsize effect on human biology. *Nature.* 2015; 523:512–513. https://doi.org/10.1038/523512a.

Cao L. et al. Design of protein-binding proteins from the target structure alone. *Nature.* 2022 May; 605(7910):551-560. doi: 10.1038/s41586-022-04654-9. Epub 2022 Mar 24. PMID: 35332283; PMCID: PMC9117152.

Check Hayden E. So similar, yet so different. *Nature.* 2007; 449:762–763. https://doi.org /10.1038/449762a.

Childs B. Sir Archibald Garrod's conception of chemical individuality: a modern appreciation. *New England Journal of Medicine.* 1970 Jan 8; 282(2):71-7. doi: 10.1056/NEJM197001082820205. PMID: 4901867.

Claussnitzer M et al. A brief history of human disease genetics. *Nature.* 2020 Jan; 577(7789):179-189. doi: 10.1038/s41586-019-1879-7. Epub 2020 Jan 8. PMID: 31915397; PMCID: PMC7405896.

Cohen J. Venter's Genome Sheds New Light on Human Variation. *Science.* 2007; 317:1311. doi: 10.1126/science.317.5843.1311.

Cooper DN et al. On the sequence-directed nature of human gene mutation: the role of genomic architecture and the local DNA sequence environment in mediating gene mutations underlying human inherited disease. *Human Mutation.* 2011 Oct; 32(10):1075-99. doi: 10.1002/humu.21557. Epub 2011 Sep 2. PMID: 21853507; PMCID: PMC3177966.

Cotton RG et al. The Human Variome Project. *Science*. 2008 Nov 7; 322(5903):861-2. doi: 10.1126/science.1167363. PMID: 18988827; PMCID: PMC2810956.

Croone, W. The Reason of the Movement of the Muscles: *Transactions, American Philosophical Society* (vol. 90, part 1) (Transactions of the American Philosophical Society, 933). ISBN 10: 087169901X ISBN 13: 9780871699015.

Curry A. Ancient DNA ties modern diseases to ancestry. *Science*. 2024 Jan 12; 383(6679):138-139. doi: 10.1126/science.adn9644. Epub 2024 Jan 11. PMID: 38207026.

Dennis C. Deaf by design. *Nature*. 2004; 431:894–896. https://doi.org/10.1038/431894a.

Dennis C. Inside Deaf culture. *Nature*. 2004. https://doi.org/10.1038/news041018-8.

Duan D, Goemans N, Takeda S, Mercuri E, Aartsma-Rus A. Duchenne muscular dystrophy. *Nature Reviews Disease Primers*. 2021 Feb 18; 7(1):13. doi: 10.1038/s41572-021-00248-3. PMID: 33602943; PMCID: PMC10557455.

Editorial. Thoughts on a legacy. *Nature Reviews Genetics*. 2022 Jul; 23(7): 385. doi: 10.1038/s41576-022-00510-8. PMID: 35672468; PMCID: PMC9172980.

Editorial. FDA Approves First Gene Therapy for Treatment of Certain Patients with Duchenne Muscular Dystrophy. *FDA Journal*. 22 June 2023. https://www.fda.gov/news-events/press-announcements/fda-approves-first-gene-therapy-treatment-certain-patients-duchenne-muscular-dystrophy.

Editorial. The true legacy of Gregor Mendel: careful, rigorous and humble science. *Nature*. 2022 Jul; 607(7919):421-422. doi: 10.1038/d41586-022-01953-z. PMID: 35854153.

Editorial. The Genes Behind the Beard. *Science*. 2009; 324:1123. doi: 10.1126/science.324_1123d.

Editorial. Is CRISPR safe? Genome editing gets its first FDA scrutiny. *Nature*. 9 November 2023; 623:234.

Editorial. First herpesvirus gene therapy. *Nature Biotechnology*. 2023; 41:739. https://doi.org /10.1038/s41587-023-01835-3.

Editorial. Gene therapy triumph. *Science*. 22 December 2017; 358:1526.

Editorial. What is the Human Variome Project? *Nature Genetics*. 2007;39:423. https://doi.org/10.1038/ng0407-423

Editorial. Marshaling the Variome. *Nature Genetics*. 2015; 47:849. https://doi.org/10.1038/ng.3377.

Eisenstein M. Closing in on a complete human genome. *Nature*. 2021 Feb; 590(7847):679-681. doi: 10.1038/d41586-021-00462-9. PMID: 33619406.

Eisenstein Michael. Fix the gene, cure the disease. *Nature*, suppl. *Nature Outlook*. 2021 Aug 26; 596(7873): S2-S4. doi: 10.1038/d41586-021-02138-wCv fjtj.

Esrick EB et al. Post-Transcriptional Genetic Silencing of *BCL11A* to Treat Sickle Cell Disease. *New England Journal of Medicine*. 2021 Jan 21; 384(3):205-215. doi: 10.1056/NEJMoa2029392. Epub 2020 Dec 5. PMID: 33283990; PMCID: PMC7962145.

Fairbanks DJ. Demystifying the mythical Mendel: a biographical review. *Heredity (Edinburgh)*. 2022 Jul; 129(1):4-11. doi: 10.1038/s41437-022-00526-0. Epub 2022 Apr 12. PMID: 35414696; PMCID: PMC9273628.

Fairbanks DJ. Mendel and Darwin: untangling a persistent enigma. *Heredity (Edinburgh)*. 2020 Feb; 124(2):263-273. doi: 10.1038/s41437-019-0289-9. Epub 2019 Dec 17. PMID: 31848463; PMCID: PMC6972880.

Firth JA, Sheldon BC. The long reach of family ties. *Science*. 2021 Jul 16; 373(6552):274-275. doi: 10.1126/science.abj5234. PMID: 34437137.

Friedman MJ. Oxidant damage mediates variant red cell resistance to malaria. *Nature*. 1979 Jul 19; 280(5719):245-7. doi: 10.1038/280245a0. PMID: 377105.

Furfaro H. Spectrum. A fragile existence. *Science*. 2018 Oct 12; 362(6411): 172-175. doi: 10.1126/science.362.6411.172. PMID: 30309943.

Garrod AE. A Contribution to the Study of Alkaptonuria. *Medico Chirurgical Transactions*. 1899; 82:367-94. doi: 10.1177/095952879908200119. PMID: 20896937; PMCID: PMC2036698.

Garrod AE. The Croonian Lectures on Inborn Errors of Metabolism (Lecture III). *Lancet*. 4 July 1908; 172(4427):1-7.

Geddes L. Height's 'missing heritability' found. *Nature*. 25 April2019; 568:444-445. doi: 10.1038/d41586-019-01157-y. PMID: 31015700.

Gershman A et al. Epigenetic patterns in a complete human genome. *Science*. 2022 Apr; 376(6588):eabj5089. doi: 10.1126/science.abj5089. Epub 2022 Apr 1. PMID: 35357915; PMCID: PMC9170183.

Ghosh S et al. Exploring Emergent Properties in Enzymatic Reaction Networks: Design and Control of Dynamic Functional Systems. *Chemical Reviews*. 2024 Mar 13; 124(5):2553-2582. doi: 10.1021/acs.chemrev.3c00681. Epub 2024 Mar 4. PMID: 38476077; PMCID: PMC10941194.

Gibbons A. Neandertal genes linked to modern diseases. *Science*. 2016 Feb 12; 351(6274):648-9. doi: 10.1126/science.351.6274.648. PMID: 26912836.

Goh K-Il, Choi IG. Exploring the human diseasome: the human disease network. *Briefings in Functional Genomics*. 2012 Nov; 11(6):533-42. doi: 10.1093/bfgp/els032. Epub 2012 Oct 12. PMID: 23063808.

Goh K-Il et al. The human disease network. *Proceedings of the National Academy of Sciences USA*. 2007 May 22; 104(21):8685-90. doi: 10.1073

/pnas.0701361104. Epub 2007 May 14. PMID: 17502601; PMCID: PMC1885563.

Gonzales LR et al. LncRNAs: the art of being influential without protein. *Trends in Plant Science*. 2024 Jul; 29(7):770-785. doi: 10.1016/j.tplants.2024.01.006. Epub 2024 Feb 16. PMID: 38368122.

Gopnik A. Van Gogh's Ear. *New Yorker*. 27 November 2009. https://www.newyorker.com/magazine/2010/01/04/van-goghs-ear.

Gray V, Thomas U, Davies K. Warrior Spirit: An Interview with Victoria Gray, Sickle Cell Pioneer. *CRISPR Journal*. 2024 Feb; 7(1):5-11. doi: 10.1089/crispr.2024.29171.vgr. PMID: 38353619.

Hall BK. How a scholarly spat shaped a century of genetic research. *Nature*. 2023 Jul; 619(7971):690-691. doi: 10.1038/d41586-023-02310-4. PMID: 37488254.

Halldorsson BV et al. The sequences of 150,119 genomes in the UK Biobank. *Nature*. 2022 Jul; 607(7920):732-740. doi: 10.1038/s41586-022-04965-x. Epub 2022 Jul 20. PMID: 35859178; PMCID: PMC9329122.

Harris JC. Lovesickness: Erasistratus Discovering the Cause of Antiochus' Disease. *Archives of General Psychiatry*. 2012; 69(6):549. doi: 10.1001/archgenpsychiatry.2012.105.

Hayden EC. Seeing deadly mutations in a new light. *Nature*. 13 October 2016; 538:154.

Herrick JB. Peculiar elongated and sickle-shaped red blood corpuscles in a case of severe anemia. 1910. *Yale Journal of Biology and Medicine*. 2001 May-Jun; 74(3):179-84. PMID: 11501714; PMCID: PMC2588723.

Hochhaus A et al. Long-term outcomes of imatinib treatment for chronic myeloid leukemia. 9 March 2017. *New England Journal of Medicine*. 2017; 376:917-927. doi: 10.1056/NEJMoa1609324.

Huerta-Sánchez E. et al. Altitude adaptation in Tibetans caused by introgression of Denisovan-like DNA. *Nature*. 2014 Aug 14; 512(7513):194-7. doi: 10.1038/nature13408. Epub 2014 Jul 2. PMID: 25043035; PMCID: PMC4134395.

Irvine DJ, Maus MV, Mooney DJ, Wong WW. The future of engineered immune cell therapies. *Science*. 2022 Nov 25; 378(6622):853-858. doi: 10.1126/science.abq6990. Epub 2022 Nov 24. PMID: 36423279; PMCID: PMC9919886.

Johannsen JW. *Elemente der exakten erblichkeitslehre*. Jena: Fischer (1909).

Kahlo F. *The wounded deer*. 1946. https://www.fridakahlo.org/the-wounded-deer.jsp.

Kaiser J. NIH megastudy analyzes first 250,000 genomes. *Science*. 2024 Feb 23; 383(6685):809. doi: 10.1126/science.ado7763. Epub 2024 Feb 22. PMID: 38386750.

Kaiser J. Rewriting DNA in the body lowers cholesterol. *Science*. 2023 Nov 17; 382(6672):751. doi: 10.1126/science.adm9506. Epub 2023 Nov 16. PMID: 37972155.

Kaiser J. United Kingdom approves first-ever CRISPR treatment, a cure for sickle cell disease and beta thalassemia. *Science*. 16 November 2023. doi: 10.1126/science.adm9989.

Kaiser J. Tweaking genes with CRISPR or viruses fixes blood disorders. *Science*. 2020 Dec 11; 370(6522):1254–1255. doi: 10.1126/science.370.6522.1254. PMID: 33303593.

Kaiser J. Gene therapy milestone looms, but field seeks better options. *Science*. 2023 May 26; 380(6647):778–779. doi: 10.1126/science.adi8792. Epub 2023 May 25. PMID: 37228210.

Kaiser J. Gene therapies that let deaf children hear bring hope – and many questions. *Science*. 26 January 2024. doi: 10.1126/science.zf53jb1.

Kaiser J. A gentler way to tweak genes: epigenome editing. *Science*. 2022 Jun 3; 376(6597):1034–1035. doi: 10.1126/science.add2703. Epub 2022 Jun 2. PMID: 35653473.

Kaiser J. A second chance. *Science*. 2017; 358:582–585. doi: 10.1126/science.358.6363.582.

Kaiser J. Gene therapy that once led to tragedy scores success. *Science*. 2021 May 21; 372(6544):776. doi: 10.1126/science.372.6544.776. PMID: 34016759.

Kaiser J. NIH's huge All of Us genes and health study releases first 100,000 genomes. *Science*. 17 March 2022. doi: 10.1126/science.adb2079.

Kaiser J. 200,000 whole genomes made available for biomedical studies. *Science*. 26 November 2021; 374(6571) 1036. doi: 10.1126/science.acx9678.

Kaiser J. Human 'knockouts' may reveal why some drugs fail. *Science*. 12 April 2017. doi: 10.1126/science.aal1041.

Keats J. Ode to a Nightingale. In *Lamia, Isabella, The Eve of St Agnes, and other poems*. London: Taylor & Hessey (1820).

Kennedy M. 'High-status' portrait of bearded woman bought by Wellcome Collection. *Guardian*. 13 December 2017.

Khamsi R. The quest for an all-inclusive human genome. *Nature*. 2022 Mar; 603(7901):378–381. doi: 10.1038/d41586-022-00726-y. PMID: 35296818.

Kitano H. Systems biology: a brief overview. *Science*. 2002 Mar 1; 295(5560):1662–4. doi: 10.1126/science.1069492. PMID: 11872829.

Knott GJ, Doudna JA. CRISPR-Cas guides the future of genetic engineering. *Science*. 2018 Aug 31; 361(6405):866–869. doi: 10.1126/science.aat5011. PMID: 30166482; PMCID: PMC6455913.

Knox WE. Sir Archibald Garrod's inborn errors of metabolism. II. Alkaptonuria. *American Journal of Human Genetics*. 1958 Jun; 10(2):95–124. PMID: 13533390; PMCID: PMC1931938.

Kolata G. Gene therapy allows an 11-year-old boy to hear for the first time. *Nature Italy*. 16 February 2024. doi: https://doi.org/10.1038/d43978-024-00032-3.

Kuchenbaecker K. Missing heritability found for height. *Nature*. 2022 Oct; 610(7933):631–632. doi: 10.1038/d41586-022-03029-4. PMID: 36224358.

Kucherlapati R. Oliver Smithies (1925-2017). *Nature*. 2017 Feb 8; 542(7640):166. doi: 10.1038/542166a. PMID: 28179651.

Lazebnik Y. Can a biologist fix a radio? Or, what I learned while studying apoptosis. *Cancer Cell*. 2002 Sep; 2(3):179-82. doi: 10.1016/s1535-6108(02)00133-2. PMID: 12242150.

Ledford H. CRISPR gene therapy shows promise against blood diseases. *Nature*. 2020 Dec; 588(7838):383. doi: 10.1038/d41586-020-03476-x. PMID: 33299166.

Ledford H. Quest to use CRISPR against disease gains ground. *Nature*. 2020 Jan; 577(7789):156. doi: 10.1038/d41586-019-03919-0. PMID: 31911695.

Ledford H. Gene therapy in mouse fetuses treats deadly disease. *Nature*. 2018 Jul; 559(7714):313-314. doi: 10.1038/d41586-018-05726-5. PMID: 30018442.

Ledford H. Success against blindness encourages gene therapy researchers. *Nature*. 2015 Oct 22; 526(7574):487-8. doi: 10.1038/526487a. PMID: 26490597.

Ledford H. FDA advisers back gene therapy for rare form of blindness. *Nature*. 2017 Oct 12; 550(7676):314. doi: 10.1038/nature.2017.22819. PMID: 29052639.

Ledford H. Gene therapy's comeback: how scientists are trying to make it safer. *Nature*. 2022 Jun; 606(7914):443-444. doi: 10.1038/d41586-022-01518-0. PMID: 35641616.

Ledford H. CRISPR 2.0: a new wave of gene editors heads for clinical trials. *Nature*. 2023 Dec; 624(7991):234-235. doi: 10.1038/d41586-023-03797-7. PMID: 38062143.

Ledford H. Last-resort cancer therapy holds back disease for more than a decade. *Nature*. 2022 Feb; 602(7896):196. doi: 10.1038/d41586-022-00241-0. PMID: 35110708.

Lee CE, Singleton KS, Wallin M, Faundez V. Rare Genetic Diseases: Nature's Experiments on Human Development. *iScience*. 2020 May 22; 23(5):101123. doi: 10.1016/j.isci.2020.101123. Epub 2020 May 1. PMID: 32422592; PMCID: PMC7229282.

Levine JD. Chronically lonely flies overeat and lose sleep. *Nature*. 2021 Sep; 597(7875):179-180. doi: 10.1038/d41586-021-02194-2. PMID: 34408302.

Li W et al. Chronic social isolation signals starvation and reduces sleep in Drosophila. *Nature*. 2021 Sep; 597(7875):239-244. doi: 10.1038/s41586-021-03837-0. Epub 2021 Aug 18. PMID: 34408325; PMCID: PMC8429171.

Liao WW et al. A draft human pangenome reference. *Nature*. 2023 May; 617(7960):312-324. doi: 10.1038/s41586-023-05896-x. Epub 2023 May 10. PMID: 37165242; PMCID: PMC10172123.

Lim WA. The emerging era of cell engineering: Harnessing the modularity of cells to program complex biological function. *Science*. 2022 Nov 25;

378(6622):848-852. doi: 10.1126/science.add9665. Epub 2022 Nov 24. PMID: 36423287.

Liu E et al. Use of CAR-Transduced Natural Killer Cells in CD19-Positive Lymphoid Tumors. *New England Journal of Medicine.* 2020 Feb 6; 382(6):545-553. doi: 10.1056/NEJMoa1910607. PMID: 32023374; PMCID: PMC7101242.

Liverpool L. First human 'pangenome' aims to catalogue genetic diversity. *Nature.* 2023 May; 617(7961):444-445. doi: 10.1038/d41586-023-01576-y. PMID: 37165229.

Long C et al. Postnatal genome editing partially restores dystrophin expression in a mouse model of muscular dystrophy. *Science.* 2016 Jan 22; 351(6271):400-3. doi: 10.1126/science.aad5725. Epub 2015 Dec 31. PMID: 26721683; PMCID: PMC4760628.

Loos RJF. 15 years of genome-wide association studies and no signs of slowing down. *Nature Communications.* 2020 Nov 19; 11(1):5900. doi: 10.1038/s41467-020-19653-5. PMID: 33214558; PMCID: PMC7677394.

Ma H et al. Correction of a pathogenic gene mutation in human embryos. *Nature.* 2017 Aug 24; 548(7668):413-419. doi: 10.1038/nature23305. Epub 2017 Aug 2. PMID: 28783728.

Maher B. Personal genomes: The case of the missing heritability. *Nature.* 2008; 456:18–21. https://doi.org/10.1038/456018a.

Manolio TA et al. Finding the missing heritability of complex diseases. *Nature.* 2009 Oct 8; 461(7265):747-53. doi: 10.1038/nature08494. PMID: 19812666; PMCID: PMC2831613.

Marris E. Parasite gives wolves what it takes to be pack leaders. *Nature.* 2022 Dec; 612(7939):202. doi: 10.1038/d41586-022-03836-9. PMID: 36424503.

Marx J. Alzheimer's disease. Play and exercise protect mouse brain from amyloid buildup. *Science.* 2005 Mar 11; 307(5715):1547. doi: 10.1126/science.307.5715.1547. Erratum in: *Science.* 2005 Apr 29; 308(5722):633. PMID: 15761131.

Massarat A, Gymrek M, McStay B, Jónsson H. Human pangenome supports analysis of complex genomic regions. *Nature.* 2023 May; 617(7960):256-258. doi: 10.1038/d41586-023-01490-3. PMID: 37165235.

Massaro G et al. Fetal gene therapy for neurodegenerative disease of infants. *Nature Medicine.* 2018; 4:1317–1323. https://doi.org/10.1038/s41591-018-0106-7.

Maugham, W.S. *Of Human Bondage.* William Heinemann, London, 1915.

McCulloch JA, Trinchieri G. Gut bacteria enable prostate cancer growth. *Science.* 2021 Oct 8; 374(6564):154-155. doi: 10.1126/science.abl7070. Epub 2021 Oct 7. PMID: 34618567.

Mendel G. Versuche über pflanzen-hybriden. *Journal of Heredity.* January 1951; 42(1):3–4. https://doi.org/10.1093/oxfordjournals.jhered.a106151.

Mendel G. Versuche über pflanzen-hybriden. *Verhandlungen des Naturforschenden Vereines, Abhandlungen, Brünn*. 1866; 4:3–47.

Mendel G. Gregor Mendel's letters to Carl Nägeli 1866–1873. *Genetics*. 35(5, 2):1–29.

Meredith W. Three authentic Beethoven locks. *William Meredith Beethoven Scholar*. https:// beethovenscholar.com/index.php/2023/03/22/three-authentic-beethoven-locks/.

Mitchell KJ. What is complex about complex disorders? *Genome Biology*. 2012; 13:237. https://doi.org/10.1186/gb-2012-13-1-237.

Motter AE, Gulbahce N, Almaas E, Barabási AL. Predicting synthetic rescues in metabolic networks. *Molecular Systems Biology*. 2008; 4:168. doi: 10.1038/msb.2008.1. Epub 2008 Feb 12. PMID: 18277384; PMCID: PMC2267730.

Musunuru K, Grandinette SA, Wang X, Hudson TR, Briseno K, Berry AM, Hacker JL, Hsu A, Silverstein RA, Hille LT, Ogul AN, Robinson-Garvin NA, Small JC, McCague S, Burke SM, Wright CM, Bick S, Indurthi V, Sharma S, Jepperson M, Vakulskas CA, Collingwood M, Keogh K, Jacobi A, Sturgeon M, Brommel C, Schmaljohn E, Kurgan G, Osborne T, Zhang H, Kinney K, Rettig G, Barbosa CJ, Semple SC, Tam YK, Lutz C, George LA, Kleinstiver BP, Liu DR, Ng K, Kassim SH, Giannikopoulos P, Alameh MG, Urnov FD, Ahrens-Nicklas RC. Patient-Specific In Vivo Gene Editing to Treat a Rare Genetic Disease. N Engl J Med. 2025 Jun 12;392(22):2235–2243. doi: 10.1056/NEJMoa2504747. Epub 2025 May 15. PMID: 40373211.

Naddaf M. Researchers welcome $3.5-million haemophilia gene therapy – but questions remain. *Nature*. 2022 Dec; 612(7940):388–389. doi: 10.1038/d41586-022-04327-7. PMID: 36474054.

Naddaf M. First trial of 'base editing' in humans lowers cholesterol – but raises safety concerns. *Nature*. 2023 Nov; 623(7988):671–672. doi: 10.1038/d41586-023-03543-z. PMID: 37957349.

Nasmyth K. The magic and meaning of Mendel's miracle. *Nature Reviews Genetics*. 2022 Jul; 23(7):447–452. doi: 10.1038/s41576-022-00497-2. Epub 2022 May 20. PMID: 35595848.

Nasser J et al. Genome-wide enhancer maps link risk variants to disease genes. *Nature*. 2021 May; 593(7858):238–243. doi: 10.1038/s41586-021-03446-x. Epub 2021 Apr 7. PMID: 33828297; PMCID: PMC9153265.

Niemi MEK et al. Common genetic variants contribute to risk of rare severe neurodevelopmental disorders. *Nature*. 2018 Oct; 562(7726): 268–271. doi: 10.1038/s41586-018-0566-4. Epub 2018 Sep 26. PMID: 30258228; PMCID: PMC6726472.

Noble W, Spires-Jones TL. Sleep well to slow Alzheimer's progression? *Science*. 2019; 363:813–814. doi: 10.1126/science.aaw5583.

Okada Y, Wang QS. A massive effort links protein-coding gene variants to health. *Nature*. 2021 Nov; 599(7886):561-563. doi: 10.1038/d41586-021-02873-0. PMID: 34697483.

Online Mendelian Inheritance in Man (OMIM). An online catalog of human genes and genetic disorders. https://www.omim.org.

Pennisi E. Of mice and men. *Science*. 2015 Jul 3; 349(6243):21-3. doi: 10.1126/science.349.6243.21. PMID: 26138961.

Piro A et al. Archibald Edward Garrod and alcaptonuria: 'Inborn errors of metabolism' revisited. *Genetics in Medicine*. 2010 Aug; 12(8):475-6. doi: 10.1097/GIM.0b013e3181e68843. PMID: 20703138.

Plotkin H. *The Nature of Knowledge*. Allen Lane (1994).

Raeke M. The boundless potential of CAR-T cell therapy, from cancer to common and chronic diseases: A Q&A with Carl June. 22 August 2023. https://www.pennmedicine.org/news/news-blog/2023/august/carl-june-on-the-boundless-potential-of-car-t-cell-therapy.

Reardon S. First gene therapy for Duchenne muscular dystrophy nears approval. *Nature*. 15 June 2023; 618:451.

Reardon S. Ancient DNA reveals origins of multiple sclerosis in Europe. *Nature*. 2024 Jan; 625(7995):431-432. doi: 10.1038/d41586-024-00024-9. PMID: 38200339.

Saçma M, Geiger H. Exercise generates immune cells in bone. *Nature*. 2021 Mar; 591(7850):371-372. doi: 10.1038/d41586-021-00419-y. PMID: 33627859.

Saleheen, D., Natarajan, P., Armean, I. et al. Human knockouts and phenotypic analysis in a cohort with a high rate of consanguinity. *Nature* 544, 235–239 (2017). https://doi.org/10.1038/nature22034.

Sankararaman S et al. The genomic landscape of Neanderthal ancestry in present-day humans. *Nature*. 2014 Mar 20; 507(7492):354-7. doi: 10.1038/nature12961. Epub 2014 Jan 29. PMID: 24476815; PMCID: PMC4072735.

Sarepta Therapeutics Press Release. Sarepta Therapeutics Announces Voluntary Pause of ELEVIDYS Shipments in the U.S. 07/21/25 7:40 PM EDT.

Savitt TL, Goldberg MF. Herrick's 1910 case report of sickle cell anemia: The rest of the story. *Journal of the American Medical Association*. 1989 Jan 13; 261(2):266-71. PMID: 2642320.

Schenck von Grafenburg J. *Observationes medicae de capite humano*. Frobeniana. 1584.

Scriver CR. Garrod's Croonian Lectures (1908) and the charter 'Inborn Errors of Metabolism': albinism, alkaptonuria, cystinuria, and pentosuria at age 100 in 2008. *Journal of Inherited Metabolic Disease*. 2008 Oct; 31(5):580-98. doi: 10.1007/s10545-008-0984-9. Epub 2008 Oct 12. PMID: 18850300.

Scull A. An ill-defined idea? *Nature*. 2005; 437:481. https://doi.org/10.1038/437481a.

Scully JL. A biographer and a bioethicist take on the CRISPR revolution. *Nature*. 11 March 2021; 591:196–197. doi: https://doi.org/10.1038/d41586-021-00579-x.

Sella G, Barton NH. Thinking About the Evolution of Complex Traits in the Era of Genome-Wide Association Studies. *Annual Review of Genomics and Human Genetics*. 2019 Aug 31; 20:461–493. doi: 10.1146/annurev-genom-083115-022316. Epub 2019 Jul 5. PMID: 31283361.

Shakespeare T. Disability or difference? *Nature Medicine*. 2005; 11:917. https://doi.org/10.1038 /nm0905-917.

Sharp N. Mutations matter even if proteins stay the same. *Nature*. 2022 Jun; 606(7915):657–659. doi: 10.1038/d41586-022-01091-6. PMID: 35676345.

Shen X, Song S, Li C, Zhang J. Synonymous mutations in representative yeast genes are mostly strongly non-neutral. *Nature*. 2022 Jun; 606(7915):725–731. doi: 10.1038/s41586-022-04823-w. Epub 2022 Jun 8. PMID: 35676473; PMCID: PMC9650438.

Sheridan C. The world's first CRISPR therapy is approved: who will receive it? *Nature Biotechnology*. 2024 Jan; 42(1):3–4. doi: 10.1038/d41587-023-00016-6. PMID: 37989785.

Silcocks M et al. National Centre for Indigenous Genomics: Indigenous Australian genomes show deep structure and rich novel variation. *Nature*. 2023 Dec; 624(7992):593–601. doi: 10.1038/s41586-023-06831-w. Epub 2023 Dec 13. PMID: 38093005; PMCID: PMC10733150.

Simonti CN et al. The phenotypic legacy of admixture between modern humans and Neandertals. *Science*. 2016 Feb 12; 351(6274):737–41. doi: 10.1126/science.aad2149. PMID: 26912863; PMCID: PMC4849557.

Smith T. First molecular explanation of disease. *Nature Structural and Molecular Biology*. 1999; 6:307. https:// doi.org/10.1038/7537.

Smithies O, Coffman T. A conversation with Oliver Smithies. *Annual Review of Physiology*. 2015; 77:1–11. doi: 10.1146/annurev-physiol-021014-071806. PMID: 25668016.

Stenn FF, Milgram JW, Lee SL, Weigand RJ, Veis A. Biochemical identification of homogentisic acid pigment in an ochronotic Egyptian mummy. *Science*. 1977 Aug 5; 197(4303):566–8. doi: 10.1126/science.327549. PMID: 327549.

Stern C, Sherwood ER. *The Origin of Genetics: a Mendel source book*. San Francisco: WH Freeman and Company (1966).

Sun BB et al. Genetic associations of protein-coding variants in human disease. *Nature*. 2022 Mar; 603(7899):95–102. doi: 10.1038/s41586-022-04394-w. Epub 2022 Feb 23. PMID: 35197637; PMCID: PMC8891017.

Suzuki K et al. *Nature*. 2024 Mar; 627(8003):347–357. doi: 10.1038/s41586-024-07019-6. Epub 2024 Feb 19. PMID: 38374256; PMCID: PMC10937372.

Tang S, Sternberg SH. Genome editing with retroelements. *Science*. 2023 Oct 27; 382(6669):370-371. doi: 10.1126/science.adi3183. Epub 2023 Oct 26. PMID: 37883564; PMCID: PMC10727090.

Trussell J. Rethinking Alexander Graham Bell's legacy. *Science*. 2021; 372:136. doi: 10.1126/science.abg8248.

Verkerk AJ et al. Identification of a gene (FMR-1) containing a CGG repeat coincident with a breakpoint cluster region exhibiting length variation in fragile X syndrome. *Cell*. 1991 May 31; 65(5):905-14. doi: 10.1016/0092-8674(91)90397-h. PMID: 1710175.

Visscher PM et al. 10 Years of GWAS Discovery: Biology, Function, and Translation. *American Journal of Human Genetics*. 2017 Jul 6; 101(1):5-22. doi: 10.1016/j.ajhg.2017.06.005. PMID: 28686856; PMCID: PMC5501872.

Visscher PM, Goddard ME. From R.A. Fisher's 1918 Paper to GWAS a Century Later. *Genetics*. 2019 Apr; 211(4):1125-1130. doi: 10.1534/genetics.118.301594. PMID: 30967441; PMCID: PMC6456325.

Waddington SN et al. Fetal gene therapy. *Journal of Inherited Metabolic Disease*. 2024 Jan; 47(1):192-210. doi: 10.1002/jimd.12659. Epub 2023 Aug 7. PMID: 37470194; PMCID: PMC10799196.

Wang JY, Doudna JA. CRISPR technology: A decade of genome editing is only the beginning. *Science*. 2023 Jan 20; 379(6629):eadd8643. doi: 10.1126/science.add8643. Epub 2023 Jan 20. PMID: 36656942.

Wang T et al. The Human Pangenome Project: a global resource to map genomic diversity. *Nature*. 2022; 604:437–446. https://doi.org/10.1038/s41586-022-04601-8.

Wang Y et al. The gut microbiota reprograms intestinal lipid metabolism through long noncoding RNA *Snhg9*. *Science*. 2023 Aug 25; 381(6660):851-857. doi: 10.1126/science.ade0522. Epub 2023 Aug 24. PMID: 37616368; PMCID: PMC10688608.

Warren M, Price M (eds). Muscular dystrophy therapy approved. *Science*. 30 June 2023; 380:1304.

Weatherall DJ. The centenary of Garrod's Croonian lectures. *Clinical Medicine (London)*. 2008 Jun; 8(3):309-11. doi: 10.7861/clinmedicine.8-3-309. PMID: 18624044; PMCID: PMC4953839.

Wesseldijk LW et al. Notes from Beethoven's genome. *Current Biology*. 2024 Mar 25; 34(6):R233-R234. doi: 10.1016/j.cub.2024.01.025. PMID: 38531312.

Winblad N, Lanner F. Biotechnology: At the heart of gene edits in human embryos. *Nature*. 2017 Aug 24; 548(7668):398-400. doi: 10.1038/nature23533. Epub 2017 Aug 2. PMID: 28783721.

Wong C. UK first to approve CRISPR treatment for diseases: what you need to know. *Nature*. 2023 Nov; 623(7988):676-677. doi: 10.1038/d41586-023-03590-6. PMID: 37974039.

Young AI, Benonisdottir S, Przeworski M, Kong A. Deconstructing the sources of genotype-phenotype associations in humans. *Science*. 2019

Sep 27; 365(6460):1396-1400. doi: 10.1126/science.aax3710. PMID: 31604265; PMCID: PMC6894903.

Zatkova A, Ranganath L, Kadasi L. Alkaptonuria: Current Perspectives. *The Application of Clinical Genetics*. 2020 Jan 23; 13:37-47. doi: 10.2147/TACG.S186773. PMID: 32158253; PMCID: PMC6986890.

CHAPTER 10

Alamos S, Shih PM. Synthetic gene circuits take root. *Science*. 2022 Aug 12; 377(6607):711-712. doi: 10.1126/science.add6805. Epub 2022 Aug 11. PMID: 35951701.

Ausländer S, Ausländer D, Müller M, Wieland M, Fussenegger M. Programmable single-cell mammalian biocomputers. *Nature*. 2012 Jul 5; 487(7405):123-7. doi: 10.1038/nature11149. PMID: 22722847.

Baker M. The next step for the synthetic genome. *Nature*. 2011; 473: 403–408. https: //doi.org/10.1038/473403a.

Ball P. Augmenting the alphabet. *Nature*. 2000. https://doi.org/10.1038/news000831-6.

Barnosky AD et al. Has the Earth's sixth mass extinction already arrived? *Nature*. 2011 Mar 3; 471(7336):51-7. doi: 10.1038/nature09678. PMID: 21368823.

Benner SA et al. Alternative Watson-Crick Synthetic Genetic Systems. *Cold Spring Harbor Perspectives in Biology*. 2016 Nov 1; 8(11):a023770. doi: 10.1101/cshperspect.a023770. PMID: 27663774; PMCID: PMC5088529.

Benner SA et al. Redesigning nucleic acids. *Pure and Applied Chemistry*. 1998 Feb; 70(2):263-6. doi: 10.1351/pac199870020263. PMID: 11542721.

Benton M. A Palaeozoic whodunnit. *Nature*. 2006; 441:27. https://doi.org/10.1038/441027a.

Berg J. Tomorrow's Earth. *Science*. 2018 Jun 29; 360(6396):1379. doi: 10.1126/science.aau5515. PMID: 29954955.

Blount BA, Ellis T. Construction of an Escherichia coli genome with fewer codons sets records. *Nature*. 2019 May; 569(7757):492-494. doi: 10.1038/d41586-019-01584-x. PMID: 31097820.

Blount BA. An antiviral molecular language barrier. *Nature*. 23 March 2023; 615:592-594. doi: https://doi.org/10.1038/d41586-023-00702-0.

Bong D, Holliger P, Yang C. Introduction to the themed collection on XNA xeno-nucleic acids. *RSC Chemical Biology*. 2022 Oct 11; 3(11):1299-1300. doi: 10.1039/d2cb90036j. PMID: 36349221; PMCID: PMC9627736.

Bourzac K. Gene therapy: Erasing sickle-cell disease. *Nature*. 2017 Sep 27; 549(7673):S28-S30. doi: 10.1038/549S28a. PMID: 28953858.

Burke AJ et al. Design and evolution of an enzyme with a non-canonical organocatalytic mechanism. *Nature*. 2019 Jun; 570(7760):219-223. doi: 10.1038/s41586-019-1262-8. Epub 2019 May 27. PMID: 31132786.

Cai H et al. Brain organoid reservoir computing for artificial intelligence. *Nature Electronics*. 2023; 6:1032–1039. https://doi.org/10.1038/s41928-023-01069-w.

Callaway E. Will these reprogrammed elephant cells ever make a mammoth? *Nature*. 2024 Mar; 627(8003):253-254. doi: 10.1038/d41586-024-00670-z. PMID: 38448533.

Callaway E et al. The discovery of Homo floresiensis: Tales of the hobbit. *Nature*. 2014 Oct 23; 514(7523):422-6. doi: 10.1038/514422a. PMID: 25341771.

Callaway E. Carbon dioxide-eating bacteria offer hope for green production. *Nature*. 5 December 2019; 576:19-20.

Cameron DE, Bashor CJ, Collins JJ. A brief history of synthetic biology. *Nature Reviews Microbiology*. 2014 May; 12(5):381-90. doi: 10.1038/nrmicro3239. Epub 2014 Apr 1. PMID: 24686414.

Campbell-Staton S. Refining the tree of life. *Science*. Oct 2021; 374:542. doi: 10.1126/science.abm0633.

Cavalli G, Heard E. Advances in epigenetics link genetics to the environment and disease. *Nature*. 2019 Jul; 571(7766):489-499. doi: 10.1038/s41586-019-1411-0. Epub 2019 Jul 24. PMID: 31341302.

Chen LG et al. A designer synthetic chromosome fragment functions in moss. *Nature Plants*. 2024 Feb; 10(2):228-239. doi: 10.1038/s41477-023-01595-7. Epub 2024 Jan 26. PMID: 38278952.

Chen Z et al. African rice cultivation linked to rising methane. *Nature Climate Change*. 2024; 14:148–151. https://doi.org/10.1038/s41558-023-01907-x.

Cho NH et al. OpenCell: Endogenous tagging for the cartography of human cellular organization. *Science*. 2022 Mar 11; 375(6585):eabi6983. doi: 10.1126/science.abi6983. Epub 2022 Mar 11. PMID: 35271311; PMCID: PMC9119736.

Christie A. *Murder on the Orient Express*. Collins (1934).

Codex Climaci Rescriptus. https://archive.org/details/Lewis1909CodexClimaciRescriptus.

Cohen SS. On biochemical variability and innovation. *Science*. 1963 Mar 15; 139(3559):1017-26. doi: 10.1126/science.139.3559.1017. PMID: 14022107.

Cooper RM et al. Engineered bacteria detect tumor DNA. *Science*. 2023 Aug 11; 381(6658):682-686. doi: 10.1126/science.adf3974. Epub 2023 Aug 10. PMID: 37561843; PMCID: PMC10852993.

Costello MJ, May RM, Stork NE. Can we name Earth's species before they go extinct? *Science*. 2013 Jan 25; 339(6118):413-6. doi: 10.1126/science.1230318. PMID: 23349283.

Curry A. Ancient DNA pioneer Svante Pääbo wins Nobel. *Science*. 2022 Oct 7; 378(6615):12. doi: 10.1126/science.adf1845. Epub 2022 Oct 6. PMID: 36201580.

Dawe RK. Charting the path to fully synthetic plant chromosomes. *Experimental Cell Research*. 2020 May 1; 390(1):111951. doi: 10.1016/j.yexcr.2020.111951. Epub 2020 Mar 6. PMID: 32151492.

Dunkelmann DL et al. Adding α,α-disubstituted and β-linked monomers to the genetic code of an organism. *Nature*. 2024 Jan; 625(7995):603-610. doi: 10.1038/s41586-023-06897-6. Epub 2024 Jan 10. PMID: 38200312; PMCID: PMC10794150.

Editorial. The legacy of Linnaeus. *Nature*. 2007 Mar 15; 446(7133):231-2. doi: 10.1038/446231b. PMID: 17361137.

Editorial. 'Dangerously fast' growth in atmospheric methane. *Nature*. 3 March 2022; 603:13. doi: https://doi.org/10.1038/d41586-022-00312-2.

Editorial. Control methane to slow global warming – fast. *Nature*. 2021 Aug; 596(7873):461. doi: 10.1038/d41586-021-02287-y. PMID: 34433951.

Editorial. Woodpecker among 23 species deemed extinct. *Science*. 8 October 2021; 374(6564): 134. doi: 10.1126/science.acx9298.

Editorial. Tiny flying saucers are actually odd new microbes. *Nature*. 12 June 2019; 570:279. doi: https://doi.org/10.1038/d41586-019-01827-x.

Editorial. A pea-sized frog. *Science*. 3 September 2010; 329:1137. doi: 10.1126/article.29987.

Editorial. Ironclad beetle takes the pressure. *Science*. 23 October 2020; 370(6515):386-387. doi: 10.1126/science.370.6515.386.

Editorial. Lilliputian lizards come to light. *Nature*. 2012; 482:442. https://doi.org/10.1038/482442a.

Editorial. Prehistoric Polly loomed large. *Nature*. 15 August 2019; 572(7796):287.

Editorial. Early penguin was a giant. *Nature*. 21/28 December 2017; 552:295.

Editorial. Scans reveal what makes beetle 'uncrushable'. *Nature*. 22 Oct 2020; 586(7830):479. doi: 10.1038/d41586-020-02840-1.

Editorial. Australia's coastal species decline. *Science*. 24 March 2023; 379:1173. doi: 10.1126/science.adh9191.

Editorial. Genome rewriting generates mouse models of human diseases. *Nature*. 2023 Nov 1. doi: 10.1038/d41586-023-03079-2. Epub ahead of print. PMID: 37914872.

Eliot TS. *The Waste Land*. Boni & Liveright (1922).

van Esse GW. The quest for optimal plant architecture. *Science*. 2022 Apr 8; 376(6589):133-134. doi: 10.1126/science.abo7429. Epub 2022 Apr 7. PMID: 35389808.

Fletcher D. Which biological systems should be engineered? *Nature*. 2018 Nov; 563(7730):177-179. doi: 10.1038/d41586-018-07291-3. PMID: 30401855.

Fodor J. Why pigs don't have wings. *London Review of Books*. 18 October 2007; 29(20):19-22.

Fredens J et al. Total synthesis of Escherichia coli with a recoded genome. *Nature*. 2019 May; 569(7757):514-518. doi: 10.1038/s41586-019-1192-5. Epub 2019 May 15. PMID: 31092918; PMCID: PMC7039709.

Galilei G. *Sidereus Nuncius*. Apud Thomum Baglionum, Republic of Venice (13 March 1610).

Gammon K. Gene therapy: editorial control. *Nature*. 2014 Nov 13; 515(7526):S11-3. doi: 10.1038/515S11a. PMID: 25390136.

Gao H. The landscape of tolerated genetic variation in humans and primates. *Science*. 2023 Jun 2; 380(6648):eabn8153. doi: 10.1126/science.abn8197. Epub 2023 Jun 2. PMID: 37262156; PMCID: PMC10713091.

Gao XJ, Chong LS, Kim MS, Elowitz MB. Programmable protein circuits in living cells. *Science*. 2018 Sep 21; 361(6408):1252-1258. doi: 10.1126/science.aat5062. PMID: 30237357; PMCID: PMC7176481.

Garrod AE. The incidence of alkaptonuria: a study in chemical individuality. *Lancet*. 13 December 1902; 160(4137):1616-1620. https://doi.org/10.1016/S0140-6736(01)41972-6.

Gaut NJ, Adamala KP. Toward artificial photosynthesis. *Science*. 2020 May 8; 368(6491):587-588. doi: 10.1126/science.abc1226. PMID: 32381709.

Głowacka K et al. Photosystem II Subunit S overexpression increases the efficiency of water use in a field-grown crop. *Nature Communications*. 2018 Mar 6;9(1):868. doi: 10.1038/s41467-018-03231-x. PMID: 29511193; PMCID: PMC5840416.

Gong F, Li Y. Fixing carbon, unnaturally. *Science*. 2016 Nov 18; 354(6314): 830-831. doi: 10.1126/science.aal1559. PMID: 27856865.

Green RE et al. A Draft Sequence of the Neandertal Genome. *Science*. 2010; 328:710-722. doi: 10.1126/science.1188021.

Grimm D. How domestic cats wiped out the Scottish wildcat. *Science*. 2023 Nov 10; 382(6671):625-626. doi: 10.1126/science.adm8227. Epub 2023 Nov 9. PMID: 37943918.

Grimstad P. The absolute originality of Georges Perec. *New Yorker*. 16 July 2019. https://www.newyorker.com/books/page-turner/the-absolute-originality-of-georges-perec.

Griffiths CJ. Back to nature. *Science*. March 2016; 351(6279):1271. doi: 10.1126/science.aaf1485.

Grome MW et al. Engineering a genomically recoded organism with one stop codon. *Nature*. 2025; 639:512–521. https://doi.org/10.1038/s41586-024-08501-x.

Gysembergh V, Williams PJ, Zingg E. New evidence for Hipparchus' Star Catalogue revealed by multispectral imaging. *Journal for the History of Astronomy*. 18 October 2022. 53:4. https://doi.org/10.1177/00218286221128289.

Hardy T. An August Midnight. In *Poems of the Past and Present*. London: Harper & Brothers (1901).

Hopkin M. Insect deaths add to extinction fears. *Nature*. 2004. https://doi.org/10.1038 /news040315-11.

Incorvaia D. Linnaeus and his life's work. *Science*. 2023; 381:611-61. doi: 10.1126/science.adj0804.

Ingram VM. Sickle-cell anemia hemoglobin: the molecular biology of the first 'molecular disease': the crucial importance of serendipity. *Genetics*. 2004 May; 167(1):1-7. doi: 10.1534/genetics.167.1.1. PMID: 15166132; PMCID: PMC1470873.

Isaacs FJ, Carr PA, Wang HH, Lajoie MJ, Sterling B, Kraal L, Tolonen AC, Gianoulis TA, Goodman DB, Reppas NB, Emig CJ, Bang D, Hwang SJ, Jewett MC, Jacobson JM, Church GM. Precise manipulation of chromosomes in vivo enables genome-wide codon replacement. *Science*. 2011 Jul 15;333(6040):348-53. doi: 10.1126/science.1205822. PMID: 21764749; PMCID: PMC5472332.

Jackson JC, Duffy SP, Hess KR, Mehl RA. Improving nature's enzyme active site with genetically encoded unnatural amino acids. *Journal of the American Chemical Society*. 2006 Aug 30; 128(34):11124-7. doi: 10.1021/ja061099y. PMID: 16925430.

Jackson K. As easy as ABC: Georges Perec once wrote a whole novel without the letter e. Kevin Jackson summarizes Oulipian life in 26 sentences. *Independent*. 06 June 1994. https://www.the-independent.com/arts-entertainment/as-easy-as-abc-georges-perec-once-wrote-a-whole-novel-without-the-letter-e-kevin-jackson-summarises-an-oulipian-life-in-26-sentences-1420994.html.

Jacobs P. Is the dire wolf back from the dead? Not exactly. *Science*. 8 April 2025. doi: 10.1126/science.zlooigi.

Jayne TS, Sanchez PA. Agricultural productivity must improve in sub-Saharan Africa. *Science*. 2021 Jun 4; 372(6546):1045-1047. doi: 10.1126/science.abf5413. PMID: 34083478.

Keasling JD. Manufacturing molecules through metabolic engineering. *Science*. 2010 Dec 3; 330(6009):1355-8. doi: 10.1126/science.1193990. PMID: 21127247.

Kidd D. *Phaenomena*, edited with introduction, translation and commentary. Cambridge (1997).

Khalil AS, Collins JJ. Synthetic biology: applications come of age. *Nature Reviews Genetics*. 2010 May; 11(5):367-79. doi: 10.1038/nrg2775. PMID: 20395970; PMCID: PMC2896386.

Khorana, H. G. (in) Caruthers, M.H. Gene synthesis with H G Khorana. Reson 17, 1143–1156 (2012). https://doi.org/10.1007/s12045-012-0131-7.

Koentges G. Evolution of anatomy and gene control. *Nature*. 2008 Feb 7; 451(7179):658-63. doi: 10.

Kwok R. Chemical biology: DNA's new alphabet. *Nature*. 2012; 491: 516–518. https://doi.org/10.1038/491516a.

Kwon D. How scientists are hacking the genetic code. *Nature*. 22 June 2023; 618:874-876. doi: https://doi.org/10.1038/d41586-023-01980-4.

Leslie M. Moss project takes first step toward first plant with an artificial genome. *Science*. 30 January 2024. https://www.science.org/content/article/moss-project-takes-step-toward-first-artificial-plant-genome.

Leslie M. 'Recoded' bacteria shrug off viral attacks. *Science*. 2022 Oct 21; 378(6617):240. doi: 10.1126/science.adf3949. Epub 2022 Oct 20. PMID: 36264809.

Li Q, Daumiller D, Bryant P. RareFold: Structure prediction and design of proteins with noncanonical amino acids. *bioRxiv*. 2025 .05.19.654846. doi: https://doi.org/10.1101/2025.05.19. 654846.

Li T et al. Developing fibrillated cellulose as a sustainable technological material. *Nature*. 2021 Feb; 590(7844):47-56. doi: 10.1038/s41586-020 -03167-7. Epub 2021 Feb 3. PMID: 33536649.

Li Y, Luo Y. Intelligent textiles are looking bright. *Science*. 2024 Apr 5; 384(6691):29-30. doi: 10.1126/science.ado5922. Epub 2024 Apr 4. PMID: 38574158.

Linnaeus. *Systema Naturae*. Leiden: Theodore Haak (1735).

de Lorenzo V, Danchin A. Synthetic biology: discovering new worlds and new words. *EMBO Reports*. 2008 Sep; 9(9):822-7. doi: 10.1038 /embor.2008.159. PMID: 18724274; PMCID: PMC2529360.

Lu Y. The gene-synthesis revolution. *New York Times*. 24 November 2021.

Lv H et al. DNA-based programmable gate arrays for general-purpose DNA computing. *Nature*. 2023 Oct; 622(7982):292-300. doi: 10.1038 /s41586-023-06484-9. Epub 2023 Sep 13. PMID: 37704731.

Lynall G. In retrospect: Gulliver's Travels. *Nature*. 2017 ; 549 :454–455. https://doi.org/10.1038/549454a.

Malyshev DA et al. A semi-synthetic organism with an expanded genetic alphabet. *Nature*. 2014 May 15; 509(7500):385-8. doi: 10.1038/nature13314. Epub 2014 May 7. PMID: 24805238; PMCID: PMC4058825.

Mangan RJ et al. Adaptive sequence divergence forged new neurodevelopmental enhancers in humans. *Cell*. 2022 Nov 23; 185(24): 4587-4603.e23. doi: 10.1016/j.cell.2022.10.016. PMID: 36423581; PMCID: PMC10013929.

Marchant J. First known map of night sky found hidden in Medieval parchment. *Nature*. 2022 Oct; 610(7933):613-614. doi: 10.1038/d41586 -022-03296-1. PMID: 36258126.

Marris E. DNA gets a fake fifth base. *Nature*. March 2005. doi: 10.1038 /news050314-8.

Morgan-Richards M, Langton-Myers SS, Trewick SA. Loss and gain of sexual reproduction in the same stick insect. *Molecular Ecology*. 6 August 2019; 28(17):3929-3941. https://doi.org/10.1111/mec.15203.

Muka S. Making and remaking life. *Science*. 2021; 373:499. doi: 10.1126 /science.abj2437.

Nadis S. Hard-hitting endeavor captures Ig Nobel. *Nature*. 2006 Oct 12; 443(7112):616-7. doi: 10.1038/443616b. PMID: 17035966.

Nelson A. Catalytic machinery of enzymes expanded. *Nature*. 2019 Jun; 570(7760):172-173. doi: 10.1038/d41586-019-01596-7. PMID: 31182827.

Nicholls H. Darwin 200: Let's make a mammoth. *Nature*. 2008 Nov 20; 456(7220):310-4. doi: 10.1038/456310a. Erratum in: *Nature*. 2009 Jan 8; 457(7226):140. PMID: 19020594.

Nicholls H. Time to sequence the 'red and the dead'. *Nature*. 2009 Apr 16; 458(7240):812-3. doi: 10.1038/458812a. PMID: 19369991.

Nicholson DJ. On being the right size, revisited. The problem with engineering metaphors in molecular biology. In *Philosophical Perspectives on the Engineering Approach in Biology*. Routledge (2020).

Nossiter A. Saleemul Huq, 71, Bangladeshi Spearhead on Climate Change, Dies. *New York Times*. 3 November 2023.

Nowell A. Knowing the Neanderthal. Review of *The Naked Neanderthal: A New Understanding of the Human Creature* by Ludovic Slimak. Pegasus (2024). *Science*. 2024 Mar; 383(6686):956. doi: 10.1126/science.adn6093. Epub 2024 Feb 29. PMID: 38422140.

Ostorov N et al. Synthetic genomes with altered genetic codes. *Current Opinion in Systems Biology*. December 2020; 24:32-40. https://doi.org/10.1016/j.coisb.2020.09.007.

Pascual U. Diverse values of nature for sustainability. *Nature*. 2023 Aug; 620(7975):813-823. doi: 10.1038/s41586-023-06406-9. Epub 2023 Aug 9. PMID: 37558877; PMCID: PMC10447232.

Patel-Tupper D et al. Multiplexed CRISPR-Cas9 mutagenesis of rice *PSBS1* noncoding sequences for transgene-free overexpression. *Science Advances*. 2024 Jun 7; 10(23):eadm7452. doi: 10.1126/sciadv.adm7452. Epub 2024 Jun 7. PMID: 38848363; PMCID: PMC11160471.

Pennisi E. Computer scan uncovers 100,000 new viruses. *Science*. 2022 Feb 4; 375(6580):483. doi: 10.1126/science.ada0854. Epub 2022 Feb 3. PMID: 35113704.

Perec, G. *La Disparition*. (Les Lettres Nouvelles). Denöel, Paris. 1969.

Pigafetta A. *Relazione del Primo Viaggio Intorno al Mondo* (The First Voyage Around the World).

Pinheiro VB et al. Synthetic genetic polymers capable of heredity and evolution. *Science*. 2012 Apr 20; 336(6079):341-4. doi: 10.1126/science.1217622. PMID: 22517858; PMCID: PMC3362463.

Priemel T et al. Microfluidic-like fabrication of metal ion-cured bioadhesives by mussels. *Science*. 2021 Oct 8; 374(6564):206-211. doi: 10.1126/science.abi9702. Epub 2021 Oct 7. PMID: 34618575.

Prüfer K. The complete genome sequence of a Neanderthal from the Altai Mountains. *Nature*. 2014 Jan 2; 505(7481):43-9. doi: 10.1038/nature12886. Epub 2013 Dec 18. PMID: 24352235; PMCID: PMC4031459.

Regaldo, A. A brain-dead man was attached to a gene-edited pig liver for three days. *MIT Technology Review*. January 18, 2024. https://www.technologyreview.com/2024/01/18/1086791/brain-dead-man-gene-edited-pig-liver/

Reif JH. Biochemistry: Scaling up DNA computation. *Science*. 2011 Jun 3; 332(6034):1156-7. doi: 10.1126/science.1208068. PMID: 21636761.

Robertson WE, Rehm FBH, Spinck M, Schumann RL, Tian R, Liu W, Gu Y, Kleefeldt AA, Day CF, Liu KC, Christova Y, Zürcher JF, Böge FL, Birnbaum J, van Bijsterveldt L, Chin JW. *Escherichia coli* with a 57-codon genetic code. *Science*. 2025 Oct 23;390(6771):eady4368. doi: 10.1126/science.ady4368. Epub 2025 Oct 23. PMID: 40743368.

Robertson WE et al. Sense codon reassignment enables viral resistance and encoded polymer synthesis. *Science*. 2021 Jun 4; 372(6546):1057–1062. doi: 10.1126/science.abg3029. PMID: 34083482; PMCID: PMC7611380.

Rusk N. Synthetic biology: Understudies of DNA and RNA. *Nature Methods*. 2012 Jun;9(6):530–1. doi: 10.1038/nmeth.2058. PMID: 22874981.

Russell B. *The Impact of Science on Society*. London: George Allen & Unwin (1952).

Sanderson K. Largest bacterium ever found is surprisingly complex. *Nature*. 2022 Jun 23. doi: 10.1038/d41586-022-01757-1. Epub ahead of print. PMID: 35750919.

Santos RG, Machovsky-Capuska GE, Andrades R. Plastic ingestion as an evolutionary trap: Toward a holistic understanding. *Science*. 2021 Jul 2; 373(6550):56–60. doi: 10.1126/science.abh0945. PMID: 34210877.

Savitsky Z. Scientists resurrect earliest star map from medieval Christian text. *Science*. 19 October 2022. https://www.science.org/content/article/scientists-resurrect-earliest-star-map-medieval-christian-text.

Schmidt G. Why 2023's heat anomaly is worrying scientists. *Nature*. 21 March 2024; 627:467. doi: https://doi.org/10.1038/d41586-024-00816-z.

Schulz K. How Carl Linnaeus set out to label all of life. *New Yorker*. 14 August 2023.

Schwille P. Bottom-up synthetic biology: engineering in a tinkerer's world. *Science*. 2011 Sep 2; 333(6047):1252–4. doi: 10.1126/science.1211701. PMID: 21885774.

Service RF. Jay Keasling profile. Rethinking mother nature's choices. *Science*. 2007 Feb 9; 315(5813):793.doi: 10.1126/science.315.5813.793 .PMID:17289982.

Service RF. Let there be dark. *Science*. 2023 Jun 9; 380(6649):1004–1007. doi: 10.1126/science.adj0814. Epub 2023 Jun 8. PMID: 37289883.

Sherwood S, Hoskins B. Clarion call from climate panel. *Science*. 2021 Aug 13; 373(6556):719. doi: 10.1126/science.abl8490. Epub 2021 Aug 10. PMID: 34376522.

Srinivasan G et al. Pyrrolysine Encoded by UAG in Archaea: Charging of a UAG-Decoding Specialized tRNA. *Science*. 2022; 296:1459–1462. doi: 10.1126/science.1069588.

Stadtman TC. Selenocysteine. *Annual Review of Biochemistry*. July 1996; 65:83–100. https://doi.org/10.1146/annurev.bi.65.070196.000503.

Stokstad E. United Nations to tackle global plastics pollution. *Science*. 2022 Feb 25; 375(6583):801–802. doi: 10.1126/science.ada1551. Epub 2022 Feb 24. PMID: 35201854.

Stokstad E. Antibody-based defense may protect plants from disease. *Science*. 2023 Mar 3; 379(6635):867. doi: 10.1126/science.adh3913. Epub 2023 Mar 2. PMID: 36862778.

Suwa G et al. The Ardipithecus ramidus skull and its implications for hominid origins. *Science*. 2009 Oct 2; 326(5949):68e1-7. PMID: 19810194.

Sweetlove L. Number of species on Earth tagged at 8.7 million. *Nature*. 23 August 2011. https://doi.org/10.1038/news.2011.498.

Swift J. Confronting the climate crisis. *Science*. 2022; 376:1277. doi: 10.1126/science.abp8954.

Tamsir A, Tabor JJ, Voigt CA. Robust multicellular computing using genetically encoded NOR gates and chemical 'wires'. *Nature*. 2011 Jan 13; 469(7329):212-5. doi: 10.1038/nature09565. Epub 2010 Dec 8. PMID: 21150903; PMCID: PMC3904220.

Thomas C. A sixth mass extinction? *Nature*. 2007; 450:349. https://doi.org/10.1038/450349a.

Thyer R, Ellefson J. Synthetic biology: New letters for life's alphabet. *Nature*. 2014 May 15; 509(7500):291-2. doi: 10.1038/nature13335. Epub 2014 May 7. PMID: 24805244.

Toomey D. *Weird Life: The search for life that is very, very different from our own*. WW Norton (2013).

de la Torre D, Chin JW. Reprogramming the genetic code. *Nature Reviews Genetics*. 2021 Mar; 22(3):169-184. doi: 10.1038/s41576-020-00307-7. Epub 2020 Dec 14. PMID: 33318706.

Tozer L. 'Biocomputer' combines lab-grown brain tissue with electronic hardware. *Nature*. 2023 Dec; 624(7992):481. doi: 10.1038/d41586-023-03975-7. PMID: 38082130.

Türk V. Protect the 'right to science' for people and the planet. *Nature*. 2023 Nov; 623(7985):9. doi: 10.1038/d41586-023-03332-8. PMID: 37914948.

Tyson M. World's first bioprocessor uses 16 human brain organoids for 'a million times less power' consumption than a digital chip. Microsoft. 2024. https://www.msn.com/en-gb/money/technology/world-s-first-bioprocessor-uses-16-human-brain-organoids-for-a-million-times-less-power-consumption-than-a-digital-chip/ar-BB1n5koX?ocid=finance-verthp-feeds.

Unger R, Moult J. Towards computing with proteins. *Proteins*. 2006 Apr 1; 63(1):53-64. doi: 10.1002/prot.20886. PMID: 16435369.

Vauquelin LN, Robiquet PJ. La découverte d'un nouveau principe végétal dans le suc des asperges. *Annales de Chimie*. 1806; 57:88–93.

Vernot B, Akey JM. Resurrecting surviving Neandertal lineages from modern human genomes. *Science*. 2014 Feb 28; 343(6174):1017-21. doi: 10.1126/science.1245938. Epub 2014 Jan 29. PMID: 24476670.

Voosen P. U.N. panel warns of warming's toll and an 'adaptation gap'. *Science*. 2022 Mar 4; 375(6584):948. doi: 10.1126/science.adb1761. Epub 2022 Mar 3. PMID: 35239376.

Walker JT et al. Genetic risk converges on regulatory networks mediating early type 2 diabetes. *Nature*. 2023 Dec; 624(7992):621–629. doi: 10.1038/s41586-023-06693-2. Epub 2023 Dec 4. PMID: 38049589.

Walsh B. New natural selection: How scientists are altering DNA to genetically engineer new forms of life. *Newsweek*. 29 June 2017. https://www.newsweek.com/2017/07/07/natural-selection-new-forms-life-scientists-altering-dna-629771.html.

Waltz E. Microbe-made jet fuel. *Nature Biotechnology*. 2024 Feb; 42(2): 163–166. doi: 10.1038/s41587-024-02136-z. PMID: 38361068.

Warne K. Organization Man. *Smithsonian*. May 2007. https://www.smithsonianmag.com/science-nature/organization-man-151908042/.

Warren M. Four new DNA letters double life's alphabet. *Nature*. 2019 Feb; 566(7745):436. doi: 10.1038/d41586-019-00650-8. PMID: 30809059.

Wellerstein A. The First Light of Trinity. *New Yorker*. 16 July 2015. https://www.newyorker.com/tech/annals-of-technology/the-first-light-of-the-trinity-atomic-test.

Wesley E et al. *Escherichia coli* with a 57-codon genetic code. *bioRxiv*. 2025.05.02.651837; doi: https://doi.org/10.1101/2025.05.02.651837.

Williams RL, Liu CC. Accelerated evolution of chosen genes. *Science*. 2024; 383:372–373. doi: 10.1126/science.adn3434.

Wilson EO. *Half-Earth: Our Planet's Fight For Life*. Liveright Publishing Corporation (2016).

Wragg Sykes R. The strange and secret lives of Neanderthals. *Nature*. 14 December 2023; 624:247–248. doi: https://doi.org/10.1038/d41586-023-03862-1.

Wu X, Wei Q, Deni S, Zhang H. China's dhole population at risk of extinction. *Science*. 2021 Apr 30; 372(6541):472. doi: 10.1126/science.abi8889. PMID: 33926944.

Zhang W et al. Mouse genome rewriting and tailoring of three important disease loci. *Nature*. 2023 Nov; 623(7986):423–431. doi: 10.1038/s41586-023-06675-4. Epub 2023 Nov 1. PMID: 37914927; PMCID: PMC10632133.

Zimmer C. How one child's sickle cell mutation helped protect the world from malaria. *New York Times*. 8 March 2018. https://www.nytimes.com/2018/03/08/health/sickle-cell-mutation.html.

Zimmer C. Scientists revive the wolf, or something close. *New York Times*. 7 April 2025.

Zürcher JF et al. Refactored genetic codes enable bidirectional genetic isolation. *Science*. 2022 Nov 4; 378(6619):516–523. doi: 10.1126/science.add8943. Epub 2022 Oct 20. PMID: 36264827; PMCID: PMC7614150.

CHAPTER 11

Austad SN, Finch CE. How ubiquitous is aging in vertebrates? *Science*. 2022 Jun 24; 376(6600):1384–1385. doi: 10.1126/science.adc9442. Epub 2022 Jun 23. PMID: 35737765.

Austen J. *Pride and Prejudice*. Whitehall: T Egerton, Military Library (1813).

Barbi E, Lagona F, Marsili M, Vaupel JW, Wachter KW. The plateau of human mortality: Demography of longevity pioneers. *Science*. 2018 Jun 29; 360(6396):1459-1461. doi: 10.1126/science.aat3119. PMID: 29954979; PMCID: PMC6457902.

Bennett NR et al. Atomically accurate de novo design of single-domain antibodies. *bioRxiv* [Preprint]. 2024 Mar 18: 2024.03.14.585103. doi: 10.1101/2024.03.14.585103. PMID: 38562682; PMCID: PMC10983868.

Berg P, Mertz JE. Personal reflections on the origins and emergence of recombinant DNA technology. *Genetics*. 2010 Jan; 184(1):9-17. doi: 10.1534/genetics.109.112144. PMID: 20061565; PMCID: PMC2815933.

Bioengineer.org. Designing Functional Genes with Genomic AI. 20 November, 2025. https://bioengineer.org/designing-functional-genes-with-genomic-ai/

Bleuler E. Die Prognose der Dementia Praecox (Schizophreniegruppe). *Allgemeine Zeitschrift für Psychiatrie und psychisch-gerichtliche Medizin*. 1908; 31:436–480.

Bleuler E. *Dementia Praecox oder Gruppe der Schizophrenien*. Leipzig und Wien (1911).

Boeke JD et al. Genome Engineering. The Genome Project-Write. *Science*. 2016 Jul 8; 353(6295):126-7. doi: 10.1126/science.aaf6850. Epub 2016 Jun 2. PMID: 27256881.

Boix CA, James BT, Park YP, Meuleman W, Kellis M. Regulatory genomic circuitry of human disease loci by integrative epigenomics. *Nature*. 2021 Feb; 590(7845):300-307. doi: 10.1038/s41586-020-03145-z. Epub 2021 Feb 3. PMID: 33536621; PMCID: PMC7875769.

Bourzac K. Writing the human genome. *Chemical & Engineering News*. 10 July 2017; 95:28. https://cen.acs.org/articles/95/i28/Writing-human-genome.html.

Brenner S, Sejnowski TJ. Understanding the human brain. *Science*. 2011 Nov 4; 334(6056):567. doi: 10.1126/science.1215674. PMID: 22053011; PMCID: PMC4757457.

Brenner S. Nature's gift to science (Nobel lecture). *ChemBioChem*. 2003 Aug 4; 4(8):683-7. doi: 10.1002/cbic.200300625. PMID: 12898617.

Bryce E. Humans 2.0: these geneticists want to create an artificial genome by synthesizing our DNA. *Wired*. 26 February 2017. https://www.wired.com/story/human-genome-synthesise-dna/.

Callaway E. Plan to synthesize human genome triggers mixed response. *Nature*. 2016 Jun 9; 534(7606):163. doi: 10.1038/nature.2016.20028. PMID: 27279189.

Callaway E. AI tools are designing entirely new proteins that could transform medicine. *Nature*. 2023 Jul; 619(7969):236-238. doi: 10.1038/d41586-023-02227-y. PMID: 37433935.

Callaway E. 'A Pandora's box': map of protein-structure families delights scientists. *Nature*. 2023 Sep; 621(7979):455. doi: 10.1038/d41586-023-02892-z. PMID: 37704851.

Callaway E. Scientists are using AI to dream up revolutionary new proteins. *Nature*. 2022 Sep; 609(7928):661-662. doi: 10.1038/d41586-022-02947-7. PMID: 36109683.

Cao L et al. Design of protein-binding proteins from the target structure alone. *Nature*. 2022 May; 605(7910):551-560. doi: 10.1038/s41586-022-04654-9. Epub 2022 Mar 24. PMID: 35332283; PMCID: PMC9117152.

Cao L et al. Design of protein-binding proteins from the target structure alone. *Nature*. 2022 May; 605(7910):551-560. doi: 10.1038/s41586-022-04654-9. Epub 2022 Mar 24. PMID: 35332283; PMCID: PMC9117152.

de Cervantes M. *El Ingenioso hidalgo Don Quixote de la Mancha*, parts I and II. Juan de la Cuesta (1617).

Church DM. A next-generation human genome sequence. *Science*. 2022; 376:34-35. doi: 10.1126/science.abo5367.

Classen AK. Tumours form without genetic mutations. *Nature*. 2024 May; 629(8012):534-535. doi: 10.1038/d41586-024-01019-2. PMID: 38658714.

Cohen J. Neanderthal brain organoids come to life. *Science*. 2018 Jun 22; 360(6395):1284. doi: 10.1126/science.360.6395.1284. PMID: 29930117.

Dahl R. *Charlie and the Chocolate Factory*. George Allen and Unwin (1967).

Dolgin E. Genome-synthesis effort shifts focus. *Nature*, 3 May 2018; 557:16-17. doi: https://doi.org/10.1038/d41586-018-05043-x.

Dolgin E. There's no limit to longevity, says study that revives human lifespan debate. *Nature*. 2018 Jul; 559(7712):14-15. doi: 10.1038/d41586-018-05582-3. PMID: 29968831.

Dunkelmann DL et al. Adding α,Symbol -disubstituted and β-linked monomers to the genetic code of an organism. *Nature*. 2024 Jan; 625(7995):603-610. doi: 10.1038/s41586-023-06897-6. Epub 2024 Jan 10. PMID: 38200312; PMCID: PMC10794150.

Editorial. Rewriting our future. *Nature Biotechnology*. 2016 Jul 12; 34(7):673. doi: 10.1038/nbt.3640. PMID: 27404859.

Editorial. How elephants dodge cancer. *Nature*. 15 October 2015; 526:297. https://doi.org/10.1038/526297b.

Editorial. Old fish, new fish: A deep dive for the clues to longevity. *Nature*. 18 November 2021; 599:351.

Eisenstein M. Closing in on a complete human genome. *Nature*. 2021 Feb; 590(7847):679-681. doi: 10.1038/d41586-021-00462-9. PMID: 33619406.

Eisenstein M. Does the human lifespan have a limit? *Nature*. 2022 Jan; 601(7893):S2-S4. doi: 10.1038/d41586-022-00070-1. PMID: 35046588.

Endy D. Should we synthesize a human genome? *MIT Libraries*. 10 May 2016. http://hdl.handle.net/1721.1/102449.

Fields S, Johnston M. Cell biology. Whither model organism research? *Science*. 2005 Mar 25; 307(5717):1885-6. doi: 10.1126/science.1108872. PMID: 15790833.

Firth, J.R. (1957), Applications of General Linguistics. *Transactions of the Philological Society*, 56: 1-14. https://doi.org/10.1111/j.1467-968X.1957.tb00568.x

FitzGerald G et al. The future of humans as model organisms. *Science*. 2018 Aug 10; 361(6402):552-553. doi: 10.1126 /science.aau7779. PMID: 30093589.

Georgirenes J. *A description of the present state of Samos, Nicaria, Patmos, and Mount Athos*. WG Moses Pitt (1678).

Gershman A et al. Epigenetic patterns in a complete human genome. *Science*. 2022 Apr; 376(6588):eabj5089. doi: 10.1126/science.abj5089. Epub 2022 Apr 1. PMID: 35357915; PMCID: PMC9170183.

Głowacka K et al. Photosystem II Subunit S overexpression increases the efficiency of water use in a field-grown crop. *Nature Communications*. 2018 Mar 6;9(1):868. doi: 10.1038/s41467-018-03231-x. PMID: 29511193; PMCID: PMC5840416.

Hawking, S. Science in the Next Millennium. Remarks by Stephen Hawking. The Second Millennium Evening at The White House March 6, 1998. https://web.math.utk.edu/~vasili/va/files/whitehouseEvening/2.hawking.mar98.html.

Hayes T. et al. Simulating 500 million years of evolution with a language model. *bioRxiv* [Preprint]. 2 July 2024. doi: https://doi.org/10.1101/2024.07.01.600583.

Hessel A. Time for another human genome project? *Huffington Post*. 14 May 2012. https://www.huffpost.com/entry/human-genome_b_1345842.

Hoshika S et al. Hachimoji DNA and RNA: A genetic system with eight building blocks. *Science*. 2019 Feb 22; 363(6429):884-887. doi: 10.1126/science.aat0971. PMID: 30792304; PMCID: PMC6413494.

Kamitaki N et al. Complement genes contribute sex-biased vulnerability in diverse disorders. *Nature*. 2020 Jun; 582(7813):577-581. doi: 10.1038/s41586-020-2277-x. Epub 2020 May 11. PMID: 32499649; PMCID: PMC7319891.

Kenyon C. Sydney Brenner (1927-2019). *Science*. 2019 May 17; 364(6441):638. doi: 10.1126/science.aax8563. PMID: 31097656.

Kini AR, Opal P. The brave new world of the synthetic genome. *Time*. 3 June 2016. https://time.com/4356581/synthetic-human-genome/.

Kipling D, Davis T, Ostler EL, Faragher RG. What can progeroid syndromes tell us about human aging? *Science*. 2004 Sep 3; 305(5689):1426-31. doi: 10.1126/science.1102587. PMID: 15353794.

Koblan LW et al. In vivo base editing rescues Hutchinson-Gilford progeria syndrome in mice. *Nature*. 2021 Jan; 589(7843):608-614. doi: 10.1038/s41586-020-03086-7. Epub 2021 Jan 6. PMID: 33408413; PMCID: PMC7872200.

Kolora SRR et al. Origins and evolution of extreme life span in Pacific Ocean rockfishes. *Science*. 2021 Nov 12; 374(6569):842-847. doi: 10.1126/science.abg5332. Epub 2021 Nov 11. PMID: 34762458; PMCID: PMC8923369.

Kupferschmidt K. Forever young? Naked mole rats may know the secret. *Science*. 2018 Feb 2; 359(6375):506-507. doi: 10.1126/science.359.6375.506. Epub 2018 Feb 1. PMID: 29420271.

Leshner AI. Behavioral science comes of age. *Science*. 2007 May 18; 316(5827):953. doi: 10.1126/science.1144897. PMID: 17510328.

Liu H et al. Single-cell DNA methylome and 3D multi-omic atlas of the adult mouse brain. *Nature*. 2023 Dec; 624(7991):366-377. doi: 10.1038/s41586-023-06805-y. Epub 2023 Dec 13. PMID: 38092913; PMCID: PMC10719113.

Lovell JT, Grimwood J. The road to accurate and complete human genomes. *Nature*. 16 Jun 2022; 606(7914):1-2. doi: 10.1038/d41586-022-01368-w.

Meng F, Ellis T. The second decade of synthetic biology: 2010-2020. *Nature Communications*. 2020 Oct 14; 11(1):5174. doi: 10.1038/s41467-020-19092-2. PMID: 33057059; PMCID: PMC7560693.

Merchant, A.T., King, S.H., Nguyen, E. et al. Semantic design of functional de novo genes from a genomic language model. *Nature* (2025). https://doi.org/10.1038/s41586-025-09749-7

Nurk S et al. The complete sequence of a human genome. *Science*. 2022 Apr; 376(6588):44-53. doi: 10.1126/science.abj6987. Epub 2022 Mar 31. PMID: 35357919; PMCID: PMC9186530.

Parreno V et al. Transient loss of Polycomb components induces an epigenetic cancer fate. *Nature*. 2024 May; 629(8012):688-696. doi: 10.1038/s41586-024-07328-w. Epub 2024 Apr 24. PMID: 38658752; PMCID: PMC11096130.

Peacock A. Can proteins be truly designed sans function? *Science*. 2020 Sep 4; 369(6508):1166-1167. doi: 10.1126/science.abd4791. PMID: 32883850.

Pennisi E. Most complete human genome yet is revealed. *Science*. 2022 Apr; 376(6588):15-16. doi: 10.1126/science.abq2802. Epub 2022 Mar 31. PMID: 35357931.

Perkel JM. Cell engineering: How to hack the genome. *Nature*. 2017 Jul 26; 547(7664):477-479. doi: 10.1038/547477a. PMID: 28748939.

Pietri P, Papaioannou T, Stefanadis C. Environment: An old clue to the secret of longevity. *Nature*. 2017 Apr 26; 544(7651):416. doi: 10.1038/544416e. PMID: 28447636.

Pollack A. Scientists talk privately about creating a synthetic human genome. *New York Times*. 13 May 2016.

Ptolemy. *Algamest*. Princeton University Press (1998). https://doi.org/10.2307/j.ctvzxx967.

Radford T, Davis N. Scientists launch proposal to create synthetic human genome. *Guardian*. 2 June 2016. https://www.theguardian.com/science/2016/jun/02/scientists-launch-proposal-to-create-synthetic-human-genome-dna.

Reardon S. A complete human genome sequence is close: how scientists filled in the gaps. *Nature*. 2021 Jun; 594(7862):158–159. doi: 10.1038/d41586-021-01506-w. PMID: 34089035.

Rivest S. Pruned to perfection. *Nature*. 1 November 2018; 563:42–43.

Robinson, N. E., Zhang, W., Ghosh, R., Gerber, B., Zhang, H., Sanfiorenzo, C., Wang, S., Di Carlo, D., Wang, K. Construction of complex and diverse DNA sequences utilizing DNA 3-Way junctions. (accepted by *Nature*). 2025.

Salinger JD. *The Catcher in the Rye*. Little, Brown and Company (1951).

Schizophrenia Working Group of the Psychiatric Genomics Consortium. Biological insights from 108 schizophrenia-associated genetic loci. *Nature*. 2014 Jul 24; 511(7510):421-7. doi: 10.1038/nature13595. Epub 2014 Jul 22. PMID: 25056061; PMCID: PMC4112379.

Sekar A et al. Schizophrenia risk from complex variation of complement component 4. *Nature*. 2016 Feb 11; 530(7589):177–83. doi: 10.1038/nature16549. Epub 2016 Jan 27. Erratum in: *Nature*. 2022 Jan; 601(7892):E4–E5. doi: 10.1038/s41586-021-04202-x. PMID: 26814963; PMCID: PMC4752392.

Service RF. Rules of the game. *Science*. 2016 Jul 22; 353(6297):338–41. doi: 10.1126/science.353.6297.338. PMID: 27463654.

Servick K. Genome writing project confronts technology hurdles. *Science*. 2017 May 19; 356(6339):673–674. doi: 10.1126/science.356.6339.673. PMID: 28522476.

Sulak M et al. *TP53* copy number expansion is associated with the evolution of increased body size and an enhanced DNA damage response in elephants. *Elife*. 2016 Sep 19; 5:e11994. doi: 10.7554/eLife.11994. Erratum in: *Elife*. 2016 Dec 20; 5:e24307. doi: 10.7554/eLife.24307. PMID: 27642012; PMCID: PMC5061548.

Thomas JH. The mind of a worm. *Science*. 1994 Jun 17; 264(5166):1698-9. doi: 10.1126/science.7911601. PMID: 7911601.

Venter, JC. *Life at the Speed of Light: From the Double Helix to the Dawn of Digital Life*. Viking Penguin, New York October 2013.

Venter JC, Glass JI, Hutchison CA 3rd, Vashee S. Synthetic chromosomes, genomes, viruses, and cells. *Cell*. 2022 Jul 21;185(15):2708-2724. doi: 10.1016/j.cell.2022.06.046. PMID: 35868275; PMCID: PMC9347161.

Vermeij WP, Hoeijmakers JHJ. Base editor repairs mutation found in the premature-ageing syndrome progeria. *Nature*. 2021 Jan; 589(7843): 522–524. doi: 10.1038/d41586-020-03573-x. PMID: 33408358.

Watson JL et al. De novo design of protein structure and function with RFdiffusion. *Nature*. 2023 Aug; 620(7976):1089–1100. doi: 10.1038/

s41586-023-06415-8. Epub 2023 Jul 11. PMID: 37433327; PMCID: PMC10468394.

Wojcik MH, Larkin K, Cipicchio M, Doupnik A, Zhao C, Cech C, Lopez D, Chandrasekar J, Leadbetter J, Mannion J, Berg K, Golkaram M, Osentowski M, Freer M, Lehmann T, Lee WM, Ormbrek E, Prindle MJ, Nabavi M, Chaturvedi A, Seberino C, Baker DN, Williams C, Toledo D, Malolepsza E, Fleharty M, Oza A, Low S, Beggs AH, Genetti CA, Strickland G, Anderson KN, Chung WK, Rehm HL, Hofherr S, Kokoris M, Lennon N. Toward Same-Day Genome Sequencing in the Critical Care Setting. N Engl J Med. 2025 Oct 15. doi: 10.1056/NEJMc2512825. Epub ahead of print. PMID: 41091060.

Yao Z et al. A high-resolution transcriptomic and spatial atlas of cell types in the whole mouse brain. *Nature*. 2023 Dec; 624(7991):317-332. doi: 10.1038/s41586-023-06812-z. Epub 2023 Dec 13. PMID: 38092916; PMCID: PMC10719114.

Young E. Now that we can read genomes, can we write them? *The Atlantic*. 10 May 2017.

Zhu ML, Schmotzer C. Writing the Genome: Are We Ready? *Clinical Chemistry*. 2017 Apr; 63(4):929-930. doi: 10.1373/clinchem.2016.270066. PMID: 28351860.

Zürcher JF et al. Continuous synthesis of E. coli genome sections and Mb-scale human DNA assembly. *Nature*. 2023 Jul; 619(7970):555-562. doi: 10.1038/s41586-023-06268-1. Epub 2023 Jun 28. PMID: 37380776; PMCID: PMC7614783.

CHAPTER 12

Adam D. How far will global population rise? Researchers can't agree. *Nature*. 2021 Sep; 597(7877):462-465. doi: 10.1038/d41586-021-02522-6. PMID: 34548645.

Alberts B. A Grand Challenge in Biology. *Science*. 2011; 333:1200. doi: 10.1126/science.1213238.

Alberts B. Is the frontier really endless? *Science*. 2010 Dec 17; 330(6011):1587. doi: 10.1126/science.1201539. PMID: 21163969.

An B et al. Engineered Living Materials For Sustainability. *Chemical Reviews*. 2023 Mar 8; 123(5):2349-2419. doi: 10.1021/acs.chemrev.2c00512. Epub 2022 Dec 13. PMID: 36512650.

Anderson JB. Clonal evolution and genome stability in a 2500-year-old fungal individual. *Proceedings of the Royal Society B: Biological Sciences*. 2018 19 Dec; 285(1893):20182233. Published online 19 December 2018. Doi. 10.1098/rspb.2018.2233.

Angrist N, Sacerdote B. The social links that shape economic prospects. *Nature*. 4 August 2022; 608:37.

Baker D, Church G. Protein design meets biosecurity. *Science*. 2024 Jan 26; 383(6681):349. doi: 10.1126/science.ado1671. Epub 2024 Jan 25. PMID: 38271530.

Ball P. Starting from scratch. *Nature*. 7 October 2004; 431:624.

Ball P. Is technology unnatural? *Nature*. 29 November 2007; 450:614.

Barabási AL, Gulbahce N, Loscalzo J. Network medicine: a network-based approach to human disease. *Nature Reviews Genetics*. 2011 Jan; 12(1):56–68. doi: 10.1038/nrg2918. PMID: 21164525; PMCID: PMC3140052.

Bioengineer.org. Designing Functional Genes with Genomic AI. 20 November, 2025. https://bioengineer.org/designing-functional-genes-with-genomic-ai/

Berkley S, Fore H. Health for all. *Science*. 2019; 364:309. doi: 10.1126/science.aax7591.

Bray D. Reasoning for results. *Nature*. 2001 Aug 30; 412(6850):863. doi: 10.1038/35091132. PMID: 11528456.

Callaway E. Synthetic biologists and conservationists open talks. *Nature*. 2013 Apr 18; 496(7445):281. doi: 10.1038/496281a. PMID: 23598317.

Callaway E. Could AI-designed proteins be weaponized? Scientists lay out safety guidelines. *Nature*. 2024 Mar; 627(8004):478. doi: 10.1038/d41586-024-00699-0. PMID: 38459345.

Calles J, Justice I, Brinkley D, Garcia A, Endy D. Fail-safe genetic codes designed to intrinsically contain engineered organisms. *Nucleic Acids Research*. 2019 Nov 4; 47(19):10439–10451. doi: 10.1093/nar/gkz745. Erratum in: *Nucleic Acids Research*. 2022 Jan 25; 50(2):1199. doi: 10.1093/nar/gkab1285. PMID: 31511890; PMCID: PMC6821295.

Ceballos G, Ehrlich PR. The misunderstood sixth mass extinction. *Science*. 2018 Jun 8; 360(6393):1080–1081. doi: 10.1126/science.aau0191. PMID: 29880679.

Chari R, Church GM. Beyond editing to writing large genomes. *Nature Reviews Genetics*. 2017 Dec; 18(12):749–760. doi: 10.1038/nrg.2017.59. Epub 2017 Aug 30. PMID: 28852223; PMCID: PMC5793211.

Check Hayden E. Promising gene therapies pose million-dollar conundrum. *Nature*. 2016 Jun 16; 534(7607):305–6. doi: 10.1038/534305a. Erratum in: *Nature*. 2016 Jun 22; 534(7608):449. doi: 10.1038/534449a. PMID: 27306167.

Chin JW. Expanding and reprogramming the genetic code. *Nature*. 2017 Oct 4; 550(7674):53–60. doi: 10.1038/nature24031. PMID: 28980641.

Cho MK, Relman DA. Genetic technologies. Synthetic 'life', ethics, national security, and public discourse. *Science*. 2010 Jul 2; 329(5987):38–9. doi: 10.1126/science.1193749. PMID: 20595601.

Cohen J. Where does embryo editing stand now? *Science*. 2022 Mar 25; 375(6587):1328–1329. doi: 10.1126/science.abq1707. Epub 2022 Mar 24. PMID: 35324278.

Collins J. Q&A: Circuit capacity. *Nature*. 2012; 483:S11. https://doi.org/10.1038/483S11a.

Community Values, Guiding Principles, and Commitments for the Responsible Development of AI for Protein Design. Responsible AI x Biodesign. 8 March, 2024. https://responsiblebiodesign.ai

Costanza R. To build a better world, stop chasing economic growth. *Nature*. 2023 Dec; 624(7992):519-521. doi: 10.1038/d41586-023-04029-8. PMID: 38123804.

Crist E. Reimagining the human. *Science*. 2018 Dec 14; 362(6420): 1242-1244. doi: 10.1126/science.aau6026. PMID: 30545872.

Doerr A. Encoding the unnatural. *Nature Methods*. 2010 May; 7(5):343. doi: 10.1038/nmeth0510-343. PMID: 20440879.

Editorial. Policing ourselves. *Nature*. 2006 May 25; 441(7092):383. doi: 10.1038/441383a. PMID: 16724020.

Editorial. The inspiring story of the Tara and its 20-year message from the corals. *Nature*. 2023 Jun; 618(7964):213. doi: 10.1038/d41586-023-01826-z. PMID: 37280289.

Editorial: Tribal gathering. *Nature*. 2014 May 8; 509(7499):133. doi: 10.1038/509133a. PMID: 24812679.

Editorial. Bottom-up biology. *Nature*. 2018 Nov; 563(7730):171. doi: 10.1038/d41586-018-07290-4. PMID: 30405231.

Editorial. Tackle sickle-cell economics. *Nature*. 5 Dec 2019; 576(7785): 7-8. doi: 10. 1038/d41586-019-03709-8.

Editorial. The best medicine for improving global health? Reduce inequality. *Nature*. 11 July 2023; 19:221.

Editorial. Hunger and famine are not accidents – they are created by the actions of people. *Nature*. 2023 Jul; 619(7968):8. doi: 10.1038/d41586-023-02207-2. PMID: 37402800.

Editorial. Research funders must join the fight for equal access to medicines. *Nature*. 2024 Feb; 626(7997):7-8. doi: 10.1038/d41586-024-00237-y. PMID: 38291143.

Editorial. Nature is our most precious asset – we must all act now to save it. *World Economic Forum*. 2 February 2021. https://www.weforum.org/agenda/2021/02/nature-is-our-most-precious -asset-form-of-wealth-and-safety-net/.

Editorial. Meanings of 'life'. *Nature*, 28 June 2007; 447(7148): 1031.

Ellis GF. Physics, complexity and causality. *Nature*. 2005 Jun 9; 435(7043): 743. doi: 10.1038/435743a. PMID: 15944681.

Elowitz M, Lim WA. Build life to understand it. *Nature*. 2010 Dec 16; 468(7326):889-90. doi: 10.1038/468889a. PMID: 21164460; PMCID: PMC3068207.

Endy D, Moront S, Alexopoulos VA, Patel R, Jain R, Bennett B. Biosecurity Really: A Strategy for Victory. Hoover Institution. 8 October 2025. https://www.hoover.org/research/biosecurity-really-strategy-victory

Ferreira FHG. Not all inequalities are alike. *Nature*. 2022 Jun; 606(7915): 646-649. doi: 10.1038/d41586-022-01682-3. PMID: 35732768.

Gilbert N. Funding battles stymie ambitious plan to protect global biodiversity. *Nature*. 2022 Mar 31. doi: 10.1038/d41586-022-00916-8. Epub ahead of print. PMID: 35361944.

Goldman MA. Biotechnology and the human soul. *Science*. 20 October 2006; 314:423.

Goldman MA. Probing the genetic future of humanity. *Science*. 2021; 371:789. doi: 10.1126/science.abg0509.

Gopalakrishnan V et al. Gut microbiome modulates response to anti-PD-1 immunotherapy in melanoma patients. *Science*. 2018; 359:97-103. doi: 10.1126/science.aan4236.

Green ED, Guyer MS, National Human Genome Research Institute. Charting a course for genomic medicine from base pairs to bedside. *Nature*. 2011 Feb 10; 470(7333):204-13. doi: 10.1038 /nature09764. PMID: 21307933.

Haldane JBS. *What is Life?* New York: Boni and Gaer (1947).

Herridge V. Before making a mammoth, ask the public. *Nature*. 2021 Oct; 598(7881):387. doi: 10.1038/d41586-021-02844-5. PMID: 34671132.

Hockfield S. Our science, our society. *Science*. 2018 Feb 2; 359(6375):499. doi: 10.1126/science.aat0957. Epub 2018 Feb 1. PMID: 29420266.

Humphreys S. How to define unjust planetary change. *Nature*. 2023 Jul; 619(7968):35-36. doi: 10.1038/d41586-023-01743-1. PMID: 37258728.

Jasechko S, Perrone D. Global groundwater wells at risk of running dry. *Science*. 2021 Apr 23; 372(6540):418-421. doi: 10.1126/science.abc2755. PMID: 33888642.

Joyce GF. Evolution. Toward an alternative biology. *Science*. 2012 Apr 20; 336(6079):307-8. doi: 10.1126/science.1221724. PMID: 22517850.

Kamitaki N et al. Complement genes contribute sex-biased vulnerability in diverse disorders. *Nature*. 2020 Jun; 582(7813):577-581. doi: 10.1038 /s41586-020-2277-x. Epub 2020 May 11. PMID: 32499649; PMCID: PMC7319891.

Kavanagh, K. World's first AI-designed viruses a step towards AI-generated life. *Nature* 646, 16 (2025). doi: https://doi.org/10.1038/d41586-025 -03055-y.

King, SH, Driscoll, CL, Li DB, Guo D, Merchant AT, Brixi G, Wilkinson ME, Hie BL. Generative design of novel bacteriophages with genome language models bioRxiv 2025.09.12.675911; doi: https://doi.org /10.1101/2025.09.12.675911

King, S. Hie B. How We Built The First AI-Generated Genomes. 17 September, 2025. Arc Institute. https://arcinstitute.org/news/hie-king -first-synthetic-phage

King BJ. Our entangled lives. *Science*. 2022; 378:840. doi: 10.1126/science .ade7859.

Knoppers BM, Greely HT. Biotechnologies nibbling at the legal 'human'. *Science*. 2019 Dec 20; 366(6472):1455-1457. doi: 10.1126/science.aaz5221. PMID: 31857473.

Koblan LW et al. In vivo base editing rescues Hutchinson-Gilford progeria syndrome in mice. *Nature*. 2021 Jan; 589(7843):608-614. doi: 10.1038/s41586-020-03086-7. Epub 2021 Jan 6. PMID: 33408413; PMCID: PMC7872200.

Lanphier E et al. Don't edit the human germ line. *Nature*. 2015; 519: 410–411. https://doi.org/10.1038/519410a.

Lenski RE. Twice as natural. *Nature*. 2001 Nov 15; 414(6861):255. doi: 10.1038/35104715. PMID: 11713507.

Ley R. The human microbiome: there is much left to do. *Nature*. 2022 Jun; 606(7914):435. doi: 10.1038/d41586-022-01610-5. PMID: 35705822.

Lin, F. AI Designs Viable Bacteriophage Genomes, Combats Antibiotic Resistance. Genetic Engineering & Biotechnology News. 17 September, 2025. https://www.genengnews.com/topics/artificial-intelligence/ai-designs-viable-bacteriophage-genomes-combats-antibiotic-resistance/

Ma S, Hong Y, Chen J, Xu J, Shen X. Single-cell nascent transcription reveals sparse genome usage and plasticity. *Cell*. 2025 Sep 26:S0092-8674(25)01034-7. doi: 10.1016/j.cell.2025.09.003. Epub ahead of print. PMID: 41015043.

Madruga RP. Linking climate and biodiversity. *Science*. 2021 Oct 29; 374(6567):511. doi: 10.1126/science.abm8739. Epub 2021 Oct 28. PMID: 34709913.

Malcom S. Strengthen the case for DEI. *Science*. 2024 Mar 29; 383(6690): 1395. doi: 10.1126/science.adp4397. Epub 2024 Mar 28. PMID: 38547281.

Maoz U. Freedom from free will. *Science*. 2023; 382:163. doi: 10.1126/science.adk1277.

Mascher M, Jayakodi M, Shim H and Stein N. Promises and challenges of crop translational genomics. *Nature*. 2024 Dec 19/26; 636:585-593. doi .org/10.1038/s41586-024-07713-5.

McCarty, N. AI-Designed Phages. Asimov Press (2025). https://doi .org/10.62211/21er-45fg.

Mills C. When is sorrow sickness? A history of depression. *Nature*. November 2020; 587(7835): 541-542. doi: 10.1038/d41586-020-03286-1.

Medawar, P. *The Hope of Progress*. Routledge. 1972. ISBN 9781032118079

Merchant, A.T., King, S.H., Nguyen, E. et al. Semantic design of functional de novo genes from a genomic language model. *Nature* (2025). https://doi.org/10.1038/s41586-025-09749-7

Muñoz JM. Achieving cognitive liberty. *Science*. 2023; 379:1097. doi: 10.1126/science.adf8306.

Napolitano R et al. Smart cities built with smart materials. *Science*. 2021; 371:1200-1201. doi: 10.1126/science.abg4254.

Noble D. Genes are not the blueprint of life. *Nature*. 8 February 2024; 626:254. doi: 10.1038/d41586-024-00327-x.

Noble D. Genes and causation. *Philosophical Transactions of the Royal Society A: Mathematical, Physical and Engineering Sciences*. 2008 Sep 13; 366(1878):3001-15. doi: 10.1098/rsta.2008.0086. PMID: 18559318.

Normile D. China sets new ethics rules for human studies. *Science*. 2023 Mar 10; 379(6636):970. doi: 10.1126/science.adh4963. Epub 2023 Mar 9. PMID: 36893237.

Nurse P. Biology must generate ideas as well as data. *Nature*. 2021 Sep; 597(7876):305. doi: 10.1038/d41586-021-02480-z. PMID: 34522015.

O'Gorman EJ. Machine learning ecological networks. *Science*. 2022 Aug 26; 377(6609):918-919. doi: 10.1126/science.add7563. Epub 2022 Aug 25. PMID: 36007050.

Paez S et al. Reference genomes for conservation. *Science*. 2022 Jul 22; 377(6604):364-366. doi: 10.1126/science.abm8127. Epub 2022 Jul 21. PMID: 35862547.

Pennisi E. Sequencing all life captivates biologists. *Science*. 2017 Mar 3; 355(6328):894-895. doi: 10.1126/science.355.6328.894. PMID: 28254891.

Peyregne S, Massilani D, Swiel Y, Boyle MJ, Iasi LNM, Suemer AP, Mesa AB, de Filippo C, Viola B, Essel E, Nagel S, Richter J, Weihmann A, Schellbach B, Zeberg H, Visagie J, Kozlikin MB, Shunkov MV, Derevianko AP, Pruefer K, Peter BM, Meyer M, Paabo S, Kelso J. A high-coverage genome from a 200,000-year-old Denisovan bioRxiv 2025.10.20.683404; doi: https://doi.org/10.1101/2025.10.20.683404

Pimm S. Edward O. Wilson (1929-2021). *Science*. 2022 Jan 28; 375(6579): 385. doi: 10.1126/science.abn9848. Epub 2022 Jan 27. PMID: 35084972.

Prüfer, K., Racimo, F., Patterson, N. et al. The complete genome sequence of a Neanderthal from the Altai Mountains. *Nature* 505, 43–49 (2014). https://doi.org/10.1038/nature12886

Ramachandran V. Blanket bans on fossil fuels hurt women and lower -income countries. *Nature*. 2022 Jul; 607(7917):9. doi: 10.1038/d41586 -022-01821-w. PMID: 35790826.

Reardon S. Global summit reveals divergent views on human gene editing. *Nature*. 2015 Dec 10; 528(7581):173. doi: 10.1038/528173a. PMID: 26659159.

Responsible AI x Biodesign. Community Values, Guiding Principles, and Commitments for the Responsible Development of AI for Protein Design. *Responsible AI x Biodesign*. 8 March, 2024. https://responsiblebiodesign.ai

Rhie A. Towards complete and error-free genome assemblies of all vertebrate species. *Nature*. 2021 Apr; 592(7856):737-746. doi: 10.1038/s41586-021-03451-0. Epub 2021 Apr 28. PMID: 33911273; PMCID: PMC8081667.

Rockström J et al. Safe and just Earth system boundaries. *Nature*. 2023 Jul; 619(7968):102–111. doi: 10.1038/s41586-023-06083-8. Epub 2023 May 31. PMID: 37258676; PMCID: PMC10322705.

Schmidt M, de Lorenzo V. Synthetic constructs in/for the environment: managing the interplay between natural and engineered Biology. *FEBS Letters*. 2012 Jul 16; 586(15):2199-2206. doi: 10.1016/j.febslet.2012.02.022. Epub 2012 Feb 23. PMID: 22710182; PMCID: PMC3396840.

Service RF. Bacteria stitch exotic building blocks into novel proteins. *Science*. 2024 Jan 19; 383(6680):247. doi: 10.1126/science.ado0968. Epub 2024 Jan 18. PMID: 38236976.

Service RF. Could chatbots help devise the next pandemic virus? *Science*. 2023 Jun 23; 380(6651):1211. doi: 10.1126/science.adj3377. Epub 2023 Jun 22. PMID: 37347875.

Service RF. Benchtop DNA synthesis raises alarms. *Science*. 2023 May 19; 380(6646):677. doi: 10.1126/science.adi7654. Epub 2023 May 18. PMID: 37200421.

Smith Hughes S. The cloning controversy. *Science*. 2015; 349:1292–1293. doi: 10.1126/science.aac9095.

Stanley M. Embrace the improbability of existence. *Science*. 2023; 381:842. doi: 10.1126/science.adj0232.

Thorp HH. To solve climate, first achieve peace. *Science*. 2022 Apr; 376(6588):7. doi: 10.1126/science.abq2761. Epub 2022 Mar 31. PMID: 35357944.

Thorp HH. The frontier is not endless for all. *Science*. 2021; 372:547. doi: 10.1126/science.abj2583.

Thorp HH. The costs of secrecy. *Science*. 2020; 367:959. doi: 10.1126 /science.abb4420.

Toews R. The next frontier for large language models in biology. *Forbes*. 16 July 2023. https://www.forbes.com/sites/robtoews/2023/07/16 /the-next-frontier-for-large-language-models-is-biology/.

Turner MS. Extremely large telescopes at risk. *Science*. 2023 Nov 24; 382(6673):857. doi: 10.1126/science.adm9964. Epub 2023 Nov 23. PMID: 37995214.

Turner D, Kropinski AM, Adriaenssens EM. A Roadmap for Genome-Based Phage Taxonomy. *Viruses*. 2021 Mar 18;13(3):506. doi: 10.3390 /v13030506. PMID: 33803862; PMCID: PMC8003253.

Wilson EO. *Half-Earth: Our Planet's Fight For Life*. Liveright Publishing Corporation (2016).

Woolfson A. Life-changing biology. *Science*. 2022; 375:982. doi: 10.1126 /science.abm9852.

Woolfson A. Synthetic life. *Daedalus*. 1 January 2008; 137(1):77–83. https:// doi.org/10.1162/daed.2008.137.1.77.

Woolfson A. The future of life. *Science*. 2024; 384:1074 . doi: 10.1126/science .adp9123.

Woolfson A. How the Human Animal Found Its Self. *Science*. 2004; 304:1248–1249. doi: 10.1126/science.1097381.

Woolfson A. So much for genes. *London Review of Books*. 8 March 2001; 23(5):7–8.

Woolfson A. Defining the Rules of Biology in the Age of Genome Writing. *GEN Biotechnology*. December 2023; 2(6):464–469. doi: 10.1089/genbio.2023.29125.awo.

Glossary

Adaptationist programme – a conceptual approach in evolutionary biology that routinely attributes adaptive functions to traits on the basis that each has been individually selected because it furnishes a selective evolutionary advantage.

Adeno-associated virus (AAV) – a harmless, laboratory-modified virus used to deliver healthy full-length genes and gene-editing entities such as CRISPR components to humans for therapeutic purposes.

AI-informed genome design – the use of AI to design synthetic genomes from scratch, or to rewrite, refactor or recode them.

AlphaFold – a computer programme developed by the company DeepMind that uses AI, in many cases but not always, to accurately predict the 3D shapes (structures) of proteins for which no experimentally determined structure currently exists from their amino acid sequence.

Amino acids – the building blocks of proteins. Proteins are made from amino acids chemically linked together into unique sequences. Their order determines the nature of the protein. Most proteins are made up from combinations of 20 different amino acids, but (rarely) some incorporate 2 additional amino acids.

Artificial Biological Intelligence (ABI) – a term invented by the author and first used in a paper published in 2025. It is the biological analogue of AGI. It comprises the ability to predict, generate, construct and boot up synthetic genomes coding for natural and unnatural species.

Artificial Biological Superintelligence (ABS) – a hypothetical form of artificial biological intelligence that surpasses the generative competence of natural biology through the use of alternative chemistries – including non-canonical amino acids – that are not utilised in nature.

Artificial General Intelligence (AGI) – a hypothetical type of AI with an intelligence equivalent to that of humans.

Artificial genomes – synthetic genomes designed and constructed on demand according to pre-specified designs.

Artificial intelligence (AI) – the ability of computing machines to simulate human intelligence and perform tasks requiring intelligence. It is facilitated by architectures such as CPUs and GPUs.

Artificial neural network – a computational model built using artificial neurons that mimic the organisation of neurons in the human brain. It comprises layers of interconnected 'nodes' that process information in a stepwise manner. It can extract patterns from complex data and make decisions and predictions based on its training.

Artivolution – a term invented by the author to describe a generative process in which species are designed and synthesised artificially according to human intent using genome design and construction methods. Unlike natural evolution, it is not dependent on 'descent' and the process of heredity.

Artificial Superintelligence (ASI) – is a hypothetical form of artificial intelligence that surpasses human intelligence in every domain. It represents a level of artificial intelligence beyond Artificial General Intelligence (AGI). ASI would outperform human minds in all endeavours, both scientific and artistic.

Authoring genomes – the creative process of writing entirely new genomes describing new functional species, much in the same way that authors write new works of literature. It relies on the ability to comprehend the generative grammar of biology and to construct genomes at scale.

Autosomal dominant disease – a disease that manifests if only a single copy of a mutant gene, inherited from the mother or father, is present.

Autosomal recessive disease – a disease that requires two abnormal copies of a mutant gene to be present for the pathological phenotype to be present.

Bacteriophage T7 – one of the best-characterised phages. A bacteriophage is a type of virus that infects and destroys bacteria. The genome of bacteriophage T7 was the first to be refactored.

Base editing – a type of gene editing that changes individual nucleotides letters in a DNA sequence. Unlike CRISPR, this is achieved without cutting the DNA.

Binary code – the language used by computer programmes. It uses combinations of just two states – 0 and 1 – to encode information. It forms the foundation for all digital information storage systems.

Biosecurity – the methods, procedures and policies put in place to protect against and prevent the generation and deployment of species that are potentially harmful to humankind and the species they rely on for their survival.

Biotechnology – an enterprise focused on transforming biological systems into technologies with multiple utilities, ranging from healthcare and agriculture to materials science.

Bioterrorism – the intentional release of hazardous biological material to cause deliberate harm and disruption to humans and the species they rely on for their survival, with a view to causing fear and achieving ideological and/or political goals.

Biowarfare – the use of synthetic biology, genome synthesis and the ability to manipulate biological systems for the purposes of warfare. This includes the design and construction of viruses and bacteria known to be harmful to humankind and the species necessary for their well-being and survival.

Bit of information – the smallest unit of information in computing; it can have a value of either 1 or 0.

Boolean logic – a logical system that enables decision-making processes according to simple binary 'yes' or 'no' rules. In an AND gate, for example, a function is triggered if both condition 1 AND condition 2 are true.

Booting up a genome – a phrase referencing the process of transplanting a synthetic genome into a recipient cell and activating it, so that it functions as the cell's operating system and determines its behaviour and functions.

C4A – the 'A' form of the complement C4 protein. It is a key part of the human innate immune system. It has been linked by GWAS to a predisposition to developing schizophrenia.

C4B – the 'B' form of the complement C4 protein, is closely related to the 'A' form. It is a key part of the human innate immune system.

CAR-T cell therapy – chimeric antigen receptor T-cell therapy. T-cells of the immune system are engineered to introduce a receptor that targets the T-cells to tumours, or in the case of autoimmune disease, to B-cells that produce pathological autoantibodies. CAR-T therapy can, in some, be curative.

Cas9 – CRISPR-associated protein 9. It is a protein that cuts DNA. It can be guided to a specific DNA sequence for use in gene editing by attaching it to CRISPR, to make a CRISPR-Cas9 construct.

Cell-as-computer – a metaphor that envisages the cell as operating like a digital computer.

Cell-as-machine – a metaphor that envisages the cell as operating like a human-made machine.

Cell therapy – a modality of clinical medical therapy in which the drug is living cells, which in some cases, such as CAR-Ts, have been genetically engineered. The cells may be derived from the patient or from a donor.

Central Processing Unit (CPU) – the core information-processing unit of a digital computer.

ChatGPT – a conversational AI chatbot. GPT is an acronym for 'Generative Pre-trained Transformer.' It uses a language model to comprehend and generate text in any written language and is trained on huge amounts of natural language data.

Chromatin – the complex of DNA and proteins found in the cell nucleus where the hereditary material is stored.

Chromosomes – thread-like structures found in the cell nucleus. They comprise very long pieces of DNA wound around proteins called histones. These are packed into more tightly folded structures. The human genome is distributed across 23 pairs of chromosomes, for a total of 46. The chromosome number varies between species. The full set of chromosomes contains the complete inventory of a cell's genetic information.

COBOL – Common Business-Oriented Language. Created in 1959, it is one of the earliest computer programming languages. This legacy software is still widely used, especially in banks and government computer systems.

Code script – a set of computer instructions written in a programming language that instructs the behaviour of a computer. It is also a metaphor for genetic sequences, which, by analogy, encode the instructions to direct the molecular activity of cells.

Codon – a triplet sequence of DNA letters (nucleotides) that specifies which amino acid a ribosome selects when synthesising the amino acid sequence of a protein according to the information in mRNA.

Complex diseases – diseases caused by small contributions from multiple genetic, environmental, lifestyle and other factors, such as the microbiome. Unlike simple, monogenic diseases, the causality cannot be attributed to a single faulty gene.

Convergence – the phenomenon whereby unrelated species independently arrive at similar evolutionary adaptive solutions to an environmental challenge.

CRISPR – an acronym for Clustered Regularly Interspaced Short Palindromic Repeats. It was discovered in bacteria, where it forms a primitive immune system. It has been repurposed to form the basis of the versatile CRISPR gene-editing tool, which – because of its simplicity, low cost and ease of use – has become the industry standard.

CRISPR-Ca9 – the synthetic molecule used for CRISPR-based gene editing. It has two components: a single guide RNA (sgRNA) and a Cas9 nuclease. The sgRNA comprises crRNA (CRISPR RNA), which guides the CRISPR to a genomic target, and a tracrRNA (trans-activating CRISPR RNA), which binds to the crRNA and helps it to recruit the Cas9.

Culturomics – a term invented by the author. It is a hypothetical – and potentially unrealisable – future area of study that may make it possible to infer, from a genome sequence, the broad kinds of cultures that a particular human species may, in principle, be capable of generating.

Debugging – the process of identifying and fixing the errors in computer software. The phrase is also used to describe the process of identifying and fixing 'glitches' in synthetic genomes – especially if they have been extensively redesigned, rewritten, recoded or refactored.

Decompressed genome – the situation where a genome is rewritten (refactored) to ensure that overlapping genes – such as those found in bacteriophage T7 – are no longer overlapping.

Deep learning – a type of artificial intelligence and machine learning that uses a multilayered neural network architecture called a transformer for parallel data processing. It can solve problems and identify complex patterns in data. 'Deep' refers to the multiple layers of the network. Deep learning has revolutionised natural language processing with applications like ChatGPT.

De-extinction – the concept of bringing extinct species back to life through whole genome synthesis of their genomes, or the genome editing of closely related species to mimic the sequence of an extinct genome: for example, transforming an elephant genome sequence into that of a woolly mammoth. It relies on the recovery of ancient reference DNA sequence information.

Deoxyribonucleic acid (DNA) – the heredity material that encodes life's genetic information. It comprises a combination of nucleotides and forms a double helical structure. Individual nucleotides comprise a sugar (deoxyribose), a nitrogenous base (A, T, C or G) and a phosphate group.

Deoxyribose – the sugar used in the nucleotide building blocks of DNA.

Dideoxy sequencing method – a method for determining the order of nucleotide bases (A, T, C and G) in DNA molecules and therefore its informational content. It was developed in 1977 by Fred Sanger in Cambridge, UK.

Dinucleotide – a DNA molecule comprising two nucleotides joined together. The first piece of synthetic DNA ever made was a dinucleotide made by Alexander Todd in Cambridge in 1955.

DNA assembly – the process by which short synthetic fragments of DNA are assembled into larger pieces of DNA. This sequential or parallel assembly process allows genomes to be constructed from small pieces of DNA. Several methods exist to do this.

DNA ligase – the enzyme that joins pieces of DNA together.

DNA polymerase – the enzyme that synthesises a new piece of DNA referencing the information in a pre-existing DNA template.

DNA sequence space – a theoretical concept that imagines every possible sequence of DNA of every possible length as being arrayed in an expansive DNA 'sequence space'.

DNA synthesis – the process by which DNA is chemically synthesised, either naturally in cells, or artificially using chemical or enzymatic synthesis methods.

Ecogenomics – a term invented by the author that describes a hypothetical future science that can predict the impact of the introduction of natural and artificial species on the dynamics of ecosystems and the species they contain.

Engineering biology – a term that addresses the science of genome design and synthesis. It treats biological systems as if they were programmable 'software' that can be reprogrammed at will to address specific outcomes and objectives.

Enzyme – a protein that selectively catalyses a specific chemical reaction, thereby enabling it to progress faster and more efficiently.

Epigenetic – references heritable chemical changes to DNA that alter gene expression but do not change the nucleotide sequences of DNA itself. They are like bookmarks that determine which parts of the DNA 'pages' may be read.

Epigenetic editing – a form of editing that changes the chemical markings on DNA while leaving the nucleotide sequence unchanged.

Epigenetic modification – a chemical modification to DNA that changes its activity without altering its nucleotide sequence. A common form of epigenetic modification is a chemical process called methylation, which silences the activity of a gene.

Epigenome – the complete collection of epigenetic marks on the DNA nucleotide sequence of a genome that influence which genes are expressed, and consequently a cell's behaviour and type.

Eukaryotic cells – the cells found in protists, fungi, plants and animals that have a nucleus and membrane-bound organelles, such as mitochondria. The (prokaryotic) cells of bacteria lack nuclei and membrane-bound organelles.

Evo – an AI foundational model that can write new DNA, RNA and protein sequences at genome scale. Although Evo has only a rudimentary ability to design whole genomes, its publication in November 2024 was a significant milestone for generative biology. The subsequent iteration of Evo, Evo 2, significantly improved its performance. The original Evo model is now known as Evo 1.

Evolutionary epistemology – as formulated by the University College London biologist and philosopher Henry Plotkin, envisages organisms as 'knowledge systems' involved in the acquisition, storage, perpetuation and transmission of information.

Extra-genetic information –the various layers of information that are not encoded genetically, but which contribute to the phenotype of an organism. These include: epigenetics, development, learning, culture, microbiome, nutrition and environment.

Foetal haemoglobin (HbF) – the form of haemoglobin used by pre-term and newborn infants. It enables the foetus to efficiently extract oxygen from the mother's blood. Following birth, the gene-encoding HbF is switched off, and HbF is replaced by adult haemoglobin (HbA).

Forward engineering – a philosophical approach that aims to comprehend biological systems through their construction, rather than by breaking them apart (reverse engineering).

Foundational models – AI machine-learning models that have been trained on vast amounts of data, and are able to perform a wide variety of general-purpose tasks. They form the core of many AI tools and can be fine-tuned to address specific tasks.

Fred's Library – a term invented by the author to describe the concept of a DNA sequence space. The philosopher Daniel Dennett referred to this as the Library of Mendel.

Fuzzy logic – unlike digital binary logic where all numerical values are either 1 or 0, in fuzzy logic, values are continuous, assuming values anywhere between 1 and 0.

Gedankenexperiment – German for a 'thought experiment', designed to explore the implications of a theory and highlight paradoxes.

Gene – traditionally taken to mean a sequence of DNA in a genome that contains instructions for making (encodes) a functional protein. We now know, however, that much of the genome is transcribed and makes RNA. The regions of DNA that encode functional RNA molecules should, therefore, also be termed (RNA-encoding) genes.

Gene editing – a technique for altering nucleotides in DNA sequences at prespecified locations. Several different technologies may be used to achieve this, including CRISPR-Cas9 and zinc finger nucleases (ZFNs).

Gene replacement therapy – a type of gene therapy where a whole (or partial) 'healthy' version of a gene is introduced into a patient's cells to compensate for the lack of function of a faulty (mutated) gene.

Gene therapy – a therapeutic strategy that aims to restore the function of a mutated gene through gene editing, gene replacement or gene silencing (inactivating a gene). Some forms of cell therapy (see CAR-T cells) are also a form of gene therapy.

Generative biology – the science of generating pre-specified unnatural DNA sequences on demand up to genome scale.

Generative grammar of biology – the hypothetical grammatical rules that naturally and eventually artificially construct biological 'utterances' much in the same way that the grammar of languages allows for the construction of coherent sentences that convey novel meaning.

Generative genomics – the new science of AI-mediated genome design and construction enabled by genome language models like Evo 2 that promises to facilitate the design and synthesis of genomes that do not exist in nature.

Genetic code – the set of rules used by living organisms to translate information encoded in their genetic material into the amino acid sequences of proteins. Each amino acid is coded by a three-nucleotide long 'codon'.

Genetic firewalling – a strategy for preventing the genetic information of synthetic organisms from mingling with that of natural organisms. The simplest way to achieve this is to recode the genome of a synthetic species.

Genome – the inventory of all the genetic material specifying an organism (or a non-living biological entity, such as a virus). Every human cell has a complete copy of the genome except egg and sperm cells, red blood cells (which lack DNA), and immune cells (where genes may be rearranged).

Genome construction/synthesis – the process of building an entire genome from scratch through the assembly of small synthetic DNA fragments.

Genome design/writing – the emerging science of intentionally redesigning and writing the sequence of a genome from scratch (typically using AI or other computational methods) to achieve a specific predefined goal.

Genome folding – the way in which DNA dynamically organises itself in three-dimensional space in the nucleus of eukaryotic cells. This likely plays a critical role in the execution of genetic 'programmes' through controlling protein- and RNA-encoding gene activity.

Genome Project-Write (GP-write) – an initiative announced in 2016 and led by George Church, Jef Boeke, Andrew Hessel and Nancy J. Kelley. It aimed to develop technologies that would facilitate the efficient synthesis of DNA at scale, including whole genomes. The project was originally named HGP-write, with the name change implemented to de-emphasise the (at the time) controversial focus on writing a human genome.

Genome recoding – rewriting the genome of an organism in a way that changes the nature of its genetic code. Codons can be reassigned so that they code for different (non-natural) amino acids.

Genomeverse – a term invented by the author to describe a theoretical mathematical space of all possible genomes.

Genome-Wide Association Study (GWAS) – a statistical method to link genetic variants in healthy and diseased human populations to specific phenotypes.

Genomic 'dark matter' – a phrase used to describe the non-coding genome, much of whose function is currently poorly understood.

Germline – in sexually reproducing organisms this refers to egg and sperm cells.

Germline cells – the cells that give rise to eggs and sperm; the cells that transmit heredity information. This contrasts with somatic cells, which do not transmit hereditary information.

Germline gene editing – the heritable editing of the genomes of egg or sperm cells. Unlike edits to somatic cells, which are not inherited, edits to germline cells are passed on to offspring and therefore comprise permanent alterations to the genetic material.

GFP – an acronym for 'green fluorescent protein', which is a naturally occurring protein found in the jellyfish *Aequorea victoria*. It exhibits green fluorescence when exposed to blue or UV light.

Gigabase (Gb) – a billion base pairs of DNA or RNA.

Graphics processing unit (GPU) – a specialised form of computer chip, originally intended for the rendering of graphics. It can rapidly perform vast numbers of parallel calculations and is ideally suited for handling complex computational tasks including the implementation of AI.

Greenhouse gases – atmospheric gases that trap heat by absorbing and emitting infrared radiation. The commonest greenhouse gases are water vapour, carbon dioxide, methane, nitrous oxide and ozone. Human activities have led to their buildup, which has resulted in global warming.

Guide RNA (gRNA) – a component of the CRISPR-Cas9 gene-editing system and guides Cas9 to a unique GPS-like nucleotide sequence 'postal' address within a genome, where it makes a precise cut (see CRISPR and CRISPR-Cas9).

Habitable exoplanet – a planet located outside our solar system that is thought to be potentially habitable.

Healthspan – the duration of an individual's life spent in good health and free from significant disease or disability.

Healthy longevity – the ability to extend lifespan while increasing healthspan in parallel.

Heredity – the transmission of genetic information in a species from one generation to another.

HGP-read – alternative (and less frequently used) names for the Human Genome Project, which emphasises that it was focused on sequencing, not writing.

Human gene kit – a phrase devised by the Cambridge biologist Sydney Brenner to describe the complete set of genes in a genome necessary to construct a particular organism.

Human genome exceptionalism – the tendency to attribute a unique and special status to biologically relevant information furnished at the genetic level, thereby elevating its importance with regard to other 'extra-genetic' sources of information that contribute to human nature.

Human Genome Project (HGP) – an international project commencing in 1990, and completed in its entirety in 2022, that successfully sequenced the human genome.

Humanised species – species in which nucleotide sequences in the genome are replaced with corresponding human nucleotide sequences in order to model human biology.

Hyperculture – a term invented by the author to describe the exponential increase in the velocity of human cultural change that has occurred as a result of human technological innovations. This unprecedented rate of change has served to accentuate the increasing disconnect between our received, ancient biology and contemporary culture and lifestyles.

Immunogenicity – the ability of a protein or other substances to evoke an immune response in the human body.

Interactome – is the network of all the biological molecules that interact with one another within a cell. While originally referencing interactions between different proteins, it can in its expanded definition include any biologically relevant molecule within a cell, including RNA.

***In utero* gene editing** – gene editing that is performed in a developing foetus before it is born. It can be either heritable (in germline cells) or somatic (non-inherited). This method may potentially be useful for individuals with known severe genetic diseases, which present early on in life.

Information space – a theoretical structure in which an extensive collection of information (of any nature) relevant to an organism's phenotype is arrayed, ordered and accessed.

Informiome – a term invented by the author to describe all the various layers of genetic and extra-genetic information that influence the phenotype of an organism.

Informiome Catalogue – a term invented by the author to describe the catalogue of all possible informiomes associated with every possible species.

Junk DNA – a (largely obsolete) term that was once used to refer to DNA nucleotide sequences that do not encode proteins, and have no known regulatory or other function. While some DNA may completely lack any known informational or structural function, it is not currently reasonable to reach such definitive conclusions. The preferred term for such sequences is non-coding DNA.

Just So Stories – references Rudyard Kipling's book *Just So Stories for Little Children*, in which the author provides fantastical explanations for how animals acquired their characteristics. The use of this phrase implies a critique of the 'adaptationist programme' which is the tendency to generate unfalsifiable adaptive narratives in an attempt to explain the nature of observed traits.

Kilobase (Kb) – a thousand base pairs of DNA or RNA.

Large language model (LLM) – an AI system that utilises a type of neural network known as a transformer, which is a deep learning architecture that processes data sequentially. LLMs are trained on massive amounts of text data to learn patterns, grammar, and meaning. LLMs are able to perform pattern recognition and complex language generative tasks.

Levinthal's paradox – formulated by Cyrus Levinthal in 1969. It highlights the fact that identifying the correctly folded state of a protein by randomly searching all possible configurations would take longer than the age of the universe, and yet in the real world proteins fold in seconds or less.

Library of Babel – a fictional expansive and alien library of books described by the Argentinian author Jorge Luis Borges in his collection of short stories *The Garden of Forking Paths*. It provides a compelling metaphor for DNA sequence space.

Library of Mendel – the name that the philosopher Daniel Dennett gave to DNA sequence space.

LMB – the MRC Laboratory of Molecular Biology, Cambridge, UK.

Long non-coding RNA (lncRNA) – RNA molecules longer than 200 nucleotides that do not code for proteins but which have been shown to have critical roles in gene regulation, genome folding and cellular function. Their sequences are specified by RNA-encoding genes.

Loopome – the complete collection of around 10,000 loops that the human genome forms when it folds up inside the nucleus. These are involved in fine-tuning gene regulation. Together they comprise a loop 'code'.

Machine learning – a form of AI that learns to discern patterns, makes decisions and predicts outcomes following training on huge datasets, without being explicitly programmed to do so. It contrasts with brute force algorithmic computing where information is pre-programmed into a computer.

Major histocompatibility complex (MHC) – a critical component of the human immune system. It encodes molecules that help the immune system to detect foreign entities, derived, for example, from viruses, bacteria or abnormal 'self' proteins produced by cancer cells. Different versions of MHC molecules are associated with susceptibility to certain diseases.

Mathematical Platonism – a philosophical perspective, which suggests that mathematical or informational structures have a timeless, abstract, mathematical reality and are therefore discovered rather than generated. DNA sequences may be viewed as having a timeless mathematical existence, with natural evolution or synthetic biology simply discovering and actualising 'pre-existing' species rather than inventing them.

Max's Library – the equivalent of DNA sequence space of Fred's library, for protein sequences. It is the hypothetical space of all possible protein sequences made from natural amino acids and their associated structures and protein sequence space.

Megabase (Mb) – a million base pairs of DNA or RNA.

Mendelian diseases – diseases caused by mutations in a single gene (monogenic) that have Mendelian patterns of inheritance within families.

Mendelian inheritance – a pattern of inheritance that follows the rules described by Mendel in 1865 and rediscovered in 1900.

Messenger RNA (mRNA) – an RNA molecule that carries genetic information from the genome (following a process called transcription) to the ribosomes, where proteins are made according to the information in the mRNA, in a process known as translation. It acts as an intermediary between information in DNA and the amino acid sequences of proteins.

Metabolic network – the complete, complex and interconnected set of biochemical reactions within a cell that transform molecules in specific ways to produce the outcomes necessary for cellular function.

Metagenomic data – refers to DNA sequences that are collected directly from environmental samples, without the need to isolate actual individual organisms. It enables whole communities of organisms to be studied simultaneously.

MHC locus – the part of the human genome (chromosome 6) where the MHC complex is located.

Microbiome – the complete collection of microorganisms in and on the human body including bacteria, viruses, fungi and their associated genomes. Microbial genes in the human body outnumber human genes by approximately 100:1. The nature of individual microbiomes has been related to states of health, disease and drug sensitivity.

Minimal genome – the minimal DNA nucleotide sequence necessary to generate and operate a functional species of a particular type. It is identified by a trial-and-error process involving sequentially stripping out DNA from a genome and determining whether a viable, self-replicating organism can still be generated if placed in an environment with sufficient nutrients to sustain it.

Mitochondria – small organelles found in the cytoplasm of eukaryotic cells that generate most of a cell's energy in the form of a molecule called ATP. Each mitochondria contains its own small genome (16,569 nucleotides) comprising 37 genes.

Model organisms – a group of non-human species including the mouse, nematode worm, fly and zebrafish that have been extensively studied and are used to model the biology of humans, with the expectation that lessons learned in one may transfer to the other given the conservation of function seen across evolutionary biology.

Molecular biology – the science of defining the structure and function of the molecular components of living organisms, and manipulating and synthesising them at the molecular level. These activities are principally focused on DNA, RNA, proteins, cells and whole organisms.

Monoclonal antibody – an antibody of a single defined specificity that is able to selectively bind a target within a molecule. The technology was invented by Köhler and Milstein, and has proved to be of great therapeutic importance, for example, in selectively targeting cancer cells.

Moonlighting – describes how some (likely 5–10% in humans) proteins perform more than one (often unrelated) biochemical or biophysical functions in a cell or organism. Many moonlighting proteins are implicated in human disease. lncRNAs can also moonlight.

Moore's Law – first mooted in 1965 by Gordon Moore, the co-founder of Intel, and based on the observation that the number of transistors on a microchip doubled approximately every two years. It reflects technological innovation in semiconductor manufacturing. It has historically led to an exponential increase in computing power and has defined the semiconductor industry for several decades.

Morphora – a term invented by the author to describe entirely novel and morphologically, biochemically and behaviourally distinct species that are synthetically generated and whose genetic information is not informed by a direct process of heredity.

Multiplexed gene editing – when more than one gene is simultaneously edited in a cell or organism.

Mycobiome – the complete collection of fungi present in or on an organism. It forms a subset of the microbiome.

Neochromosome – an unnatural synthetic chromosome constructed by humans to implement new genetic programmes in the cells of species to alter the biology of existing or synthetic organisms.

Network medicine – a term invented by the author to describe a new and hypothetical form of medical intervention that attempts to modify the dynamics and behaviour of a metabolic network to achieve a specific biological outcome such as curing a disease.

Neural network – a computing architecture and type of machine-learning model comprising layers of interconnected nodes (neurons) that process information. It is designed to mimic the structure and function of the human brain. Following training, it can learn to perform tasks such as pattern recognition.

Non-canonical monomers – chemical building blocks that are not amino acids and whose sequences could, in principle, be coded by the genetic code if the system was engineered appropriately. The use of such building blocks would allow biology to transcend the limits of protein-base life.

Non-computable – a number, problem or informational structure that cannot be computed, calculated or 'solved' algorithmically by implementing a process of computation. Also known as 'undecidable'.

Nonillion – a 1 followed by 30 zeros.

Non-coding genome – the 99% of the human genome that does not encode proteins.

Non-linear behaviour – occurs when the output of a network system is not proportional to the input. Small changes may therefore have large effects, while large changes may have small effects. Non-linearity arises, amongst other things, as a result of feedback loops and cooperativity within networks.

Non-natural amino acids – all natural proteins are built from a core 'canonical' set of 20 standard amino acids. There are, however, thousands of unnatural amino acids that could be accommodated within protein sequences and potentially introduce novel properties into them. These can be incorporated into proteins by recoding an organism's genome by and assigning existing codons.

Non-playable characters (NPCs) – characters in video games that cannot be controlled by the player. Their behaviour is pre-programmed.

Nuclein – the term given to the heredity substance by Friedrich Miescher, who first discovered it in 1869 in an extract prepared from white blood cells.

Nucleosome – a structural unit of DNA packing that comprises around 147 base pairs of DNA wound around a core of histone proteins. It is the basic structural unit of chromatin.

Nucleotide – the building-block molecule of DNA and RNA. Each nucleotide is made up of a phosphate group, a sugar, and a base. In DNA the sugar in the nucleotide is deoxyribose, whereas in RNA, it is ribose. The 4 bases in DNA are A, T, C and G. In RNA, T is replaced by U, which stands for uracil.

Nucleotide base – (see Nucleotide)

Nucleotide sequence – the way in which nucleotides are joined together into information-containing combinatorial sequences that encode the hereditary information of life.

Nucleus – a membrane-bound organelle found in eukaryotic cells that houses a cell's genome. Note that prokaryotic organisms like bacteria do not have a nucleus and the genome is free-floating rather than encapsulated.

Oligogenic editing – the targeted editing of a small number of genes. It references a number of genes somewhere between monogenic (one gene) and polygenic (multiple genes).

Oligonucleotides – short nucleotide sequences, typically 2 to 3100 nucleotides long (although they can be longer). They have many uses in molecular biology, and form the feedstock for genome assembly technologies.

Operating system (OS) – the software that manages a computer's hardware and software resources. By analogy, a genome can be viewed as an OS that operates a cell.

Order for free – a term used by Stuart Kauffman to reference the spontaneous, emergent, self-organising behaviours in physical and chemical systems that generate non-programmed order. The non-programmed hexagonal symmetry of a snowflake provides a good example. Kauffman suggests that genes have learned how to entrain and exploit these wells of naturally occurring order.

Organoid – is a three-dimensional cluster of cells – a 'mini-organ' – derived from stem cells and grown in tissue culture fluid. They may partially or fully mimic the structure and function of a body organ.

Orthogonal components – components of machines that function entirely independently of one another and whose functions do not overlap with those of other components. In machines made from orthogonal parts, broken parts may readily be replaced without interfering with the function of other parts.

Orthogonality – the property of being orthogonal, namely having an activity and function that is independent of all other components.

Orthologue – a gene found in different species that has a common evolutionary origin and typically retains the same function.

Petabyte (PB) – a unit of information storage equivalent to 1,000 terabytes (or 10 to the power of 15 bytes).

Phenotype – the overall nature of an organism, which is rooted in the (genetic) information encoded in its genome, but is also determined by other (extra-genetic) aspects of its informiome, including epigenetics, culture and environment.

Φ-X174 – a bacteriophage (a virus that infects bacteria, in particular *E. coli*). It was of key importance in the history of molecular biology and contributed to many fundamental discoveries.

Phosphoramidite method – introduced in 1981, it became the industry gold standard for oligonucleotide synthesis.

Phosphotriester chemistry – one of the earliest chemical synthesis methods for making DNA. It modifies phosphate groups with an ester bond, and nucleotides are added sequentially.

Pleiotropic effect – is when a single gene impacts multiple different traits.

Polygenic diseases – these result from small contributions from multiple genes.

Programmable biology – the notion that once the rules of biology have been fully comprehended and suitably versatile technologies are available to efficiently write genomes of a predetermined sequence on demand, it will be possible to programme biology like a piece of computer software.

Programmable engineered cell – a hypothetical cell with a highly streamlined, refactored and recoded genome that is engineered to readily receive additional genetic code script to introduce new functions on demand.

Protein – molecules comprised of sequences of amino acids. These amino acid chains 'fold' to form a three-dimensional structure, which can be inferred using AI such as AlphaFold, or more reliably, using an experimental method such as X-ray crystallography. The structure of a protein determines its function.

Protein Data Bank (PDB)/Research Collaboratory for Structural Bioinformatics Data Bank (RCSB PDB) – the key global repository for experimentally determined 3D protein structures.

Protein design – a more generalised form of protein engineering that includes the generation of novel proteins that do not exist in nature.

Protein engineering – the activity of optimising the design of protein sequences and the 3D structures they assume on a rational predictive basis or using AI to achieve a desired functional outcome.

Protein folding – the process by which a protein adopts its three-dimensional configuration.

Protein structure – the shape (assay of atoms in space) that a protein adopts once it has 'folded'.

Quaternary code – unlike the digital binary code of computer software that uses just two symbols, 1 or 0, the code of DNA and RNA uses four symbols: A, T, C and G (DNA), and A, U, C and G (RNA). DNA and RNA may thus be considered as quaternary information systems.

Quintillion – 10 to the power of 18 (a million trillion).

Recoding – the process by which the genetic code of an organism is modified to reassign codons and allow for the incorporation of unnatural amino acids into the proteins synthesised by the cell.

Recombinant DNA – a biotechnology method that allows DNA from a variety of sources, including different species, to be combined into hybrid sequences that are not found in nature.

Refactoring – the process of rationalising computer code (or the sequence of a genome) to deconvolute it, simplify it and fix bugs and vulnerabilities, without changing its overall function and performance.

Reverse engineering – a philosophical approach to understanding biological systems that involves breaking them apart. It was pioneered by Francis Bacon in 1620.

Rewilding – the concept of letting land impacted by human activities revert to its natural state. This may involve reintroducing species and restoring their habitats.

Rewind the tape of life – a thought experiment described by the paleontologist Stephen Jay Gould in his 1989 book *Wonderful Life*. He argued that life is highly historically contingent, and speculated that if the history of life could be rewound, and the play button pressed again, the history of life on Earth would be different on each replay.

RFdiffusion – a deep learning/neural network AI model developed by David Baker at the University of Washington for the generation of novel protein designs.

Ribonucleic acid (RNA) – a single-stranded nucleic acid molecule that performs multiple informational and structural functions in cells.

Ribose – the sugar used by RNA.

Ribosomal RNA (rRNA) – a type of RNA that forms the structures of ribosomes, which are the molecular machines in eukaryotic cells that synthesise proteins.

RNA polymerase II – the enzyme that transcribes DNA into messenger RNA and also makes several other non-coding RNA molecules such as lncRNA.

RNA Renaissance – a term that captures the fact that we have, until now, focused on protein-encoding genes in the human genome at the expense of RNA-encoding genes located in the non-coding regions of the human genome that play key regulatory roles and form an indispensable component of the human genome's Operating System.

RNAome (RNome) – the inventory of all the different types of RNA molecules (transcriptome) and their modified versions, which are made by a cell. It is the RNA equivalent of the genome.

Rules of biology – the hypothetical grammatical rules by which biological systems generate the structures and functions of living things. Once these rules have been deciphered, biology will be transformed into a predictive engineering material.

Sanger dideoxy sequencing method – developed by Fred Sanger in 1977. It was the main method used by the Human Genome Project to sequence the human genome.

Sc2.0 – *Saccharomyces cerevisiae* 2.0. It refers to the synthetic yeast project, an international consortium led by Jef Boeke at the NYU Langone Medical Center, focused on constructing a streamlined synthetic yeast genome.

Sidewinder – a method developed in 2025 at Caltech for the construction of DNA at scale. It enables DNA of any complexity to be assembled rapidly, accurately and at low cost. It uniquely uses 'page numbers' to assemble DNA, and assembles DNA without needing to reference the sequence information of the DNA sequence to be assembled.

Small world network (SMN) – a complex real-world network found in many types of systems – including metabolic – in which each node can be reached from any other node by travelling a relatively short distance.

Software – a set of programmatic instructions designed to perform a specific task. The nucleotide sequences of genomes may be thought of as representing DNA programmes that describe how to build and operate a living organism.

Software bug – an error, fault or flaw in a piece of computer software source code, which causes it to behave in an unanticipated manner.

Somatic cells – all the cells in humans, other than egg and sperm germ cells. Modifications to the genomes of somatic cells are not heritable.

Source code – a collection of instructions written by a programmer in a human-readable computer programming language. DNA nucleotide sequences that perform specific functions, such as directing the synthesis of a protein, may be viewed as comprising genetic source code.

Spaghetti code – a pejorative term in computer science that refers to tangled, unstructured, messy, hard-to-understand and difficult-to-maintain pieces of computer code. The same term may be applied to nucleotide sequence code that has been 'written' by evolution.

Species – historically defined as a group of organisms that share common morphological and biochemical features and are capable of breeding to produce fertile offspring via a process of heredity. The existence of species categories prevents the intermingling of DNA between species. If new species can be made synthetically without referencing heredity we may need to adopt a new terminology.

Species Catalogue – a term invented by the author to describe a version of Fred's Library that goes beyond mathematical sequence spaces cataloguing only DNA, and extends to the mathematical sequence spaces describing all possible chemistries capable of generating living species. It is the exhaustive, hypothetical mathematical space containing all possible forms of potentially living things. Once charted, it will comprise an exhaustive map of all biological possibility.

Syn57 – a recoded *E. coli* made at the LMB in Cambridge, UK, which uses 57 codons instead of the available 64, thereby freeing up 7 codons that can be used to code for unnatural amino acids.

Syn61 – a laboratory-synthesised and genomically recoded *E. coli* that was constructed at the LMB in Cambridge, UK. It uses 61 codons instead of the available 64, thereby freeing up 3 codons that can be used to code for unnatural amino acids. It was the first fully synthetic bacterial genome built from scratch.

SynHG – the Synthetic Human Genome project, announced in June 2025 and led by Jason Chin at the LMB in Cambridge and Ellison Institute of Technology in Oxford.

Synthetic biology – a term that addresses the science of genome design and synthesis. It treats biological systems as programmable 'software' that can be recoded at will to address specific outcomes and objectives.

Synthetic rescue – a potential strategy for indirectly treating monogenic diseases by 'rewiring' the metabolic networks in which faulty genes are embedded.

Tappan Zee – a term referencing the Tappan Zee Bridge in New York State that was invented by the author to describe the situation where a feature of a biological system represents a maladaptive vestige of a now irrelevant past, rather than a specific adaptation. In the absence of the ability to naturally engineer biology, biology is unable to completely re-write and re-imagine itself, meaning that sub-optimal features may get locked into biological systems indefinitely.

T-cells – cells of the human immune system that eliminate abnormal cells such as cancer cells. They also help B-cells to make antibodies and prevent autoimmunity. They can be genetically engineered to make CAR-T cells.

Technical debt – the 'mess' in computer code that accumulates when software is not routinely and regularly subjected to refactoring.

Telomere-to-telomere consortium (T2T) – established to address the fact that despite the completion of the draft human genome, no individual chromosome had been completed end-to-end and hundreds of gaps existed. The consortium succeeded in producing the first complete human genome sequence with no gaps.

Terminal deoxynucleotidyl transferase (TdT) – a specialised DNA polymerase that can write relatively short sequences of DNA in a template-independent manner. Engineered versions have been used to synthesise artificial DNA.

Topologically associated domains – large regions (domains) of the genome that interact with sequences located within the domain more frequently than those located outside it.

Trans-activating CRISPR RNA (tracrRNA) – a critical part of the CRISPR-Cas9 gene editing machinery. It works with CRISPR RNA (crRNA) by binding to it and guiding Cas9 to DNA sequences of interest.

Transcription – the process by which information in the nucleotide sequence of a gene is copied into a complementary messenger RNA molecule (mRNA). This serves as a template for making a protein in the process of translation occurring in ribosomes.

Transcription factor – a protein that binds to DNA nucleotide sequences and controls the extent to which a gene is transcribed.

Transfer RNA (tRNA) – brings specific amino acids to the ribosome according to the codon sequences comprising the mRNA molecule, thereby ensuring that a gene sequence is translated into the correct corresponding protein sequence.

Underground metabolism – the low-level interactions (side reactions) between the molecular components of a metabolic network that are not related to their core function and primary pathway. These may generate unintended byproducts.

Universal Turing Machine – a theoretical machine imagined by the mathematician Alan Turing. It became the foundational concept for computer science and forms the theoretical basis for a programmable computer.

Uracil (U) – one of the four nucleotides bases used to make the nucleotide sequences of RNA. It replaces thymine (T), which is used in DNA but not in RNA.

Virome – the complete inventory of the viruses in an organism. It forms a subset of the microbiome.

Whole genome sequencing (WGS) – determining the complete nucleotide sequence of a genome.

Xenobiology – a field of synthetic biology focused on constructing novel biological systems using unnatural chemistries and recoded genomes.

Xeno nucleic acids (XNAs) – synthetic analogues of natural nucleic acids (DNA and RNA) with distinct properties that may, potentially, provide the basis of alternative genetic materials and heredity systems.

X-linked inheritance – a form of Mendelian inheritance related to traits or disease conditions determined by mutant genes located on the X chromosome.

X-ray crystallography – an experimental method used to determine the molecular structures of protein molecules through the analysis of X-ray diffraction patterns. These reveal the positions of individual atoms at varying degrees of resolution.

Zettabyte (ZB) – a unit of information storage equivalent to one sextillion (10 to the power of 21) bytes.

Index

A Note on the Author

Adrian Woolfson is the co-founder of Genyro, a California-based biotechnology company specialising in AI-augmented synthetic genome design and construction. Born in London, he studied medicine at Balliol College, Oxford, molecular genetics at Gonville and Caius College, Cambridge, and was formerly the Charles and Katherine Darwin Research Fellow at Darwin College, Cambridge, where he worked in the Division of Protein and Nucleic Acid Chemistry at the MRC Laboratory of Molecular Biology. He is the author of the critically acclaimed *Life Without Genes: The History and Future of Genomes* and *An Intelligent Person's Guide to Genetics*. He has authored over 160 scientific papers, book chapters, reviews, and patents, contributed to multiple publications including the *Washington Post*, *Nature*, the *Spectator*, the *Literary Review*, *Prospect* magazine, the *Times Literary Supplement*, the *London Review of Books* and the *Financial Times*, and is a regular contributor to the *Wall Street Journal* and *Science* magazine. He currently lives in San Francisco.

A Note on the Type

The text of this book is set in Bembo, which was first used in 1495 by the Venetian printer Aldus Manutius for Cardinal Bembo's *De Aetna*. The original types were cut for Manutius by Francesco Griffo. Bembo was one of the types used by Claude Garamond (1480–1561) as a model for his Romain de l'Université, and so it was a forerunner of what became the standard European type for the following two centuries. Its modern form follows the original types and was designed for Monotype in 1929.